WILD FLOWERS OF THE CANARY ISLANDS

TO DON ENRIQUE

“Pro loco et tempore”

Wild Flowers of the Canary Islands

by

David Bramwell MSc., PhD.,
and
Zoë I. Bramwell BSc.

Stanley Thornes (Publishers) Ltd.

First published in 1974 by
Stanley Thornes (Publishers) Ltd.
Educa House
Old Station Drive
Leckhampton
CHELTENHAM GL53 0DN

First Paperback Edition 1984

British Library Cataloguing in Publication Data

Bramwell, David
Wild Flowers of the Canary Islands
1. Wild Flowers 1. -Canary Islands
I. Title II. Bramwell Zoe
13/09649 QK42.2

ISBN 0-85950-227-9

Printed in Great Britain by Colour Reproductions Ltd, Billericay, Essex, and The Pitman Press, Bath.

Bound at The Pitman Press.

Table of Contents

Preface

The mountains, forests and coasts of the Canary Islands harbour a wealth of wild flowers, a large proportion of which are peculiar to the archipelago or shared with the neighbouring Madeira Islands. This endemic element in the flora is strikingly evident to even the most casual observer. The native 'cardon' and 'tabaiba' (*Euphorbia* species) form the dominant vegetation of many of the uncultivated regions of the dry, lower zone of the islands whilst the upper slopes are clad with dense forests of Canary laurel and its relatives or of the local Canary pine.

Despite this richness of endemic plants it remains difficult for anyone other than a dedicated, well-trained botanist to identify and name the majority of the native wild flowers. A knowledge of rare Mid-Ninteenth Century Floras and old botanical journals mostly written in Latin, French or German and possessed by only a handful of major botanical institutes, or the time and patience to work through large numbers of dried specimens preserved in National Herbaria such as Kew, Paris or Florence are primary requisites unavailable to the majority of people. Such tasks often spoil rather than enhance the enjoyment of wild plants in their natural habitat for all but the keenest professionals.

This book, it is hoped, will meet the current need for a simple, illustrated guide to the endemic wild flowers of the Canary Islands for use by naturalists, whether amateur or professional, horticulturally-minded visitors and tourists who wish to explore the hills and mountains of these magnificent islands away from the main towns and roadsides. In view of the total size of the Canarian flora (almost 2000 species) it has not been possible to include all the plants to be encountered as there has been an overwhelming need to keep the book to a manageable size and to make it available quickly at a reasonable price.

A number of non-endemic, mainly Mediterranean, species are found in the islands but these should present little problem to anyone familiar with the recent epidemic of popular guides to the wild flowers of the Mediterranean region. As the endemic plants dominate most regions of the islands away from the immediate vicinities of towns and cultivated fields we have deliberately chosen to concentrate on this part of the flora. In some cases, however, we have allowed ourselves a wide interpretation of the term endemic and included some species found in North Africa or the West Mediterranean region when they form an important part of native plant communities in the Canaries. We have also included most of the species shared with the other Atlantic Islands, Madeira, the Azores and the Cape Verde Islands, but not found elsewhere. This choice of species to be included has necessarily been a personal one and we hope it will prove to be satisfactory.

The main feature of this book lies in the illustrations which should suffice, along with the keys and short descriptions provided, to identify most of the plants concerned especially if used in conjunction with information provided about localities. The descriptions and keys have been kept as simple as possible and a fairly extensive glossary of botanical terms is provided as an aid to their use. We have attempted to keep things basic enough for anyone interested in, but not necessarily familiar with the technical details of, plants whilst retaining enough scientific detail for the book to be of use to the professional botanist until a more complete scientific Flora of the islands is available.

A very good illustrated guide to the subtropical and tropical trees planted along roadsides and in parks and squares in the Canaries is provided by Gunther Kunkel in his book 'Arboles Exoticos' which is published in Las Palmas and is on sale locally.

In an area such as the Canary Islands, where many of the endemic plants are very restricted in their range and are often limited to only one or two localities, information about distribution can be an important aid to identification. We have, therefore, given localities for most of the plants especially when these coincide with local beauty spots or areas likely to be visited even by the only mildly adventurous. This, of course, raises the very serious problem of endangered species and conservation. We believe that the time has come to stop collecting plants like postage stamps or cheese labels and to appreciate them in their natural surroundings so we strongly recommend that any species mentioned in this book as being rare or restricted in distribution should not be collected or damaged in any way. Further, we hope that this book will help to stimulate interest in the endemic flora and to further the cause of conservation in the Canary Islands which, though strenuously pursued by a small group of local botanists and others, is still very much in its infancy.

Acknowledgements

During the preparation of this book we have been fortunate to have had the assistance of numerous friends and colleagues both in the Canary Islands and at home. In the first place we must mention Eric R. Sventenius who was unfortunately killed in a road accident during the summer of 1973. Eric Sventenius was unrivalled in his knowledge of the plants of the Canary Islands and during the last five years introduced us to most of them. Without his help and generosity this book would not have been possible.

Dr. C. J. Humphries and Miss Angela Aldridge prepared the initial accounts of the genera *Argyranthemum* and *Sonchus* respectively and the following people helped with the preparation of other groups: Miss E. A. Leadlay (*Senecio*), Miss D. H. Davis and Mr. A. J. Scott (*Lotus*), Miss B. Petty (*Descurainia*, *Bencomia*) and Miss V. W. Smith (*Kickxia*). Colour photographs have been loaned by several people particularly Dr. C. J. Humphries, Dr. R. Melville, Dr. Per Sunding, Dr. W. T. Stearn, Mr. G. D. Rowley, Dr. D. M. Moore, Miss Angela Aldridge and Dr. C. N. Page. Mr. D. Gibson provided photographs of *Lavatera acerifolia*. To all these people we are extremely grateful.

Our many visits to the Canary Islands have been facilitated by a number of local botanists and naturalists and several of these have been invaluable companions in the field. We are greatly indebted to the following people: Professor W. Wildpret de la Torre, Snr. D. Arnoldo Santos Guerra, Snr. D. Eduardo Barquin Diez, Snr. D. Ventura Bravo, Snr. D. Jose Luis Vega Mora, Snra. Da. C. Kercher and Mr. Gunther Kunkel. We are also grateful to the staff of the Jardín Canario, Tafira and to Snr. D. Carlos Gonzalez Martín of the Jardín de Aclimatación, Puerto de la Cruz for permission to photograph plants in cultivation.

Various colleagues have helped during the preparation of the manuscript with advice and comments and we would like especially to thank Dr. D. M. Moore, Miss E. A. Leadlay and Professor V. H. Heywood. We are extremely grateful to Mrs. Abigail Gillett, Mrs. F. Ord-Hume and Miss Christine Reid for typing the manuscript.

We would like to extend our thanks to our publishers for their help and encouragement over the preparation and publication of the book.

Last but not least we would like to express our gratitude to the British Council, The Consejo Superior de Investigaciones Cientificas, Madrid and Professor E. F. Galiano (Sevilla) who made possible a 10 month visit to the Canaries in 1968–69 when many of the photographs were taken and line drawings prepared.

History of Botanical Exploration

1750–1800

Though a handful of Canary Island plants were known to the 18th Century Swedish botanist Linnaeus, 'the father of modern botany' (The Dragon tree *Dracaena draco*, the Canary bell-flower *Canarina canariensis*, etc.), the earliest real botanical exploration of the islands was carried out by the first plant collector sent out by the Royal Botanic Gardens, Kew, Francis Masson, who went to the islands on at least two occasions in the 1770's. Masson collected many endemic plants some of which he presented to Linnaeus and which were later described and named by the great botanist's son. The major part of Masson's collections including living plants and seeds were, however, kept at Kew and were named by William Aiton who was senior gardener there.

Towards the end of the 18th Century the Canaries were visited by the German explorer and botanist Alexander von Humboldt who first described the major vegetation zones on the islands. At the turn of the Century the islands were also explored by the French botanist P. M. A. Broussonet who was French Consul on Tenerife.

1801–1900

In 1815 the islands were studied by a Norwegian, Christen Smith, and his German companion Leopold von Buch who later wrote the first catalogue of the plants of the islands following Smith's untimely death whilst plant-collecting in the Congo.

Following this period a major landmark in Canarian botany was reached. This was the preparation and publication, over a period of more than 20 years of Webb & Berthelot's *Histoire Naturelle des Iles*

Canaries. Phillip Barker Webb, an English botanist and philanthropist, visited the Canaries in the course of a proposed expedition to Brazil and so liked the islands that he abandoned his Brazilian journey and remained in the Canaries in the company of a French naturalist Sabin Berthelot. On his return to his home in Paris several years later Webb undertook the preparation, financed largely out of his own pocket, of the outstanding treatise on the natural history of the islands. In the course of this work he collected together a large herbarium and library of the Canarian flora which is now housed at the University of Florence. Webb & Berthelot's work, which extends to 3 tomes comprising 10 volumes, remains the only one in which descriptions and illustrations of Canarian plants are provided in sufficient detail for identification.

During the second half of the 19th Century several botanists from Europe made use of Webb and Berthelot's groundwork and continued the botanical exploration of the islands. The Spanish botanist Ramon Masferrer y Arquimbo wrote a catalogue of the plants of Tenerife while the Swiss Herman Christ and German Carle Bolle worked on the still poorly known eastern islands of Lanzarote and Fuerteventura as well as on the western group. In addition Christ worked through Webb's unpublished papers and wrote a long account of his post-*Histoire Naturelle* studies of the Canarian flora.

1901–1973

It was not, however, until the early part of the 20th Century that the writing of a second Flora of the Canaries was attempted. An English botanist the Rev. R. P. Murray worked extensively in the islands between 1890 and 1904 but unfortunately died suddenly before he could prepare his proposed Flora. It was left to two French botanists, C. J. Pitard and L. Proust, to prepare and publish an extensive account of the flora in 1908. This work was not a Flora in the modern sense but an annotated catalogue giving distributions of each species but lacking keys or descriptions as aids to identification of the plants.

In 1913 two botanists from Kew, Thomas Sprague and John Hutchinson, visited the Canaries and on their return wrote a series of articles on plants from the islands. Later R. Lloyd Praeger from Dublin studied the succulent flora of the Canaries paying particular attention to the *Sempervivum* group (Houseleeks) which he monographed in 1932. A German doctor O. Burchard who lived for many years on Tenerife contributed a valuable study on the ecology and distribution of the endemic flora of the islands in 1928. More recent studies of Canarian plants have been carried out by L. Ceballos and

F. Ortuño who surveyed the forest floras of the Western Canaries, Kornelius Lems who, in 1960, wrote an important check-list outlining the then current state of knowledge of the flora and Johannes Lid whose *Contributions to the Canary Islands Flora* is a valuable guide to the distribution of many species.

A notable modern achievement has been the establishment at Tafira on Gran Canaria of the *Jardín Canario Viera y Clavijo*. This botanical garden contains a splendid collection of Canarian endemic plants and should be visited by anyone interested in seeing very rare plants growing in virtually natural surroundings. The garden is the fruit of almost twenty years labour by its first director the late Dr. Eric R. Sventenius to whom we have dedicated this book. Eric Sventenius has also contributed many major publications on Canarian plants during the past thirty years during which he was resident in the islands and he has been the main force behind the revival of interest in the Canary Islands and their natural history over the past decade.

Climate

The climate of the Canary Islands is basically a Mediterranean one but is influenced by a series of modifying factors. These are the close proximity of the eastern islands to the coast of North Africa and, therefore, the Sahara Desert, the oceanic position of the western islands on the edge of the North East Trade Wind system and finally the high altitude of the peaks of the western islands.

RAINFALL

Generally the Canaries have hot, dry summers and warm, wet winters but locally there are often considerable deviations from this basic pattern. The North East Trade Winds bring in moisture from the sea which, when forced to rise by the mountain barriers of the western islands, is cooled and forms a zone of precipitation at about 800 to 1500 m. (Fig. I.) This causes a more or less persistent cloud-layer at this level on the north side of all the western islands and has a great influence on the natural vegetation.

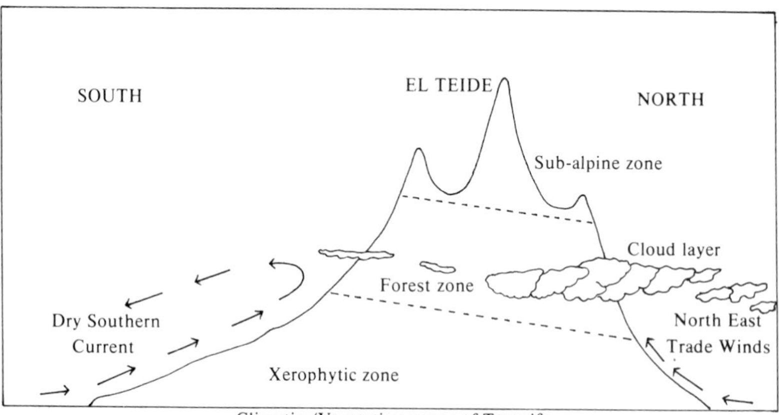

Climatic /Vegetation zones of Tenerife

Fig. I

The southern sectors of these islands are in a rain shadow and receive much less precipitation. They are generally without a dense forest zone at mid-altitude level and are much more xerophytic in nature. The climate is more humid and the vegetation more luxuriant the further west one goes in the archipelago due to the increasingly oceanic position of the western islands and the greater strength of the Trade Winds. The extreme dryness of the eastern islands and the south of Gran Canaria is partially due to the hot dry Saharan winds, the Harmatan or Levante, which sometimes reach the eastern Canaries blowing for up to a week at a time with a strong desiccating effect on the vegetation.

The eastern islands are too low to intercept the Trade Wind moisture except at their highest points (Jandia on Fuerteventura, Famara on Lanzarote) where small areas of more luxuriant vegetation are developed.

TEMPERATURE

The coolest month tends to be February with a mean temperature of just below 18°C. The hottest period is from July to September with a mean of between 20° and 25°C for the western islands. Few data are available for Lanzarote and Fuerteventura, the two semi-desert islands, but summer means of up to 35°C have been recorded. The Trade Wind effect is responsible for differences of over 10°C between the north and south coasts of the larger islands and snow lies on the upper slopes of Tenerife for several months in the winter. The diurnal temperature range in Las Cañadas is often as much as 25°C.

Vegetation Zones and Plant Communities

Diversity of climate coupled with high altitude have a profound effect on the distribution of vegetation types in the islands. There is a striking altitudinal zonation which is summarized in Fig. II. The following notes are intended to give a brief guide to each of the dominant types of vegetation for the main zones.

Vegetation	Altitude (m.)	Climate	Habit
Subalpine scrub	2600	Subcontinental; cold winter, hot dry summer	Perennial herbs
Montane scrub	1900–2500		Small-leaved shrubs
Pine savanna	800–1900	Dry Mediterranean	Needle-leaved trees
Tree heath and evergreen forest	400–1300	Wet Mediterranean	Broad-leaved trees, shrubby heaths
Juniper scrub	S. slopes 400–600	Mediterranean	Small-leaved shrubs
Semi-desert succulent scrub	0–700	Hot, dry Mediterranean	Candalabra and spiny shrubs, therophytes

Fig. II

1. XEROPHYTIC SCRUB ZONE

In this zone, which occupies the lower slopes of all the islands, several different plant communities are found. Stem and leaf succulents predominate with species of *Euphorbia*, *Aeonium* and the family Compositae most common. Some of the cliff areas, particularly those composed of older Tertiary basalts and phonolites, harbour relict local endemics which often have their greatest affinities with species in the East and South African regions or even in South America.

At its upper limit the xerophytic zone merges into a forest scrub zone of *Erica arborea* (tree heath) with *Juniperus phoenicea* (juniper) on some south slopes giving a type of vegetation very reminiscent of parts of the Mediterranean region.

2. EVERGREEN FOREST ZONE

The evergreen forest zone is dominated by four species of laurel (bay), *Laurus azorica*, *Apollonias barbusana*, *Ocotea foetens* and *Persea indica*. These are relicts of a now virtually extinct Tertiary Mediterranean flora which occupied Southern Europe and North Africa about 15–40 million years ago (see p. 9).

Unfortunately only relatively small areas of evergreen forest have survived the ravages of Man in the Canaries. The forest has been cut for fuel, building and agricultural purposes to such an extent that on Gran Canaria only about one percent of the original forest area still survives and even on Tenerife the figure is rather less than ten percent and this is diminishing rapidly. In many areas over-vigorous de-forestation has led to severe soil erosion.

The following trees and shrubs form an important part of the laurel forest community:

Trees	Shrubs and Herbs
Laurus azorica	*Geranium canariense*
Persea indica	*Cedronella canariensis*
Apollonias barbusana	*Hypericum grandifolium*
Ocotea foetens	*Woodwardia radicans*
Arbutus canariensis	*Viburnum rigidum*
Pleiomeris canariensis	*Ranunculus cortusifolius*
Heberdenia bahamensis	*Isoplexis canariensis*
Prunus lusitanica	*Rubus ulmifolius*
Ilex platyphylla	*Convolvulus canariensis*
Visnea mocanera	*Canarina canariensis*
Salix canariensis	*Rhamnus glandulosa*

The laurel forest is generally confined to the wetter north coast areas of the western islands but is occasionally found in climatically favourable places on south slopes for example at Barranco de Badajoz above Guimar on Tenerife.

3. PINE FOREST ZONE 1200–2000 m

Natural Canarian pine forest, with the endemic *Pinus canariensis*, is found on the islands of La Palma, Tenerife, Gran Canaria and Hierro. It is generally an open, savanna-like formation with few herbs and shrubs in the ground-layer and large areas of bare ground covered only by pine needles. The most common shrubs are *Adenocarpus foliolosus*, *Cistus symphytifolius*, *Daphne gnidium* and *Micromeria* species. *Lotus* species, and *Asphodelus microcarpus* are frequent herbs. Occasionally local endemics are found in the pine forest, these include *Micromeria pineolens* on Gran Canaria and *Lactucosonchus webbii* on La Palma. On all the western islands *Pinus canariensis* has been extensively planted as part of a reforestation programme carried out in the early 1950's because of its economic importance for timber and the vital role played by forest regions in the water economy of the islands.

4. MONTANE ZONE ABOVE 1900 m

The vegetation of the highest peaks of Tenerife, La Palma and to some extent Gran Canaria support an open scrub community with shrubs of the family Leguminosae dominant. Many rare, endemic species are to be found in this area such as *Echium wildpretii*, *Pterocephalus lasiospermus*, *Spartocytisus supranubius*, *Plantago webbii* and *Silene nocteolens*.

The Las Cañadas National Park has the richest montane flora and many of the endemic species can be found within easy reach of the Parador Nacional del Teide which is a convenient stopping place in the park.

The upper slopes of the Pico de Teide have a unique community which consists entirely of the small, endemic alpine violet *Viola cheiranthifolia*. This species is locally very abundant but has recently become depleted in numbers due to the construction of a cable railway linking Las Cañadas with the summit of the peak.

Origin of the Canary Islands Flora

The geological origin of the Canary Islands themselves has been the subject of considerable controversy for many years. Originally it was suggested that they were the peaks of high mountains of the sunken continent of Atlantis and that the old, now unfortunately extinct, inhabitants of the islands, the Guanches, were the last survivors of the race of great warriors said to have occupied the lost continent.

More recently two alternative theories have been presented, firstly that the islands were once part of the African continent and at some stage in the history of the west part of Africa they split off and drifted westwards, and secondly, that the islands arose from the sea-bed as independent volcanoes. Modern geological and oceanographical studies suggest that the true story really lies somewhere between these two possibilities.

The two eastern islands, Lanzarote and Fuerteventura, were once probably part of North Africa, perhaps breaking away when the continents of Africa and South America split apart millions of years ago. Gran Canaria and the western group of islands, Tenerife, La Palma, Hierro and Gomera, however, appear to have had a purely volcanic origin.

The endemic flora of the islands reflects their considerable age. Fossils of leaves and fruits found in many places in the Mediterranean region and South Russia (Barcelona, Rhone Valley, S. Italy, Godanski Pass, etc.) (Fig. III) are of plants identical to species now found only in the Canary Islands and Madeira. These fossils, of plants such as the Dragon Tree, the Canarian laurels and many of the Canarian ferns, date from the Miocene and Pliocene periods of the Tertiary Epoch and are up to 20 million years old.

During this period the Mediterranean region formed part of the basin of an ancient ocean, the Tethys Sea, which separated Europe from Africa. On the margins of this subtropical sea the vegetation must have been very similar in composition and appearance to the laurel forest communities of the present-day Canary Islands.

Many living Canarian plants have their nearest relatives in South or East Africa or even in South America and thus the flora appears to be an ancient survivor of a bygone age.

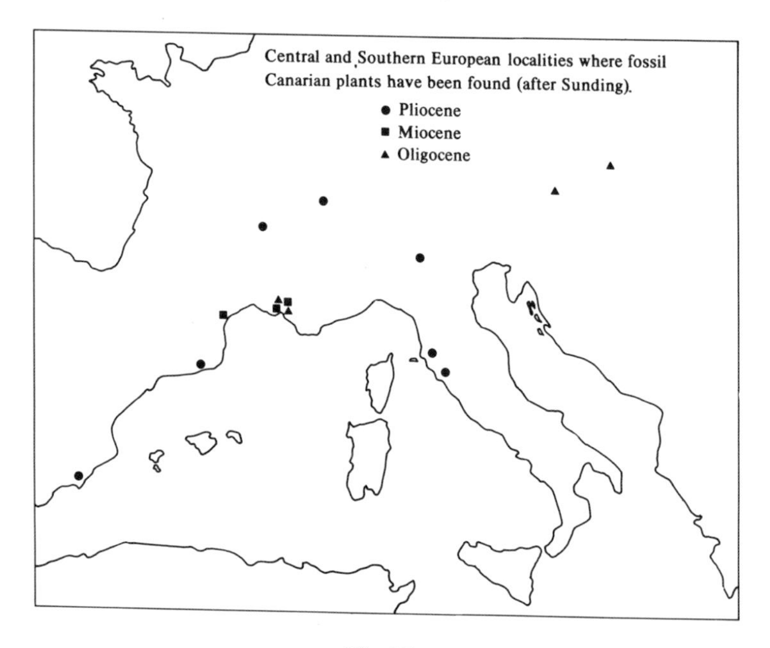

Fig. III

The question may be asked—how did these plants survive in the Canaries? and furthermore—what happened to their contemporaries on the continents of Europe and Africa which caused them to be almost completely wiped out? The answers are complex. Taking the second question first, the main cause of the changes in the Tethyan-Tertiary flora at the end of the Tertiary period was the onset of the Ice Ages. Southern expansion of the Polar Ice Cap with the consequent cooling of the climate led to a southwards migration of the subtropical flora and at almost the same time the area of North Africa now occupied by the Sahara Desert began to dry out. This left the subtropical flora with nowhere to go, restricted by increasing cold to the north and aridity to the south it was more or less wiped out and is now found in a very modified form at a few places in the south of Spain and western Portugal.

The Atlantic Islands, however, were buffered from both types of climatic catastrophe, the northern cold and the Saharan dessication, by their oceanic position which appears to have afforded them considerable protection from major changes. Thus the subtropical flora was able to survive with the loss only of some of the more tropical elements. The high altitude of most of the islands probably served as further insurance against climatic changes by enabling plant communities to migrate altitudinally in order to escape the ravages of varying conditions and also, therefore, may have contributed to the survival of this remarkable flora to the present-day.

Plant Folklore

Many of the endemic plants have long featured in the legends and folklore of the Canary Islands. Sabin Berthelot in 1840 tells of the natives of Tenerife, in periods of drought, removing the bark of the 'cardon' (*Euphorbia canariensis*) and with it the poisonous milky latex and sucking the inner flesh of the stem to quench their thirst. We cannot, however, recommend this practice in view of the fact that the latex was once used by the ancients as a purgative and emetic (a treatment now discontinued because of its violent action and its apparent induction of severe delirium), was used by the Guanches as a fish poison and in addition can cause blindness if splashed in the eyes.

The legendary Canarian Dragon Tree (*Dracaena draco*) yields a red gum, 'dragon's blood', which was highly prized in the Middle Ages by alchemists and physicians for its medicinal and mystical powers. The Dragon Tree is said to achieve great age. The famous tree at Orotava which perished in a hurricane in 1867 was said by the explorer Alexander Von Humboldt to be over six-thousand years old. Lindley and Moore in their *Treasury of Botany* write 'The famous Dragon Tree of Orotava was a giant amongst the plants . . . with an antiquity that must at least be greater than that of the pyramids'. This particular tree was examined by Fenzi in the 1860's and he reported in the *Gardener's Chronicle and Agricultural Gazette* that it was seventy-eight feet in circumference and over seventy-five feet high.

One of the most famous legends of the Canary Islands tells of the 'rain tree' of Hierro. The tree, also known as the 'holy tree' or 'garoe', at the time of the Spanish conquest of the Canaries in the 15th Century distilled sufficient water from the sea mists to meet the needs of all the inhabitants who, prudently, covered it with dry grass so as to conceal it from the invaders. The secret was, however, betrayed by a young girl who fell in love with one of the Spanish soldiery and revenge was exacted by the islanders for her duplicity. The girl, it is said, was condemned to death, the only known instance of capital punishment in the history of the island. An account of the conquest of the Canaries written in the 1630's suggests that the tree was a large 'til' (*Ocotea foetens*), the leaves of which condensed the mountain mists and caused water to drip into two large cisterns which were placed beneath. The tree was destroyed in a storm in 1612 but the site is known and the remnants of the cisterns preserved.

Areas of Botanical Interest

TENERIFE

Tenerife is a roughly triangular island about 80 km long and 60 km wide at its widest point. A long ridge traverses the centre of the island

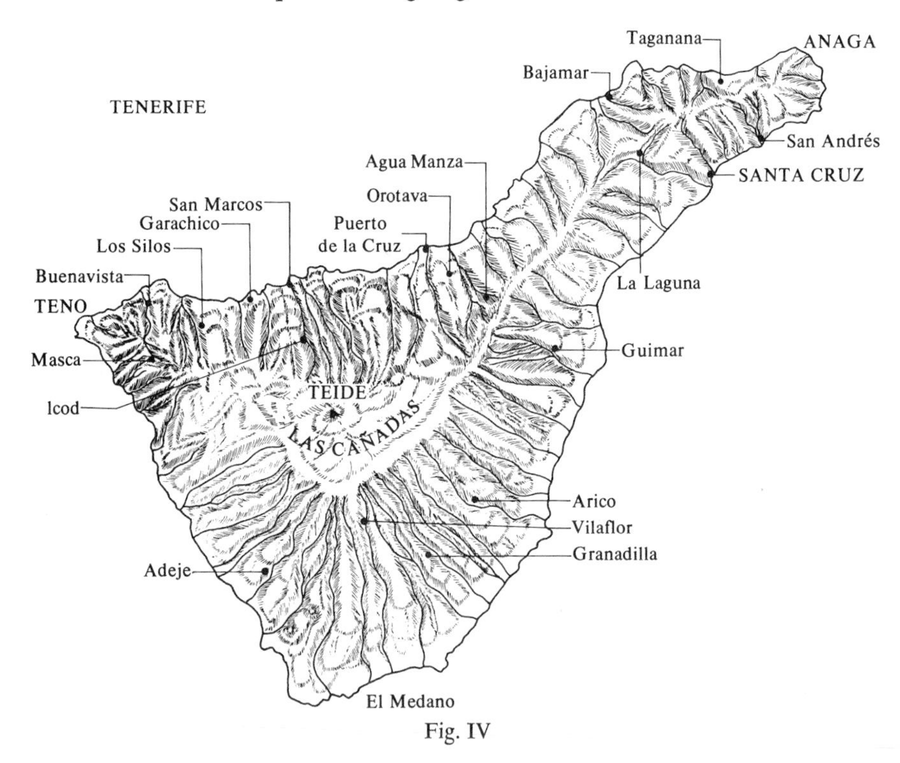

Fig. IV

from end to end with deep valleys dissecting the flanks. The north coast is steep with high sea-cliffs whereas the south side has a coastal plain which is wide and flat in the region of Granadilla. The island is dominated by the central volcanic peak of El Teide (3707 m) with its vast crater of Las Cañadas at 2000 m, and newer cone of the Pico rising from the centre of this old crater. In the extreme west and east regions, Teno and Anaga, are two mountain blocks of old Tertiary basalt. These have characteristic V-shaped valleys and razor-backed ridges with forested crests. (Fig. IV.)

Fig. V

1. Montañas de Teno

This area consists of a block of old deeply dissected mountains with deep valleys and abrupt, vertical cliffs. To the north, in the Buenavista and Teno Bajo region the coasts are flat but elsewhere the mountain spurs drop precipitously to the sea.

Of the northern cliffs of the area, those at Los Silos (Barranco de Cuevas Negras) and El Fraile are outstandingly rich botanically. The former shelter such rarities as *Lavatera phoenicea*, *Dorycnium spectabile* and *Marcetella moquiniana* as well as more common species like *Sonchus radicatus*, *Canarina canariensis* and *Hypericum reflexum*. The shrubs, *Maytenus canariensis* and *Picconia excelsa* occur on north-facing cliffs and the dry lower slopes provide the only known locality for the hemi-parasitic *Osyris quadripartita* in the Canaries. The cliffs of El Fraile between Buenavista del Norte and Punta de Teno are part of one of the richest areas in the Canaries with over 300 species of flowering plants in an area of a few square kilometres. El Fraile, an ancient basalt cliff with almost vertical faces, abounds with endemic species. *Centaurea canariensis*, *Limonium fruticans*, *Tolpis crassiuscula*, *Vieraea laevigata* and *Argyranthemum coronopi-olium* are all rare plants which have their largest populations here. Several species of *Euphorbia*, *E. bourgaeana*, *E. aphylla* and *E. canariensis* dominate the cliff communities while crevices and ledges house succulent plants such as *Monanthes silensis* and *Aeonium tabuliforme*.

To the west, beyond El Fraile, is the coastal platform of Punta de Teno. *Euphorbia balsamifera* and *E. canariensis* are the most common plants along with *Ceropegia dichotoma* and *Launaea arborescens*. Notable endemics include the pink-flowered *Parolinia intermedia*, the almost tree-like *Sonchus arboreus*, *Justicia hyssopifolia* and the extremely rare *Sideritis nervosa*.

The southern sector of the Teno mountains is dominated by the vast chasm of the barranco of Masca. The ridge between Santiago del Teide and Masca provides a convenient starting point for exploring this area. The track passes large populations of the red-flowered *Euphorbia atropurpurea* and the spiny, white-flowered *Echium aculeatum*. On the west side of the ridge are high cliffs with *Echium virescens*, enormous *Aeonium urbicum* plants and, in small, shady crevices, the tiny *Sonchus tuberifer*. In Spring the dry slopes have two small plants of special interest, *Orchis canariensis* with pale pink flowers and the yellow and purple-flowered *Romulea columnae*. The descent to the village of Masca is a long one, passing slopes covered with the local *Lotus mascaensis*, *Aeonium sedifolium* and white-flowered *Retama monosperma*. Below the village the valley narrows with steep cliffs on either side and the area is rich in local endemics, *Teline osyroides*, *Crambe laevigata*, *Convolvulus perraudieri* and *Sonchus capillaris* are common. Typical Teno-area plants at Masca include *Vieraea laevigata*, *Polycarpaea carnosa* and the pink-grey *Dicheranthus plocamoides*. A word of warning is necessary here, if intending to visit Masca and the area below the village allow a full day for the trip, take a packed lunch and save enough energy for the steep climb back.

Beyond the Teno mountains the Valle de Santiago del Teide, with its lava-flows and cinder fields of *Adenocarpus foliolosus* and the attractive *Cheiranthus scoparius* (the form known as *C. cinereus*), leads to the village of Tamaimo with nearby cliffs such as El Retamal and Hoya de Malpais. These cliffs have many plants of special interest like *Sonchus fauces-orci*, *Argyranthemum foeniculaceum* and red-flowered *Teucrium heterophyllum*. Further south, near the town of Adeje is the deep, narrow gorge of Barranco del Infierno where there are small enclaves of Teno plants, *Tolpis crassiuscula*, *Sonchus fauces-orci*, *Senecio echinatus* and *Marcetella moquiniana* as well as the locally endemic *Sideritis infernalis* and an extremely rare shrub *Sideroxylon marmulano*.

2. Sierra de Anaga

The Anaga mountains, a range of jagged, razor-backed hills reaching almost 1000 m, run from the proximity of the town of La Laguna eastwards to the furthest north-east point of the island, Punta de Anaga. The crests are still wooded in many places with dense groves of laurel forest beginning at Las Mercedes and extending almost to Chamorga. The main trees are *Laurus azorica*, *Prunus lusitanica*, *Rhamnus glandulosa* and *Ilex canariensis*. Many interesting plants can be easily found in this area. The white-flowered shrub *Senecio appendiculatus* and *Isoplexis canariensis* with its superb bright orange-red flowers can be found in the deepest forest with *Convolvulus canariensis* and *Crambe strigosa*. In Summer the yellow flowers of *Ixanthus viscosus* and *Teline canariensis* predominate but the best time to see the woodland flora is in late Spring. Forest cliffs house several endemic plants such as *Bencomia caudata*, *Silene lagunensis* with white flowers and the *Sedum*-like *Monanthes anagensis* with reddish leaves.

Probably the best way to see the Anaga region is to follow the road from La Laguna to Las Mercedes and then on through the woods to Pico Ingles and the forest house at Vueltas de Taganana, then pass via the highest ridge, still following the road, to El Bailadero and down the south slopes to San Andrés and Santa Cruz. The excursion can be made very easily using local taxis or bus services.

The coasts of the Anaga region are botanically very rich. Roque de las Animas by Taganana has several local endemics including the large, white-flowered *Echium simplex*, *Lugoa revoluta*, a pale yellow *Centaurea* species, *C. tagananensis* and a tiny leguminous shrub *Teline linifolia* subspecies *teneriffae*. The coast at Bajamar and Punta del Hidalgo is equally interesting with the rare *Pterocephalus virens* on sea-cliffs and in steep valleys *Ceropegia dichotoma*, *Ruta pinnata*, *Pimpinella anagodendron* and the tiny *Monanthes praegeri*. Several succulent *Aeonium* species are common including *A. lindleyi*, *A. cuneatum* and *A. ciliatum*. On the south slopes of Anaga in the valleys of San Andrés and Igueste there are areas with many endemics. In the higher part, about 600 m, *Solanum vespertilio* with mauve flowers and *Sideritis dendrochahorra*, the tallest of the many Canarian species of this genus, are locally frequent, and several cliff plants such as *Salvia broussonetii*, *Sonchus tectifolius* and *Polycarpaea carnosa* are abundant in coastal regions.

3. Ladera de Guimar

One of the most worthwhile visits to the south of Tenerife can be made very easily from Santa Cruz. This is to the high cliffs known as the Ladera de Guimar. The old south road from the town of Guimar crosses these cliffs at about 400 m where numerous characteristic plants abound. Two species of *Sonchus*, *S. gummifer* and the fine-leaved *S. microcarpus* occur on the steepest faces with the arborescent *Crambe arborea* and pretty, pink-flowered *Senecio heritieri*. *Euphorbia atropurpurea* and *Lavatera acerifolia* can both be found on dry slopes with the peculiar succulent-leaved *Campylanthus salsoloides* and floriferous *Pterocephalus dumetorum* at its only Teneriffean locality. Cliff-crevices shelter the locally endemic *Monanthes adenoscepes* and *Micromeria teneriffae*. In the high part of the Ladera tracing the Barranco de Badajoz up towards Izana and the Cumbres de Pedro Gil it is possible to follow a water-canal close to the cliffs where the blue-pink *Echium virescens* and hairy-leaved *Greenovia aizoon* are common. Eventually the track leads to a small area of relict laurel forest with *Arbutus canariensis* and several other typical forest plants.

4. El Medano

In the extreme south of the island are several dry coastal areas with an almost Saharan flora. These include El Medano and Los Cristianos. Both areas have low dune formations with *Launaea arborescens*, the fleshy, yellowish *Zygophyllum fontanesii* and the dense dome-shaped shrub *Euphorbia balsamifera* dominant. In dry rocky areas a special community of *Lotus sessilifolius* and *Polycarpaea nivea* develops and in areas of shifting sand it is possible to find *Euphorbia paralias* which occurs in similar situations on the coasts of Britain. At El Medano the littoral zone has groves of *Tamarix canariensis* with *Heliotropium erosum* a common weed. Here the area is overshadowed by the high, red, cinder cone known as Montaña Roja where the brown-flowered *Ceropegia fusca* is common and several species such as *Kickxia urbanii* and *Herniaria canariensis* whose affinities lie with the eastern islands have their main Tenerife stations.

5. Agua Mansa

The village of Agua Mansa is situated in the Orotava Valley on the road to Las Cañadas. Below the village is a zone of farmland mainly given over to potato cultivation but above begins the pine forest which extends from about 1200 m to almost 2000 m at El Cabezón. *Pinus canariensis*, *Erica arborea* and *Myrica faya* are the most common shrubs and trees on the roadside slopes with the grey-leaved *Andryala pinnatifida* and *Micromeria herphyllimorpha* frequent in the ground-layer. The shallow barranco running from the road towards the giant cliffs of Los Organos has an interesting flora. *Monanthes brachycaulon*, *Greenovia aurea* and *Aeonium spathulatum* are common on the rocks and cliffs with the maiden-hair fern *Asplenium trichomanes* also frequent in crevices. Shrubby species include a white woolly-leaved *Sideritis* (*S. candicans*) and *Adenocarpus foliolosus*. The steep slopes up to the cliffs of Los Organos appear to have once harboured a small pocket of laurel forest. *Viburnum rigidum*, common in the Anaga forests, is found here whilst on the cliffs themselves *Laurus azorica* occurs as a shrub in large crevices. The slopes are now planted with chestnut (*Castanea sativa*) and two leguminous shrubs are common, *Teline canariensis* with yellow flowers and *Chamaecytisus proliferus* with white. Amongst them scrambles the local endemic white-flowered *Vicia scandens*. The vertical faces of Los Organos are not particularly rich in species but *Greenovia aurea* and *Echium virescens* are to be found in abundance along with a local form of *Argyranthemum canariense*, the widespread *Lobularia intermedia* and *Crambe strigosa*, a further reminder that the area was once laurel woodland.

6. Las Cañadas de Teide

The vast panorama of Las Cañadas comprises an ancient volcanic crater with the south wall more or less intact but the north and west sides destroyed by more recent surges of volcanic activity. In the centre stands the huge often snow-covered cone of Pico de Teide with its two adjacent, subordinate peaks of Montaña Blanca and Pico Viejo. The flora of Las Cañadas consists mainly of endemic plants many of them known only from this particular area of Tenerife. The strongly scented, white-flowered *Spartocytisus supranubius* and the sticky, yellow-flowered *Adenocarpus viscosus* are the most common shrubs over wide areas of pumice and volcanic debris in this almost lunar landscape. In flat areas and sheltered depressions such as Cañada de las Arenas Negras, Llano de Ucanca and Cañada Blanca *Pterocephalus lasiospermus*, *Nepeta teydea* and *Micromeria julianoides* are frequent whilst *Scrophularia glabrata* and *Tolpis webbii* can be found almost everywhere. In a number of areas, particularly near

El Portillo and the Parador Nacional del Teide three very colourful species are abundant, *Cheiranthus scoparius* with purplish flowers, *Descurainia bourgaeana* with masses of yellow flowers and *Argyranthemum teneriffae* (*A. anethifolium*) which in late Spring and Summer is covered with white flowers. At the Parador Nacional it is possible to see one of the most spectacular Canarian plants of all, the giant, red-flowered *Echium wildpretii* which can also be found sporadically along cliff walls of the crater from La Fortaleza to Ucanca. A second, very rare, *Echium* species is also found in the area near Las Arenas Negras. This is *E. auberianum* which is one of the few Canarian endemic species not growing into a woody shrub. Places such as La Fortaleza, and the southern cliffs at Montaña de Diego Hernandez, Topo de la Grieta and Boca de Tauce, have very restricted local communities with species such as *Pimpinella cumbrae*, *Senecio palmensis*, *Aeonium smithii*, *Rhamnus integrifolia* and *Silene nocteolens*. Other local rarities include yellow-flowered *Centaurea arguta*, the very attractive *Cistus osbeckifolius* with large pink flowers, the spiny *Carlina xeranthemoides* and two almost extinct species *Juniperus cedrus* (now fortunately replanted in some areas) and *Bencomia exstipulata*. The higher reaches of the Montaña Blanca and Pico de Teide support only a single species of flowering plant, a pansy-like violet, *Viola cheiranthifolia* which is not known from any other part of the world.

GRAN CANARIA

The island of Gran Canaria is roughly circular, resembling an inverted saucer with a flattened central dome. The slopes of the dome are dissected by a series of radial barrancos arising from the centre of the island and reaching down to the coasts. The largest of these valleys are in the south and west, Tirajana, Tejeda, Mogan, San Nicolas, Agaete, Fataga, those of the north and east being generally smaller and less spectacular. Subsidiary cones and volcanic monoliths are abundant, Montaña de Galdar, Caldera de Bandama and Roque Nublo being amongst the most prominent. Coastal plains are found only in the east and south and tend to be no more than a few kilometres wide. A small area near the mouth of the barranco of Fataga in the extreme south was once occupied by an area of marsh and brackish lagoons but this area has now been drained to create the tourist centre of Maspalomas. (Fig. VI.)

7. Los Tiles de Moya

The formerly quite extensive laurel woods of the north coast of Gran Canaria have, over several hundred years, been so savagely exploited that, at the present time, only a very small area remains.

Even this is in a secondary or even tertiary state of degradation but still presents the visitor to the island with the chance to see at least some typical forest endemics. There are two valleys which still shelter laurel woodland, these are Los Tiles de Moya and Barranco de la Virgen. The former, as it is easily accessible, is well worth visiting and,

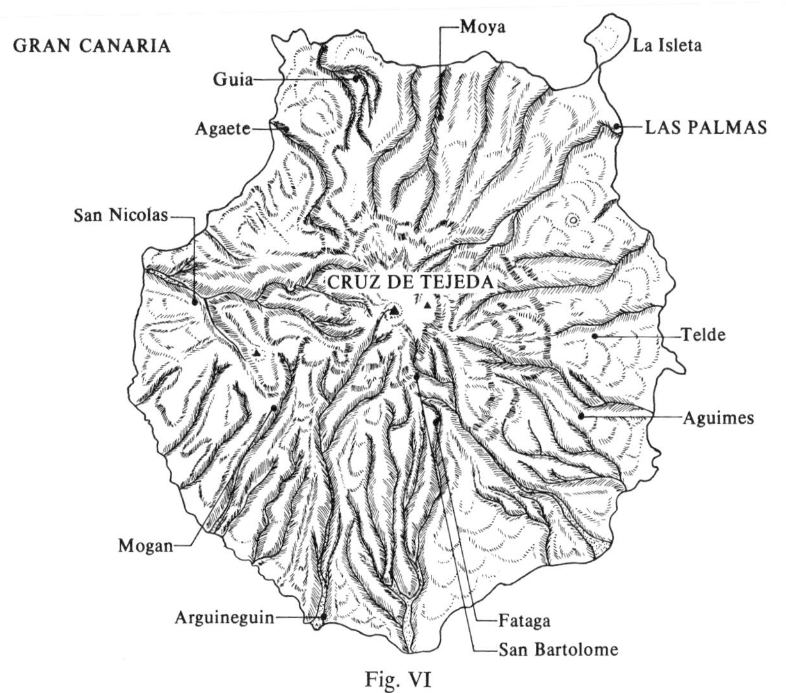

Fig. VI

at the time of writing, has just been placed under the protection of the Cabildo Insular as a nature reserve. The track to Los Tiles passes a dry stream-bed with *Salix canariensis* in abundance along the water-course. A pink or white-flowered Cineraria (*Senecio webbii*) is very common and walls and bushes are sometimes covered with *Convolvulus canariensis* notable for its attractive pale mauve flowers. The small forest area has several tree species, *Ocotea foetens* ('til'), *Ilex canariensis* and *Laurus azorica* are predominant. Amongst the trees in the herb layer *Canarina canariensis* is quite common with *Sonchus congestus*, *Semele androgyna* subsp. *gayae* with the flowers borne near the centre of the cladodes, *Hypericum canariensis* and *Echium strictum*. In the higher part of the valley it is possible to encounter several rare species such as *Bencomia caudata*, *Isoplexis chalcantha*, *Ixanthus viscosus* and *Bystropogon canariensis*.

8. Valle de Agaete, Guayedra, Andén Verde

The west side of Gran Canaria resembles the Teno region of Tenerife both in its rugged mountain scenery and in its rich endemic flora. The valley of Agaete runs from the west coast into the Tamadaba massif and has natural health-springs in the higher part. A small spur

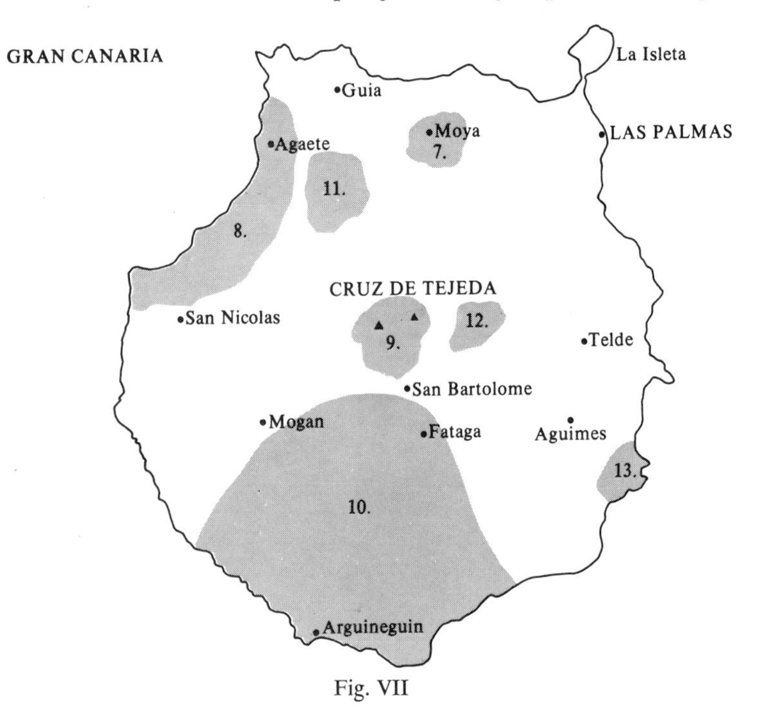

Fig. VII

arising from the main valley is known as Las Berrezales and is the site of the medicinal springs. The high cliffs beyond which rise starkly towards the high forests of Pinar de Tamadaba house many interesting plant species. The small, yellow-flowered *Descurainia artemisoides* is frequent along the side of the narrow track as well as on the cliffs which give refuge to two of the rarest and most treasured plants of the island, the palm-like *Dendriopoterium menendezii* and an almost arborescent *Centaurea*, pink-flowered *C. arbutifolia*. Also on the cliffs one can find *Crambe pritzelii* with small white flowers and an attractive yellow composite *Prenanthes pendula*. Along the bed of the valley *Dracunculus canariensis* with its cream-coloured *Arum*-like flowers abounds.

The west coast road from Agaete to San Nicolas passes along dark craggy cliffs which drop steeply to the sea. The barranco of Guayedra and its immediate area is famous for several rare plants. The lower slopes have *Tanacetum ferulaceum*, *Sonchus brachylobus* and a local form of *Argyranthemum frutescens* whereas the high, inaccessible cliff-faces are the home of one of the rarest Canarian endemics *Sventenia bupleuroides*. Continuing south past the small hamlet of Tirma towards Paso del Herreros one encounters the cliffs of Andén Verde with several local endemics, *Lotus callis-viridis* hanging from the cliffs or forming dense carpets with its bright yellow flowers, *Argyranthemum lidii* and *Lyperia canariensis* which turns up in several places on the west side of the island. A yellow-flowered umbellifer, *Seseli webbii*, has its only Gran Canaria station here and *Echium bonnetii*, *Reseda crystallina* and *Drusa glandulosa* may be found along the roadsides. On the seaward side of Andén Verde huge clumps of *Euphorbia canariensis* protect a few plants of the very rare, tall *Sonchus canariensis* and on windswept slopes it is possible to find a succulent-leaved form of the small annual crucifer *Erucastrum canariensis*.

9. Cruz de Tejeda

Cruz de Tejeda, occupying an almost central position on the island, acts as a route centre for the mountain roads from north, south, east and west but it is an area of special interest for botanists in its own right. The altitude, almost 1600 m, gives the area a montane flora with *Cheiranthus scoparius*, *Argyranthemum canariense* and the curious, houseleek-like, *Aeonium simsii* very common. A yellow-flowered broom *Teline microphylla* is very widespread in the mountain regions and a second legume *Chamaecytisus proliferus* with silvery leaves and white flowers is very common near Cruz de Tejeda. A large buttercup *Ranunculus cortusifolius* can be found in damp places along roadsides. The road to Artenara yields several botanical treasures including the very rare *Hypericum coadunatum* which occurs on wet cliffs and *Umbilicus heylandianus*. On the route to Santa Brigida and Tafira are endemics such as *Aeonium undulatum*, *A. manriqueorum*, *Orchis canariensis* and *Adenocarpus foliolosus*.

10. Caldera de Tirajana, Fataga, Mogan

The southern sector of Gran Canaria can be reached most easily from the main south road by turning beyond the airport onto the road to Aguimes and then to San Bartolome de Tirajana. A more spectacular route, however, is that from Cruz de Tejeda to San Bartolome via Paso de la Plata. Along this route *Echium decaisnei* with pyramidal inflorescences of white and pale blue flowers is common with several species of *Micromeria* and *Cistus symphytifolius*. Paso de la Plata is the classical locality for the very rare *Tanacetum ptarmaciflorum* and the more common, narrow-leaved *Echium onosmifolium*. Cliffs in this area have *Sonchus platylepis* with enormous flower-heads and the filiform-leaved *Descurainia preauxiana*. Lower down the valley towards San Bartolome the road passes dry slopes with clumps of a small, hairy-leaved *Lotus*, *L. holosericeus*. Between San Bartolome and Termisas is the small town of Santa Lucia de Tirajana where, on cliffs to the east of the town, it is possible to find a very rare mauve-flowered species *Solanum lidii* along with *Campylanthus salsoloides*, *Tanacetum ferulaceum* with much narrower leaf-lobes than at its Guayedra station, *Asteriscus stenophyllus* and a local, white-leaved form of *Salvia canariensis*. The two *Echium* species, *E. decaisnei* and *E. onosmifolium* are both common along with hybrids between them. On cliffs near Termisas one meets another rarity *Kickxia pendula*, and the more common, spiny *Carlina canariensis*. From San Bartolome it is possible to reach the south coast by way of the barranco of Fataga. This deep valley has dense palm-groves of *Phoenix canariensis* in the bed while the enormous cliffs shelter numerous interesting plants. *Prenanthes pendula* is frequent with a large-flowered *Micromeria*, *M. helianthimifolia*, *Convolvulus glandulosus* and *Ruta oreojasme*. *Limonium rumicifolium*, one of the most beautiful of Canarian plants is found here. The dry slopes have populations of *Salvia canariensis*, *Cneorum pulverulentum*, and a few plants of *Parolinia ornata*, more common further south at Arguineguin and Mogan. Two species of *Argyranthemum* are fairly common, these are the tiny-headed, fine-leaved *A. filifolium* and a strange, slender local form of *A. canariense*. In the lower part of the valley *Ceropegia fusca* occurs on very dry cliffs.

Along the south coast *Euphorbia balsamifera* covers the dry slopes with *Launaea arborescens* and *Argyranthemum filifolium*. The valley of Arguineguin is the home of large populations of *Parolinia ornata*, *Phagnalon purpurascens* and the shrubby, virgate *Convolvulus scoparius*. In the valley-bed are extensive *Tamarix* groves and on high cliffs at the head of the barranco one of the very few colonies of wild Dragon Trees to be found on Gran Canaria. In the south-west of the island is a deep, extensive valley known as the barranco of Mogan carrying a rough road via the Degollada de Tazartico to San Nicolas. The lower part of Mogan valley is dry with xerophytic scrub vegetation. A small, pink *Echium*, *E. triste* is frequent with a dense, compact *Asteriscus* species related to *A. stenophyllus* and several other rare species can be found. The higher mountains are clothed with large colonies of *Euphorbia obtusifolia* and *Echium decaisnei* with occasional pockets of rarities such as *Argyranthemum escarrei*.

11. Pinar de Tamadaba

The Tamadaba Massif, in the north-west of the island, is an area of pine-woodland dominated by *Pinus canariensis*. On entering the forest from the Artenara road the sparseness of the undergrowth is immediately striking. *Cistus symphytifolius* and *Asphodelus microcarpus* are the most common species amongst the pines and *Lotus spartioides* borders the road. Towards the centre of the forest near the Casa Forestal three species of *Micromeria* are found. The most notable is the tall shrubby *M. pineolens* with large pink flowers. *M. benthamii* and the densely white-hairy *M. lanata* are frequent while a tiny fern *Ophioglossum lusitanicum* and *Polycarpaea aristata* can be found in wet areas. On its western margin the Tamadaba massif drops steeply down to the coast with huge, vertical cliffs. The upper part of these is accessible from the forest area and several very rare species occur there. *Isoplexis isabelliana* with deep orange-red flowers is common only in this area. *Argyranthemum jacobifolium*, *Scrophularia calliantha* with the biggest flowers in the genus, *Phyllis nobila* and the rosaceous *Dendriopoterium menendezii* make the climb down well worthwhile.

12. Rincón de Tenteniguada

This area is one of the classical localities for Canarian plants. The narrow gorge and steep cliffs have a number of rare endemics such as *Echium callithyrsum* with blue flowers, *Tinguarra montana*, *Scrophularia calliantha* and *Bencomia brachystachya*. *Monanthes brachycaulon* is frequent on cliffs and *Aeonium spathulatum*, *A. undulatum* and *Aichryson porphyrogennetos* also occur in the valley. *Crambe pritzelii* and *Senecio webbii* are locally abundant.

13. Punta de Arinaga

The south-east coast of Gran Canaria has several areas of dry, windswept volcanic rubble and sand bordering the sea. These include Punta de Melenara and Arinaga. The latter is accessible by road from the south 'autopista' and a unique plant community occurs there with most plants forming dense prostrate mats or hummocks. The spiny *Convolvulus caput-medusae* and *Atractylis preauxiana* are local rarities with two yellow-flowered plants also present, *Lotus leptophyllus* and *Kickxia urbanii*. Typical Canarian halophytes include *Polycarpaea nivea*, *Gymnocarpos decander*, *Limonium pectinatum* and *Frankenia laevis*.

One of the most important features of the island of Gran Canaria is the *Jardin Canario Botanical Garden* at Tafira. Here a large collection of Canarian endemic species is maintained in almost natural conditions. Making full use of natural cliffs most of the Canarian species of *Sonchus*, *Centaurea* and *Limonium* have been collected together and set in magnificent surroundings. Particularly interesting are a cliff with a large colony of *Sventenia bupleuroides* and *Crambe laevigata* with *Isoplexis* and *Scrophularia* close by, a splendid monument to famous Canarian botanists such as Webb, Berthelot, Pitard and Masferrer with a long avenue dominated by giant specimens of *Bencomia*, *Marcetella* and *Dendriopoterium*, and a large area of laurel forest in the lower part of the garden with groves of *Arbutus canariensis* and the laurel species. On the walls of the administration building there is a colony of *Lotus berthelotii* with deep red flowers, *Convolvulus lopez-socasii* and *Semele androgyna*. Any botanist visiting the islands should endeavour to see the garden in Spring when most of the endemics are in full flower. It can be seen on a day excursion from Tenerife by air but spend a couple of days there if possible.

LA PALMA

La Palma is a pear-shaped, mountainous island with a large central crater, La Gran Caldera de Tabouriente, the outer rim of which forms the highest point of the island, 2483 m at Roque de los Muchachos. The northern half of the island consists of the outer slopes of the Caldera and is dissected by deep valleys and spurs with high coastal cliffs. The southern part has a narrow central ridge, Cumbre Vieja, but is without the deeply dissected valleys of the north. Many areas of the southern part of the island show signs of recent volcanic activity with cinder cones and lava streams particularly in the extreme south at Fuencaliente where the Volcan de Teneguia erupted as recently as 1971 and along the west coast above Puerto Naos where an eruption took place in 1949. (Fig. VIII.)

14. Cubo de la Galga, Los Tilos, Barlovento

The north-east coast of La Palma is still extensively forested though recent exploitation has cleared large areas at La Galga and Barlovento and in the last year or so at Cumbre Nueva. Cubo de la Galga is reached by a narrow track along a water canal from the small village of La Galga. It passes a long forest cliff where *Silene pogonocalyx*, *Aeonium palmense*, *Cryptotaenia elegans* and *Senecio papyraceus* are abundant. The forest area is dominated by old trees of *Ocotea foetens*. A second track, now extended to a forest road climbs above the village into the mountains. Several forest endemics can be found here. *Echium pininana*, reduced to a few examples hides away deep in the heart of the woodland while *Gonospermum canariense* with its bright yellow flowers stands out as brightly as its local name

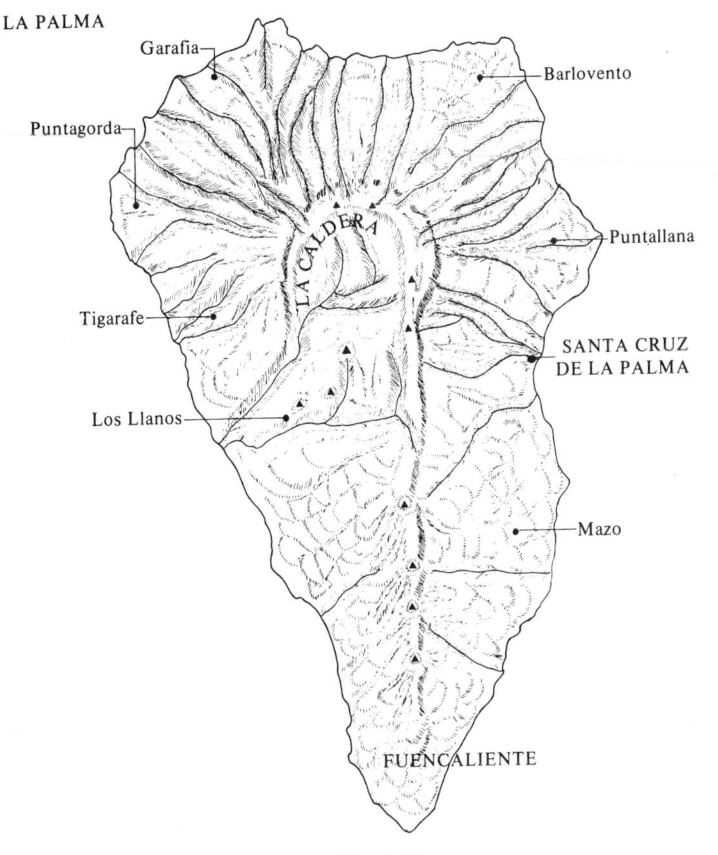

Fig. VIII

'faro' (lighthouse) suggests. Further north, above the town of Los Sauces is the barranco of Los Tilos reached by a motorable track. The main valley supports dense forest on both sides with *Woodwardia radicans* in shady places. A ruined hotel with a supply of drinking water and, on occasions, a small bar makes a good base for exploring the forest. High cliffs have several La Palma endemics such as *Polycarpaea smithii*, *Crambe gigantea*, *Argyranthemum webbii* and the gentian-blue *Echium webbii*. Several shrubs can be found at the edges of the woodland, *Sonchus palmensis* with a huge inflorescence of yellow flowers, *Teline stenopetala*, the Canarian ivy, *Hedera canariensis* and its fellow liana *Semele androgyna* are abundant and the most frequent tree is *Ocotea foetens* with *Laurus azorica* also common. A dark, wet gorge leads off the main valley, it is aptly called Barranco del Agua. Large examples of *Crambe gigantea* and a very rare, robust form of *Bencomia sphaerocarpa* can be found with a little searching but the latter now appears to be almost extinct at least in this area. *Euphorbia mellifera*, a tree species, used to occur in this locality but the species now seems to be confined to Cumbre Nueva where it is endangered by forest clearing.

Further north, on the north side of the island are more rich forests. These can be found in the mountains above Barlovento. *Echium pininana*, *Geranium canariense*, *Scrophularia langeana* and *Ixanthus viscosus* can be located in this area but the zone is still relatively poorly known and further along the coast a very rare plant *Lactucosonchus webbii* was recently rediscovered after over 100 years.

15. Fuencaliente

Situated near the southern tip of La Palma, the village of Fuencaliente lies between pine forest and coastal zones. Between the town and the ultimate south point of the island is a vast area of recent volcanic origin with lava rock, cinder fields, black basaltic sand and ash cones. The land is famous for its vineyards and rich, sweet wines. A road winds down from Fuencaliente to the south coast passing dry slopes rich in Canarian plants. *Sonchus hierrense* and a white-flowered form of *Aeonium ciliatum* are common on walls whilst the dry field-edges have large colonies of a white-stemmed, stick-like *Ceropegia*, *C. hians* and the curious composite *Phagnalon umbelliforme*. *Lotus hillebrandii*, *Argyranthemum haouarytheum* and white-flowered *Echium brevirame* all have local, dwarf forms in the windswept lava and the tiny prostrate plants of *Euphorbia balsamifera* found in the coastal regions seem to survive in the most inhospitable conditions. In the area between Fuencaliente and a large cone known as Volcan de San Antonio is an outcrop of Tertiary basalt called Roque de Teneguia which is the home of a rare purple-flowered *Centaurea*, *C. junoniana*. The dry desolate area around the rock is poor in plant species with just *Echium brevirame*, *Artemisia canariensis* and a few plants of *Lotus hillebrandii* and *Kickxia spartioides*. The rock itself, however, has a very interesting series of Guanche inscriptions in the form of strange carved spirals and circles and it is hoped that the construction of a nearby water canal will not endanger either the *Centaurea* or the inscriptions.

16. La Cumbrecita

A road from El Paso on the west side of Cumbre Nueva runs into the Gran Caldera to a high vantage point known as La Cumbrecita from which spectacular views of the interior of the giant crater can be seen. The mountain is densely forested with *Pinus canariensis* and the road passes by colonies of a tall form of *Argyrathemum haouarytheum* and *Sideritis dendrochahorra*. Cliffs on either side of the small Casa Forestal situated in the saddle area of La Cumbrecita have an interesting flora but require a fairly energetic, steep climb to reach them. *Teline linifolia* with large, silvery leaves, *Aichryson bollei* and *Greenovia aurea* are common on the cliffs *with Echium webbii*, *Plantago arborescens* and *Pimpinella dendrotragium* which are rather more rare. The slopes below the cliffs on the east side are the only known habitat in the Canaries for *Orchis mascula*, a widespread Mediterranean and European species of pink-flowered orchid.

17. Barranco de las Angustias

This large valley leads out from the Caldera de Tabouriente on the

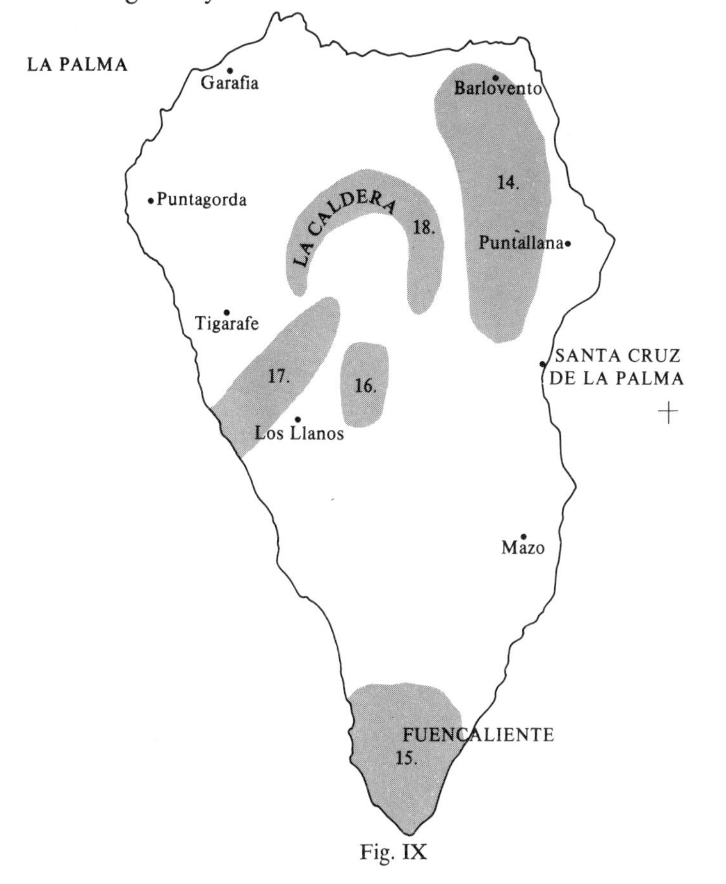

Fig. IX

west side. A track goes along the south side of the valley via a wide water-canal from the town of Los Llanos near the mouth of the barranco right into the centre of the Caldera. If one follows this track, it has to be on foot, along the canal one passes through a very rich area from a botanist's point of view. Of special interest is the magnificient red-flowered *Aeonium nobile* with its almost orbicular, fleshy, yellowish leaves which is only found in a few places in the northern half of the island. The track passes close to cliffs supporting colonies of *Senecio palmensis*, a pendulous plant with yellow flowers, *Dorycnium eriophthalmum* with white flowers and reddish-brown pods, *Pimpinella dendrotragium*, here quite common, and a very robust form of the Canarian cocksfoot *Dactylis smithii*. On the slopes below the track is a large population of *Echium* hybrids with a variety of pink, blue and white flowers, the parent species are *E. brevirame* and *E. webbii* and the hybrid has been given the name *E. bond-spraguei*. On the sides of the canal several shrubs are quite common including a tall, shrubby, white-flowered *Convolvulus*, *C. floridus*, *Spartocytisus filipes* a white broom-like plant with black pods and the very narrow-leaved *Bupleurum salicifolium*. At the water's edge *Aeonium palmense* abounds and towards the end of the canal is a small colony of *Convolvulus fruticulosus*, a species otherwise confined to Tenerife.

18. Roque de los Muchachos, Cumbres de Garafia

The highest mountains of La Palma are those formed by the rim of the Caldera. Several peaks reach over 2300 m, as high as Las Cañadas on Tenerife, and so there is a well developed subalpine flora like that of Tenerife. Several species such as *Nepeta teydea*, *Adenocarpus viscosus* and *Juniperus cedrus* occur on both islands while a number of local endemics may also be found including *Echium gentianoides*, *Viola palmensis*, *Cerastium sventenii*, *Tolpis calderae* and *Pterocephalus porphyranthus*. Several tracks from the north of the island lead up to the high peaks of Topo Alto de Los Corralejos, Pico del Cedro, Roque de los Muchachos and Cueva de la Tamagantera but in all cases the journey requires a very long day.

GOMERA

The island of La Gomera closely resembles a miniature Gran Canaria. It rises to almost 1450 m in the centre and the flanks of its flattish dome are dissected by 38 deep, radial barrancos which open narrowly to the sea between precipitous cliffs. The south is hot and dry but there is no coastal plain and the north is reminiscent of the north coast of Tenerife. (Fig. X.)

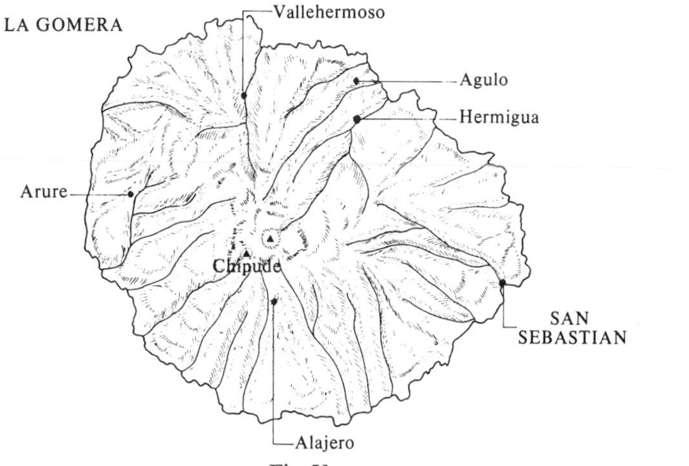

Fig. X

19. Barranco de la Villa

The largest valley on the island runs from the centre to the east coast and contains, at its mouth, the capital town San Sebastian. A road runs along the north side linking San Sebastian with the northern villages of Hermigua and Vallehermoso. On the roadside cliffs are numerous endemic plants. A woolly-leaved *Sideritis* with pendulous inflorescences (*S. gomeraea*) is abundant while a second species *S. lotsyi* is rather more rare. *Dicheranthus plocamoides*, also found on Tenerife at Teno, is frequent on rocks with the solitary rosettes of *Greenovia diplocycla*. On the dry slopes a small-flowered form of *Argyranthemum frutescens* intermingles with the succulent *Senecio kleinia* and several *Euphorbia* species while it is possible to find at least four species of *Aeonium* amongst dry rocks, *A. subplanum* with flat green rosettes, *A. decorum* with reddish leaves and two grey-leaved species *A. castello-paivae* and *A. gomerense*. In the nearby Barranco de la Laja is yet another *Aeonium*, *A. saundersii*. In Spring both cliffs and slopes may be golden yellow with the flowers of the Gomeran endemic *Lotus emeroides*.

20. Roque de Agando

This basaltic monolith situated at the head of Barranco de la Laja is best reached on foot from Tagamiche. The track passes through *Erica arborea* heath and descends the cliffs of Riscos de Tagamiche where three local endemics can be found, a tall, blue *Echium*, *E. acanthocarpum*, *Argyranthemum callichrysum* with yellow flowers and *Crambe gomeraea*. In the area at the base of Roque Agando are large groups of the tall, yellow-flowered *Aeonium rubrolineatum* and a very rare *Sideritis* species, probably *S. marmorea* Bolle. Wet rocky areas harbour colonies of the orchid *Neotinea intacta* and two yellow leguminous shrubs are common, *Adenocarpus foliolosus* with very narrow leaflets and long petioles and *Teline linifolia* in yet another of its many forms, this time with small, oblong leaves. One can return to San Sebastian by taking a relatively easy track down the Barranco of la Laja.

21. Riscos de Agulo, Roque Cano de Vallehermoso

These two areas of basalt cliff are considered together because of the similarity of their floras. Both are relatively easily reached from the north road of the island though quite a vigorous climb is needed to reach the base of Roque Cano. The Agulo cliffs have two endemic *Sonchus* species, *S. gonzalez-padronii* which becomes quite frequent on rocks and cliffs further to the west and *S. regis-jubae* which appears to be restricted to the north between Agulo and Vallehermoso. *Silene bourgaei* and *Limonium brassicifolium*, however, appear to be more or less confined to Agulo. *Convolvulus subauriculatus* occurs at both localities while *Euphorbia bravoana* is represented at Agulo by very few plants, being more common in the Barranco de Majona to the east.

Roque Cano, in the barranco of Vallehermoso, has sheer cliff-faces on all sides. The slopes below the rock are covered with small shrubs of *Juniperus phoenicea* and with the curious endemic grass *Brachypodium arbuscula*. The faces of the south side of Roque Cano have several interesting plants on them, notably *Sonchus ortunoi*, *Senecio hermosae*, a local endemic, *Dicheranthus plocamoides* and a small sticky *Aeonium*, *A. viscatum*. Several shrubs occur on the slopes of the north side, *Teline linifolia* with large leaves, *Sonchus regis-jubae* and the blue-flowered *Globularia salicina*. The coastal region at Puerto de Vallehermoso has several interesting communities with *Euphorbia aphylla* and *E. balsamifera* predominant.

22. Forests of El Cedro

The central region of Gomera has some of the finest laurel woodland in the Canaries. The forests of El Cedro at the head of the Hermigua Valley is probably the richest of all. *Laurus azorica*, *Ilex canariensis*, *Myrica faya* and *Salix canariensis* are the dominant trees and streams, wet banks and cliffs with lush vegetation are very common here whereas in most other Canarian forests exploitation of ground-water has resulted in the drying out of the ground-layer and the almost complete absence of running water. *Hypericum grandifolium*, *Cedronella canariensis*, *Gesnouinia arborea* and *Scrophularia langeana* form the main part of the shrub layer with, in some areas,

Teline stenopetala and *Bystropogon canariensis*. The ferns *Woodwardia radicans*, *Athyrium umbrosum* and *Asplenium onopteris* are abundant along stream-sides with *Aichryson punctatum* which often grows here with its 'feet in water'. Very old laurel trees with enormous trunks are common in the forest and in Summer the ground is often covered with a carpet of reddish-brown leaves of Viñatigo (*Persea indica*). El Cedro can easily be reached from San Sebastian or Hermigua and makes a wonderful days' excursion for a naturalist.

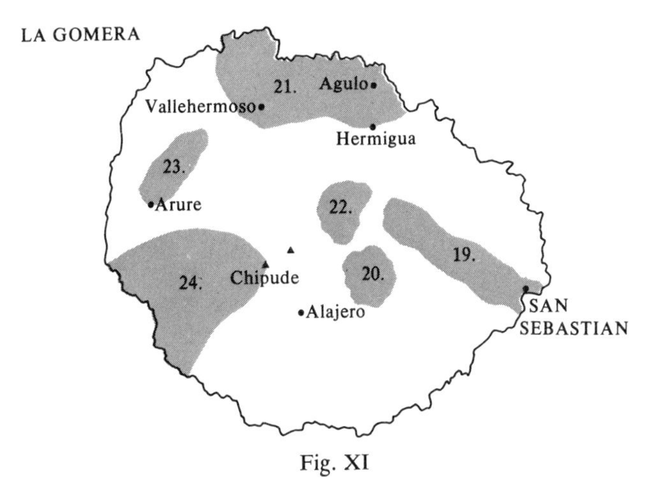

Fig. XI

23. Chorros de Epina, Arure

From Vallehermoso the north coast road goes on to Vallegranrey on the west side of the island. About half an hour's drive along the road is a small track which leads to a spring known as the Chorros de Epina. The spring is on the lower edge of a rich forest region below some of the most interesting cliffs on Gomera. The tall nettle-like shrub *Gesnouinia arborea* and a shrubby heather *Erica scoparia* can be found at the edge of the track and near the stream are a few plants of a very rare species *Euphorbia lambii* with large yellowish bracts surrounding the flowers. Beyond the Chorros a rough track passes the base of a steep slope below a series of very high cliffs. If one climbs the slope to the base of the cliffs and then follows the cliffs towards the west coast some of the rarest botanical treasures of the island can be found often in abundance. *Sideritis nutans* with a nodding spike of whitish flowers, *Pimpinella junionae*, *Sonchus gonzalez-padronii* and *S. filifolius* can be found rooted in cliff crevices with a strange fern with kidney-shaped leaves, *Adiantum reniforme*. *Aeonium subplanum* with its flat rosettes is common and in the higher parts of the cliffs

Aeonium rubrolineatum, a tall shrub with yellow flowers, is locally frequent. Two other succulent species, *Greenovia diplocycla* and *Monanthes laxiflora* cover areas of bare rock and in shady places below the cliffs a white-flowered *Senecio*, *S. appendiculatus* is abundant.

Returning to the Vallegranrey road which winds on its way towards the village of Arure one passes through an area of forest where *Geranium canariense*, *Hedera canariensis* and *Andryala pinnatifida* abound but where, unfortunately many of the larger trees were cut down a number of years ago to feed a small charcoal manufacturing industry. This is an area of mountain mists and many trees have strange green lichens hanging from the branches and trunks covered with moss especially on their north sides.

24. Vallegranrey, Barranco de Argaga, Chipude

The area in the southwest of Gomera with its deep valleys and cliffs is rich in endemic plants but these are often very difficult to find. The deep barranco of Vallegranrey has very high cliffs which are almost inaccessible and largely unexplored. On the cliffs along the road several species are common, *Descurainia millefolia*, a large form of *Aeonium decorum* with deep red leaves, *Sonchus filifolius* with fine leaf-lobes and small yellow flowers and, in a few places, *Sideritis nutans* with strongly scented leaves. In the bed of the valley there are large groves of palms and tamarisk. To the south of Vallegranrey and accessible only from the coast there is a small valley known as Barranco de Argaga. In the dry coastal area at the mouth of the valley there are small populations of *Echium triste* and on the sea-cliffs the very rare *Limonium dendroides*. Higher up the valley one can find a second very local endemic, the pink-flowered *Parolinia schizogynoides* amongst dry scrub vegetation of *Lavandula pinnata*, *Messerschmidia fruticosa*, *Kickxia scoparia* and *Micromeria densiflora*.

Further to the south is the mountain village of Chipude dominated by a huge table-mountain known as the Fortaleza where several Gomera endemics occur. A visit is well worthwhile. It is almost impossible to get further away from tourism, hotels and concrete anywhere in the Canaries. The life-style of the local people has hardly changed this Century and it is possible to buy small articles of pottery made in exactly the same manner as that of the original inhabitants of the islands, the Guanches. The cliffs of the Fortaleza are rich in plants such as *Sideritis lotsyi*, *Aeonium urbicum*, *Bystropogon plumosus*, *Pimpinella junionae*, *Paronychia gomerensis* and *Crambe gomeraea*.

HIERRO

This almost semicircular island is possibly a fragment of an old volcano. It consists of a high, flattish table-land in the centre with steep cliffs on all sides reaching their maximum development in the huge basin of El Golfo on the west side. Hierro is the smallest island in the Canaries group but reaches over 1500 m at its highest point. (Fig. XII.)

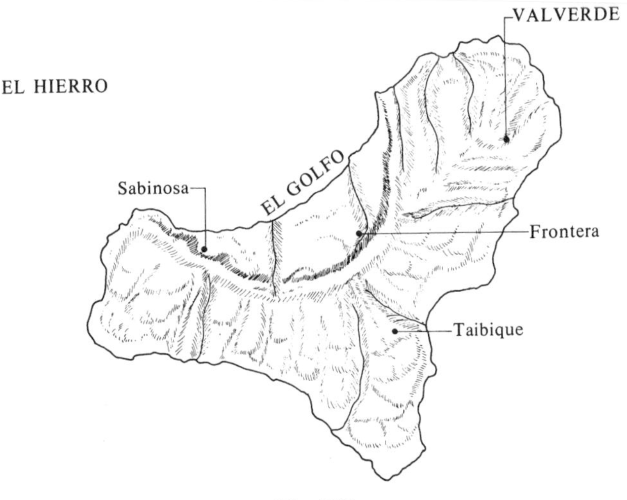

Fig. XII

25. El Golfo

The upper region of El Golfo consists of steep cliffs, densely clothed with pine and laurel forest. There is a track from Jinamar to Frontera which passes through the forest and makes an interesting excursion. A particularly rich area surrounds a spring known as Fuente de Tinco where several rare endemics are located. *Bencomia sphaerocarpa*, an unusually large form of *Crambe strigosa* and small groups of *Sideritis canariensis* can be found but are rather rare. On the path from Jinamar the forest is dense and often very wet. *Echium strictum* with deep blue flowers and a local form of *Aeonium holochrysum* are common. *Sonchus hierrensis* with large yellow heads and *Teline stenopetala* both occur on the forest cliffs with *Tinguarra montana* and on the forest floor *Geranium canariense* and *Myosotis latifolia*, here often with pink flowers, abound. *Visnea mocanera*, locally called 'mocan' is a dominant tree and in the lower part of the forest tree-like examples of *Euphorbia regis-jubae* can be found. The lower zone to the north west of Frontera has several areas of interest. The cliffs at Las Casitas have a local endemic *Sonchus gandogeri* and the pink *Echium hierrense*. Walls are covered with *Aichryson parla-*

Fig. XIII

torei with golden flowers and *Argyranthemum hierrense* which has white, daisy-like flowers, is common. Between Frontera and the town of Sabinosa the roadside cliffs are covered with *Senecio murrayi*, sometimes with pink and sometimes with white flowers, and a large cabbage-like form of *Aeonium palmense*. Rarities on these cliffs include *Silene sabinosae* with pink flowers and *Centaurea durannii*.

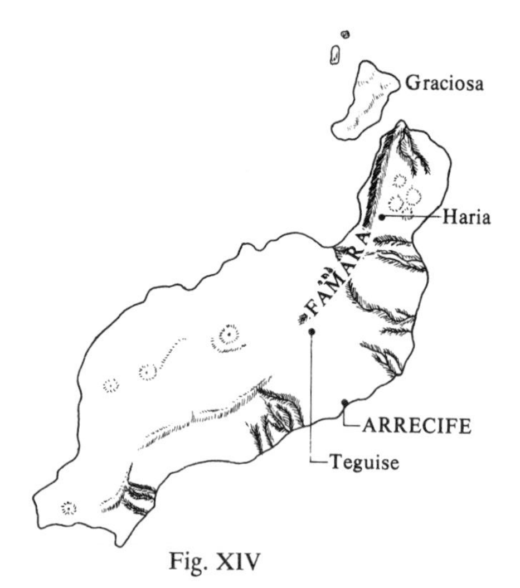

Fig. XIV

LANZAROTE

The island of Lanzarote is less mountainous than the western islands reaching only about 700 m in the north part which is dominated by the ridge of Famara. South of Famara is an area of sandy plain with areas of dunes and cinder cones. From the centre to the extreme south the island is composed of a low mountain area with volcanic peaks and craters. In the extreme south-east and south there is a narrow coastal plain of sand and lava with occasional cones and cinder heaps. The coasts in this area are steep cliffs. (Fig. XIV.)

26. Riscos de Famara

Virtually all the endemic plants of Lanzarote are concentrated in the mountain region of Famara centred on the town of Haria. The base of the cliffs reaching down to the Playa de Famara is now explorable as a road has recently been built. Near the sea are small colonies of *Pulicaria canariensis* and *Asteriscus schultzii*, two rare species found only on Lanzarote and Fuerteventura. The cliffs themselves have a local species of *Reichardia*, *R. famarae* and a succulent-leaved endemic species of *Kickxia*. *Aichryson tortuosum* is common in crevices and in the upper parts two *Aeonium* species *A. lancerottense* and *A. balsamiferum* can be found, the former with pink and the latter with yellow flowers. The high cliffs shelter a spectacular *Limonium*, *L. bourgaei* and the very rare *Echium decaisnei* subsp. *purpuriense*

Fig. XV

and one of the few yellow-flowered *Argyranthemum* species *A. ochroleucum* can also be found. Two endemic species of *Helichrysum* and several grasses including *Dactylis smithii* also occur.

On the top of the Famara cliffs the peaks known as Peñitas de Chache also have a number of these rare endemics. *Ferula lancerottensis*, a giant plant with yellow flowers is frequent while a small *Sedum* creeps over rockfaces intermingled with the local *Thymus origanoides*. *Lavandula pinnata* is represented by a particularly broad-leaved form and an annual *Echium*, *E. pitardii* is abundant.

The town of Haria is set between the hills in an oasis of palms, *Phoenix canariensis*. Beyond the town, to the north, the coastal cliffs overlook the islet of Graciosa from a cliff-top known as Mirador del Rio which is the classical locality for a second species of *Limonium*, *L. puberulum*.

FUERTEVENTURA

The northern part of Fuerteventura consists of extensive hilly plains with deep depressions. There are extensive areas of sand-dunes on the north coast. The central and southern areas as far south as Gran Tarajal consist of a central plain surrounded by isolated hills to the east and a line of low hills to the west. The area south of Gran Tarajal forms a narrow peninsula connected to the main area of the island by a narrow sandy isthmus. The peninsula of Jandia is an old, high volcanic ridge with steep slopes on either side dropping abruptly, especially on the north side, to long sandy beaches. (Fig. XVI.)

Fig. XVI

27. La Oliva, North Coast Dunes

The town of La Oliva is set at the foot of a large volcanic cone of black ash and cinder. Amongst the rock and lava of the area are several interesting plants, a small blue *Echium*, *E. bonnetii*, a scrambling, twining *Kickxia* with bright, yellow flowers (*K. heterophylla*) and a small umbelliferous plant *Ruthea herbanica*. The succulent shrub *Senecio kleinia* is common with spiny *Asparagus pastorianus* and amongst the stones a real treasure, the small, square-stemmed succulent *Caralluma burchardii*.

The north coast of the island is covered by an extensive dune formation from Tostón in the west to Corralejo in the east. Many halophyte species occur in the region including *Traganum moquinii*. *Lotus lancerottensis* and *Polycarpaea nivea* are common with the yellow *Reseda crystallina* and several *Beta* species. Near Corralejo the dunes are the classical locality for *Androcymbium psammophilum*, a small, white-flowered *Crocus*-like plant.

28. Jandia, Pico de la Zarza

The southern peninsula is the highest part of the island and also the most botanically rich. South of the village of Morro Jable the coastal

Fig. XVII

region is very desolate but is the home of two very rare plants, the cactus-like *Euphorbia handiensis* and an extremely interesting species *Pulicaria burchardii* which is known only from a small area of a few square yards with about six plants. The high peaks such as Pico de la

Zarza are rather tedious to climb but the summits shelter a wealth of rarities such as *Argyranthemum winteri*, *Echium handiense* with blue flowers, *Bupleurum handiense*, *Sideritis massoniana* and several species which can also be found at Famara on Lanzarote. The extremely attractive *Asteriscus sericeus* is common on the summit as is a small pink-flowered chickweed-like plant *Minuartia platyphylla*.

Conservation

MAN'S INFLUENCE ON THE FLORA

Inevitably the flora and vegetation of small islands are extremely vulnerable to the pressures of Man's progress. In the Canary Islands large areas have been under cultivation for several hundred years with the earliest major crop sugar cane being replaced successively by vines, ice plant (*Mesembryanthemum crystallinum*) for the extraction of soda, cochineal cactus, tomatoes and potatoes and bananas. At present the declining banana industry is slowly giving way to the production of flowers such as *Strelitzia*, carnations and chrysanthemums and market garden products such as cucumbers and several exotic fruits on a very large scale.

The forest zones have long been exploited for timber but never so rapidly or so intensively as at the present time when power-saws and good forest roads make cutting and transport simple. Many areas in the cloud forest regions are now turned over to potato production and a large acreage of woodland has been cleared for this purpose. Mountain streams, springs and underground water supplies have all been tapped to provide for the needs of both the agricultural and tourist industries and in some areas water has become a scarce and valuable commodity as the recent construction of several sea-water distillation plants demonstrates.

The effect on the flora and vegetation has been considerable. Already a number of endemic plants appear to be extinct and many more are endangered but the real danger for the immediate future seems to be that whole plant communities and unique vegetation formations can easily be destroyed if positive steps are not taken to conserve them.

THE NEED FOR CONSERVATION

The rapidly growing, international, conservation and ecology movement with its pleas for consideration of the needs of future generations presents a strong moral case for the conservation of nature and natural resources. The forest zones of the Canaries are the primary link in the water supply chain, the forests condense moisture from the Trade Wind clouds as anyone who has been to Agua Mansa on a wet day will testify. The steep slopes of the islands result in a rapid run-off of surface water and when the dense vegetation is cleared soil erosion rapidly follows as can be seen in Sierra Anaga and parts of northern Gran Canaria. From this fact alone there is a strong case for preserving the forest regions.

As we hope this small book shows, the plants of the Canaries offer a great deal to visitors and local inhabitants alike. They can provide considerable pleasure for tourists travelling round any of the islands and are of tremendous scientific importance.

Several areas have already been designated as nature reserves, Los Tiles de Moya and parts of Cuesta de Silva on Gran Canaria, or national parks, Las Cañadas de Teide on Tenerife, but there is an urgent need for many more of these. It is almost impossible to conserve individual species except on a short term basis and collections of endemic plants such as that at the Jardin Canario are probably the best means of achieving this but whole communities such as the laurel forests of Anaga, El Cedro, northern La Palma and El Golfo on Hierro or the rich coastal vegetation of Punta de Teno, Andén Verde or Fuencaliente and the cliff vegetation of many areas could still be conserved if steps can be taken in time. Conservation, however, cannot deprive local people of their livelihood and can only succeed by means of a policy of education. We hope very much that this book will play a part in the latter process by bringing the valuable assets of the flora of the Canaries to general notice.

Many of the plants described as being rare are endangered species and we recommend them only as subjects for photography and not collection. In many cases even the collection of seed can help to exterminate a rare species and many of them cannot be grown satisfactorily away from their natural habitats. Collection of dried specimens should always be done with thought and consideration for the well-being of the local colony of the species and whole plants should not be taken unless the group of individuals is a large one. Most of the Canarian succulents grow well from small cuttings so there is no need to take whole plants for propagation.

We would strongly urge anyone who benefits from the use of this book to support local conservation as much as possible and also to enjoy their exploration of the Canary Islands.

Some Suggestions for Further Reading

Many of the older guide-books and accounts of travel in the Canary Islands make fascinating reading particularly for anyone who knows the islands well. Of the modern guides those published by *Planeta* for Tenerife and Gran Canaria are amongst the best. We recommend the following books to anyone wishing to find out more about the Canaries. Those marked with an asterisk are of special interest.

HISTORICAL AND GENERAL

Bannerman, D. A., 1922.* THE CANARY ISLANDS, THEIR HISTORY, NATURAL HISTORY AND SCENERY. Gurney & Jackson.

Du Cane, E. & F., 1911. THE CANARY ISLANDS. A. & C. Black.

Edwardes, C., 1888.* RIDES AND STUDIES IN THE CANARY ISLANDS. T. Fisher Unwin.

Espinosa, A. de, 1907. THE GUANCHES OF TENERIFE WITH THE SPANISH CONQUEST AND SETTLEMENT. *Hakluyt Soc.* ser. 2, **21.**

Salmer Brown A., 1932.* BROWN'S MADEIRA, CANARY ISLANDS AND AZORES. 14th Edition. Simpkin, Marshall, Hamilton Kent & Co., Ltd.

Stone, O. M., 1889. TENERIFE AND ITS SIX SATELLITES.

BOTANICAL

Bramwell, D., 1971. The Vegetation of Punta de Teno, Tenerife. CUADERNOS DE BOTANICA CANARIA **11:** 4–37.

Ceballos L. & Ortuno F., 1951* VEGETACION Y FLORA FORESTAL DE LAS CANARIAS OCCIDENTALES. Inst. Forestal, Madrid.

Kunkel G., 1969. ARBOLES EXÓTICOS 1. Cabildo Insular, Las Palmas.

Lid J., 1968. CONTRIBUTIONS TO THE FLORA OF THE CANARY ISLANDS. Norske Vidensk.-Acad., Oslo.

Line Drawings

Fig. 1a. Notholaena marantae. 1b. Woodwardia radicans.
1c. Asplenium hemionitis.

Fig. 2. Salix canariensis.

Fig. 3. Myrica faya.

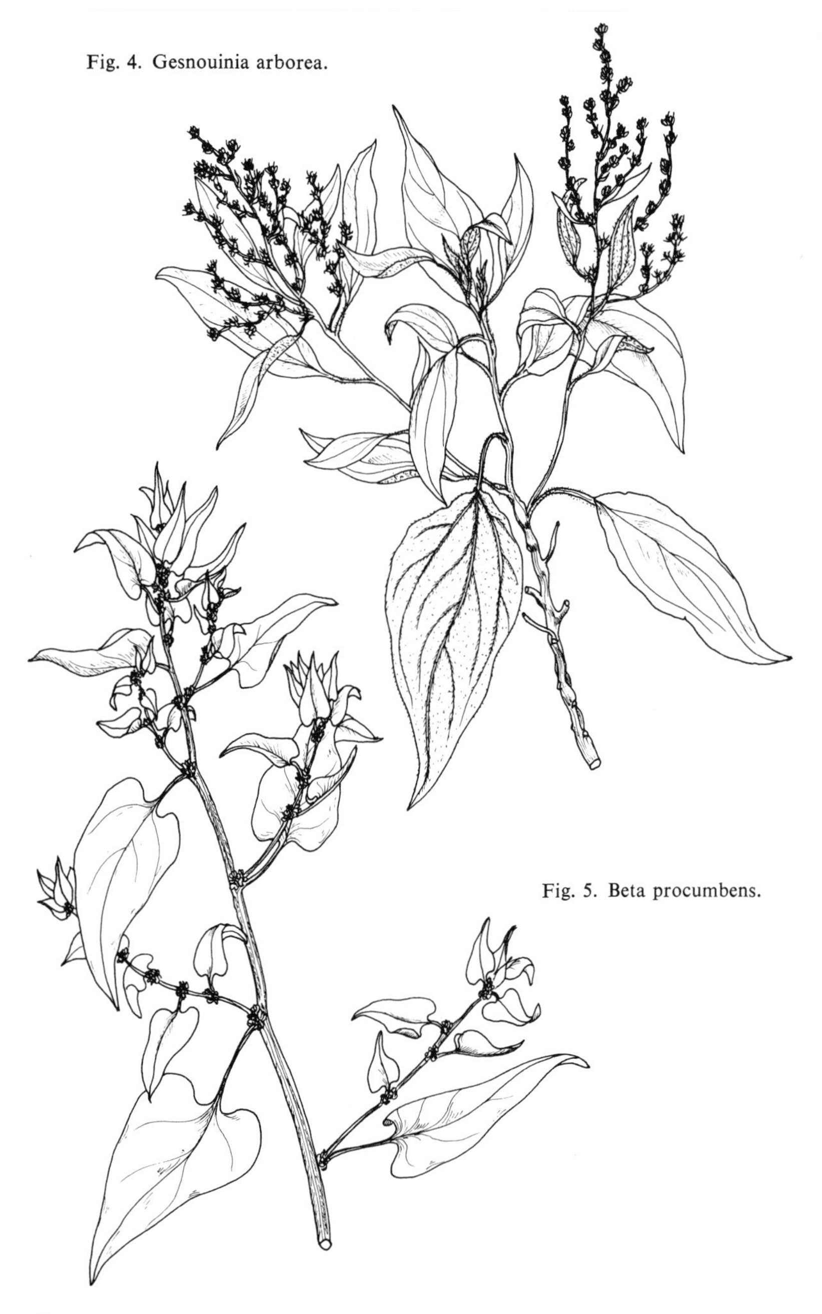

Fig. 4. Gesnouinia arborea.

Fig. 5. Beta procumbens.

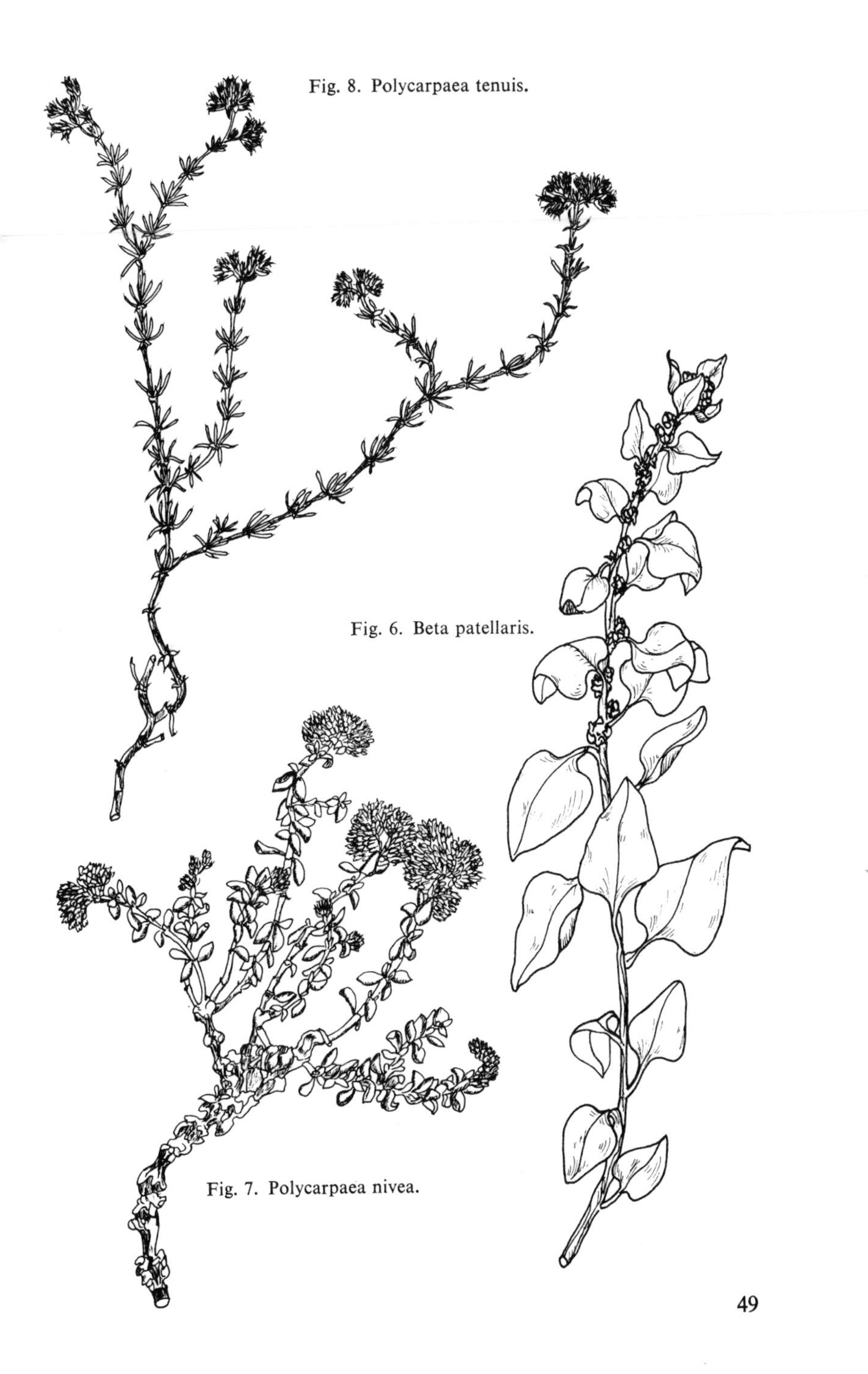

Fig. 8. Polycarpaea tenuis.

Fig. 6. Beta patellaris.

Fig. 7. Polycarpaea nivea.

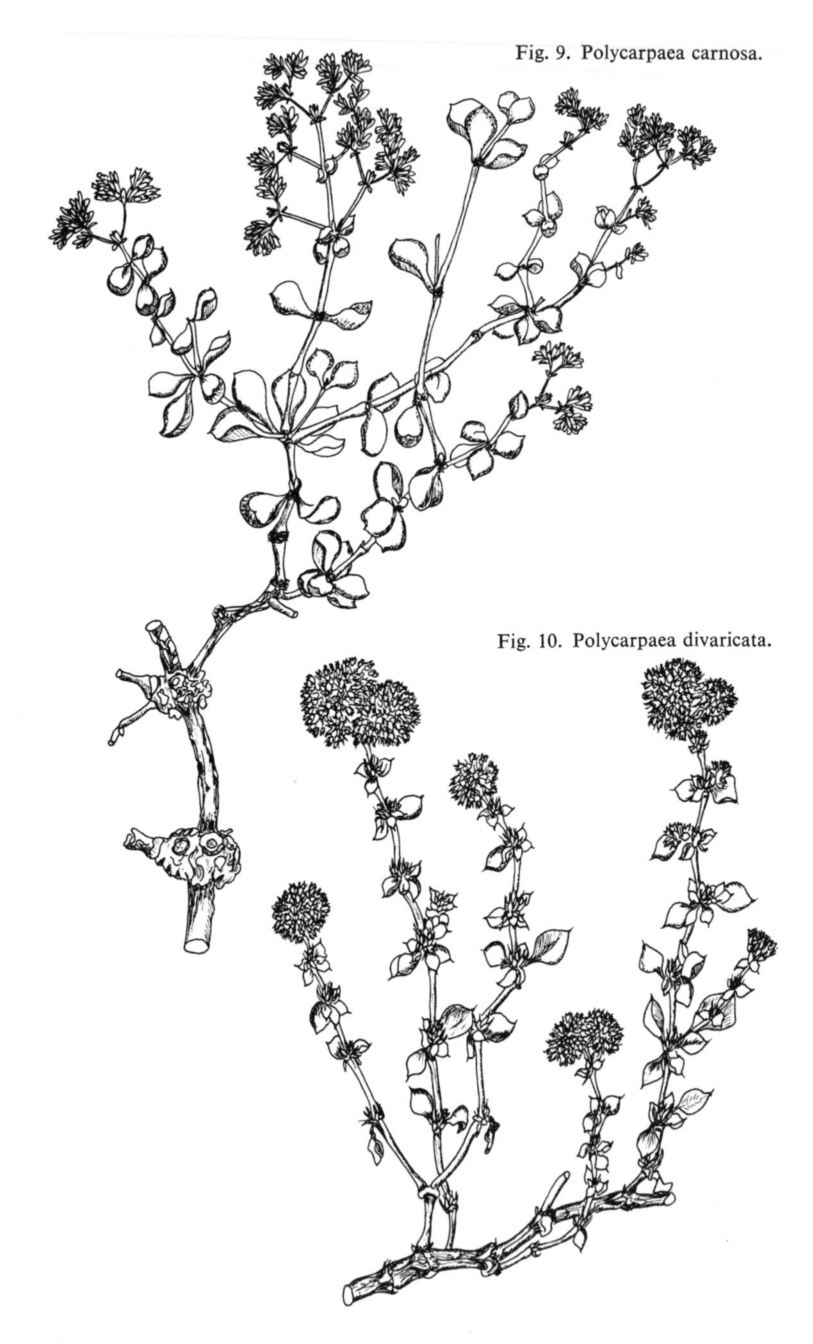

Fig. 9. Polycarpaea carnosa.

Fig. 10. Polycarpaea divaricata.

Fig. 12. Paronychia canariensis.

Fig. 11. Silene nocteolens.

Fig. 13. Ranunculus cortusifolius.

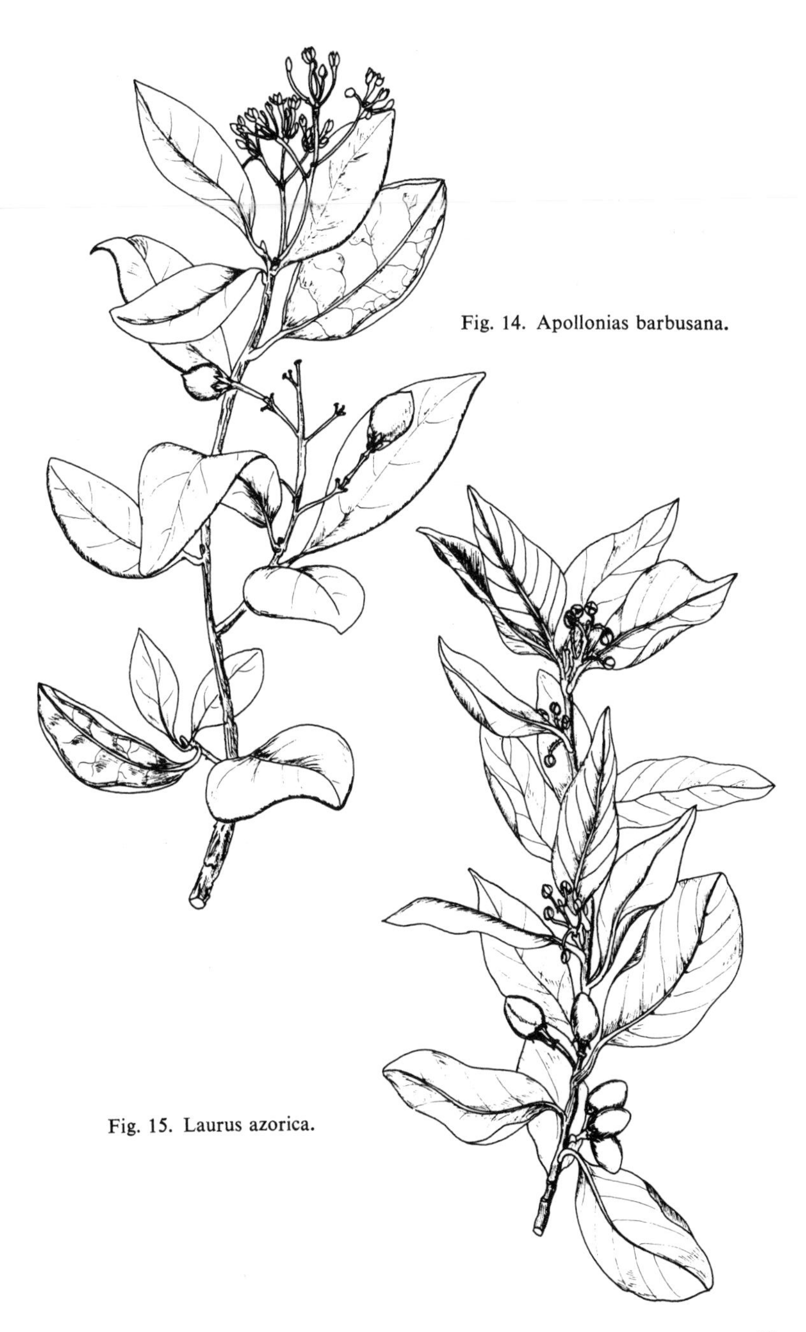

Fig. 14. Apollonias barbusana.

Fig. 15. Laurus azorica.

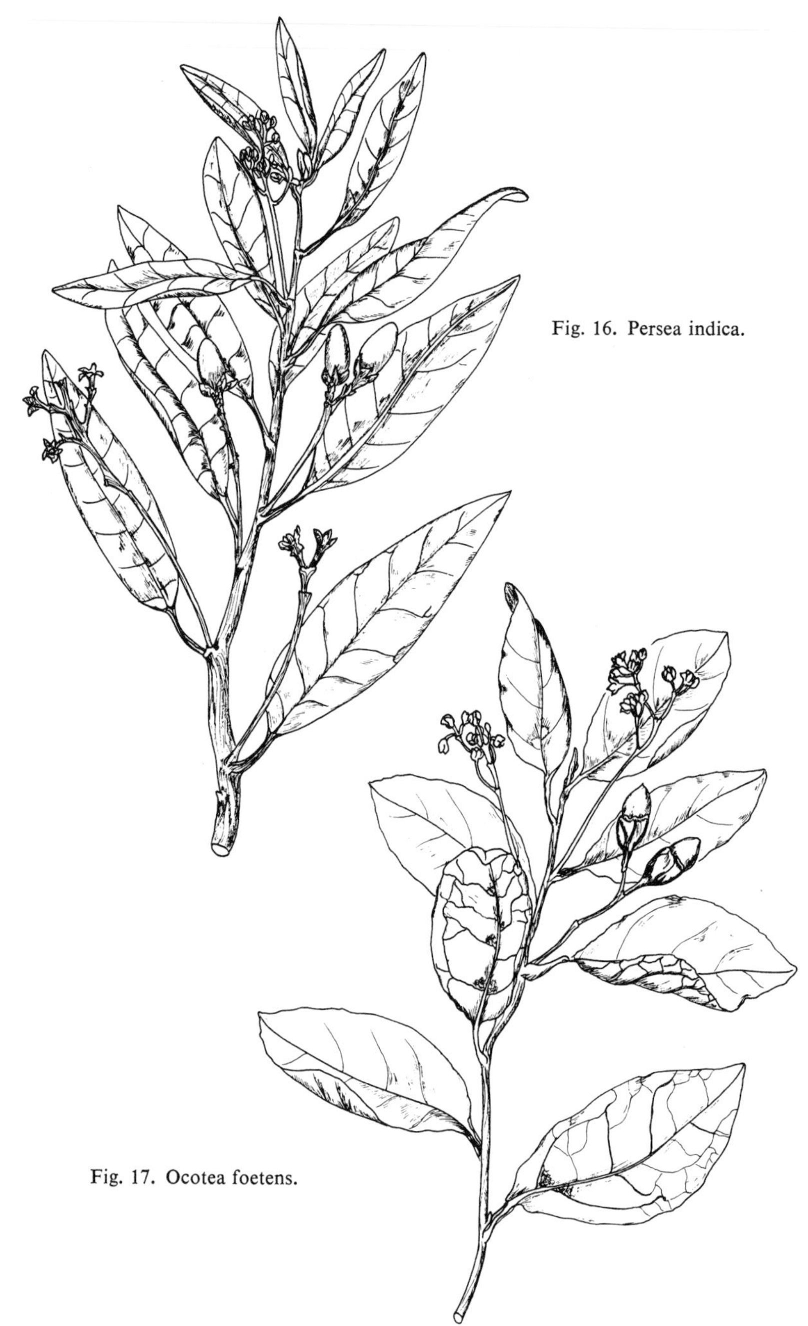

Fig. 16. Persea indica.

Fig. 17. Ocotea foetens.

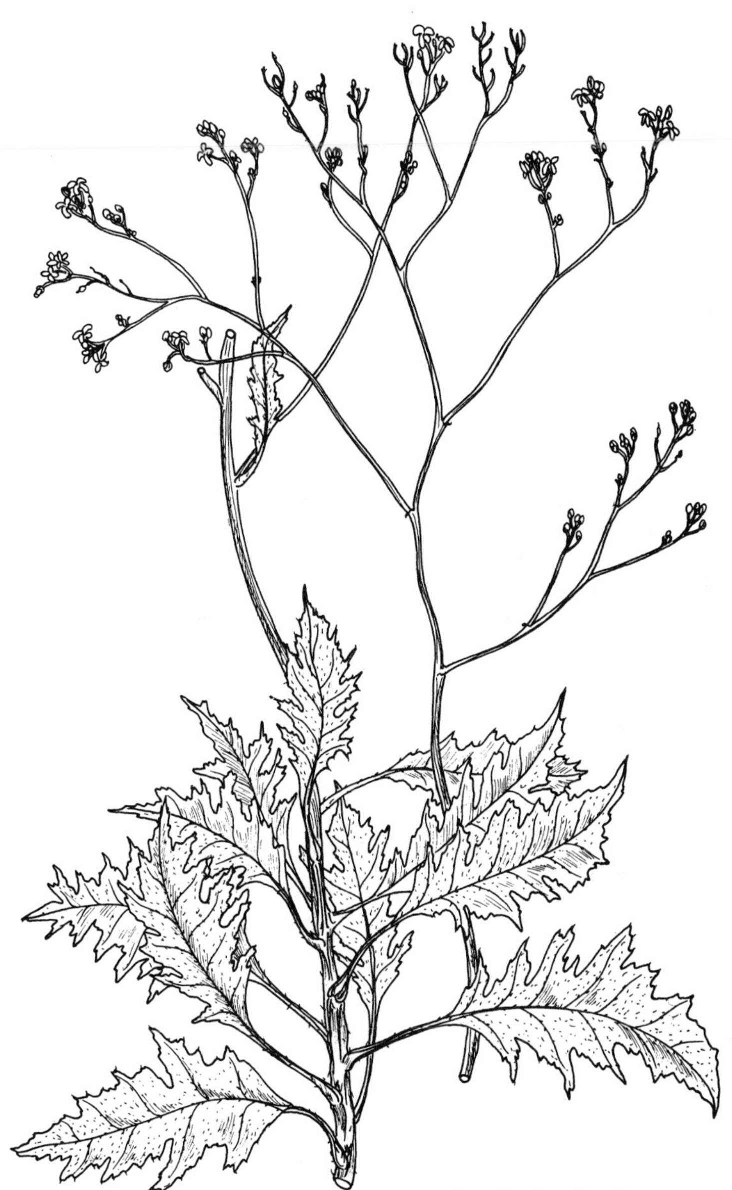

Fig. 18. Crambe arborea.

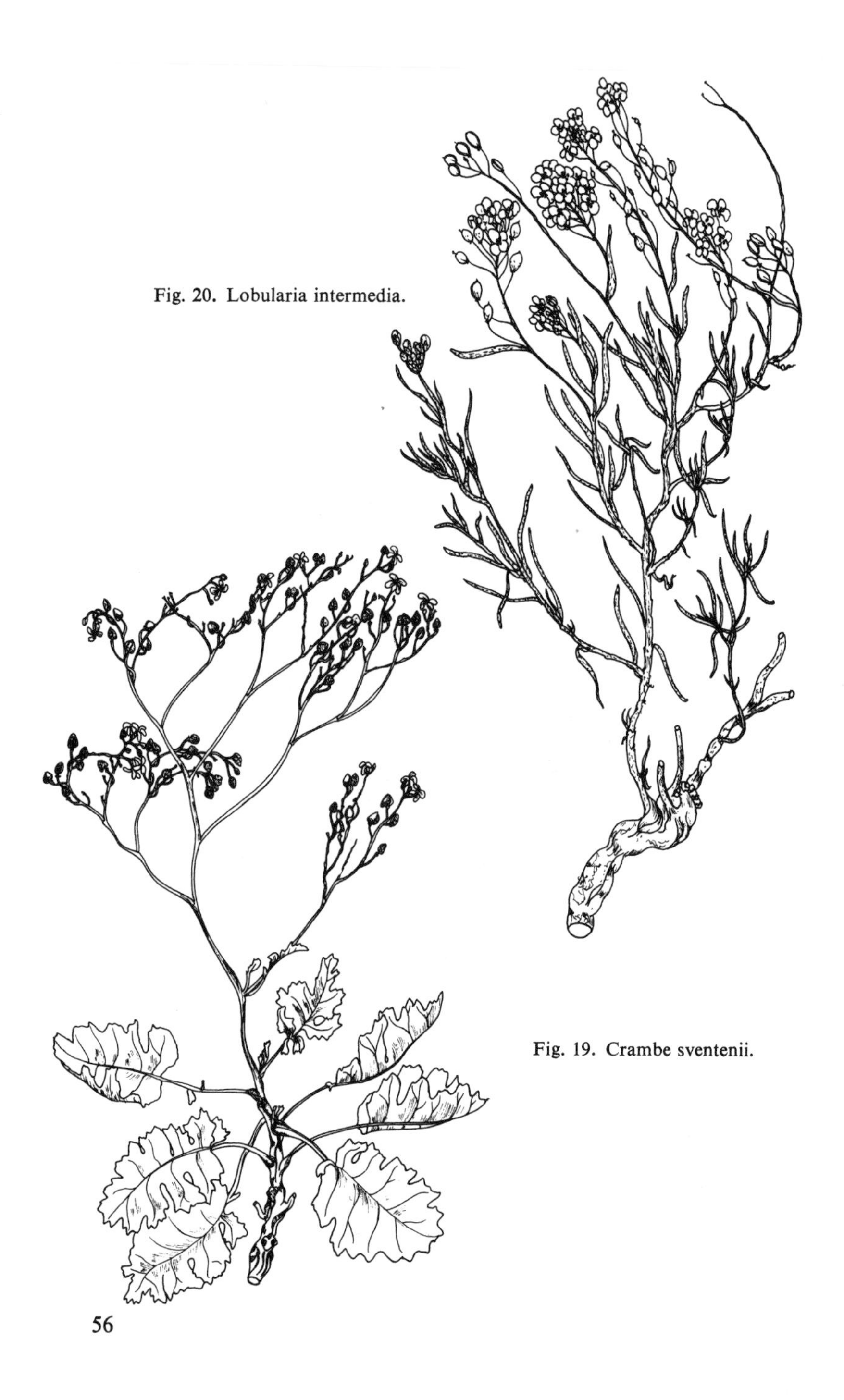

Fig. 20. Lobularia intermedia.

Fig. 19. Crambe sventenii.

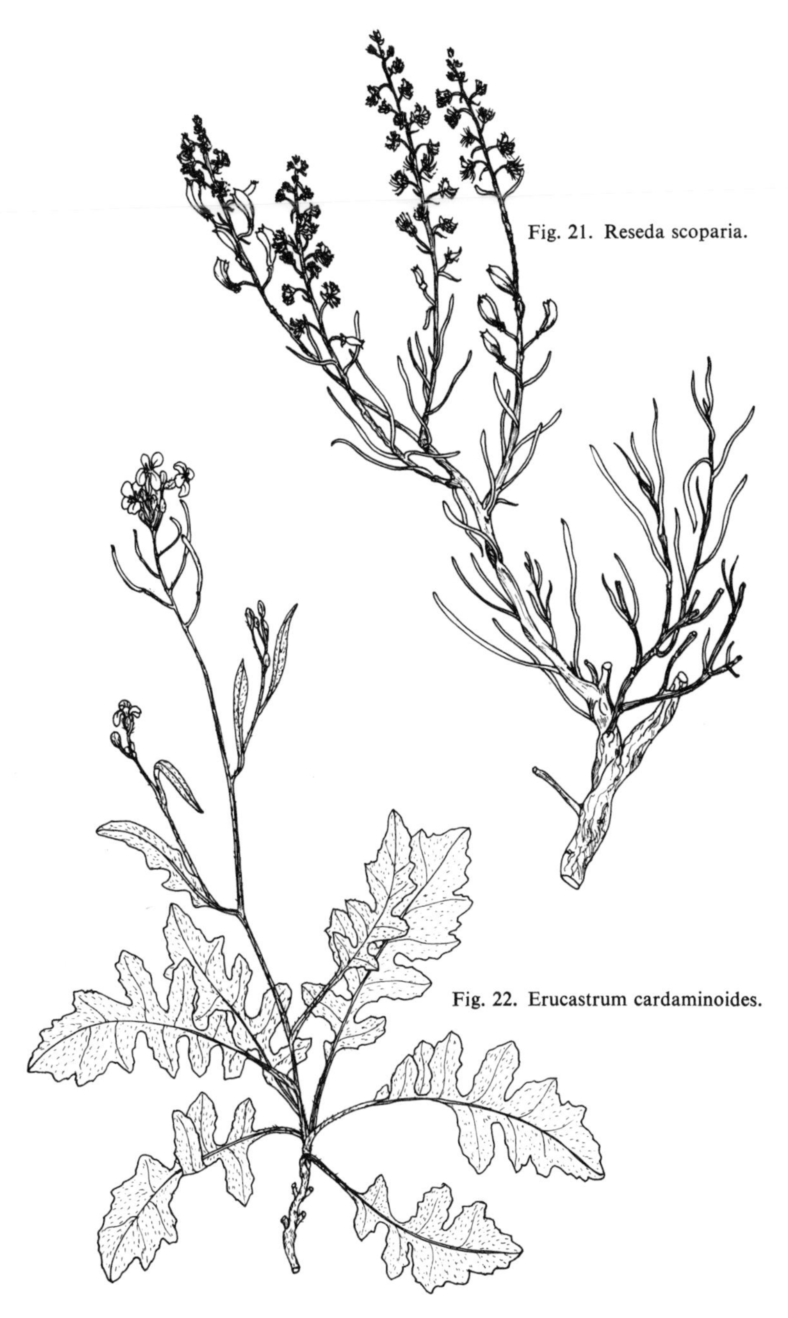

Fig. 21. Reseda scoparia.

Fig. 22. Erucastrum cardaminoides.

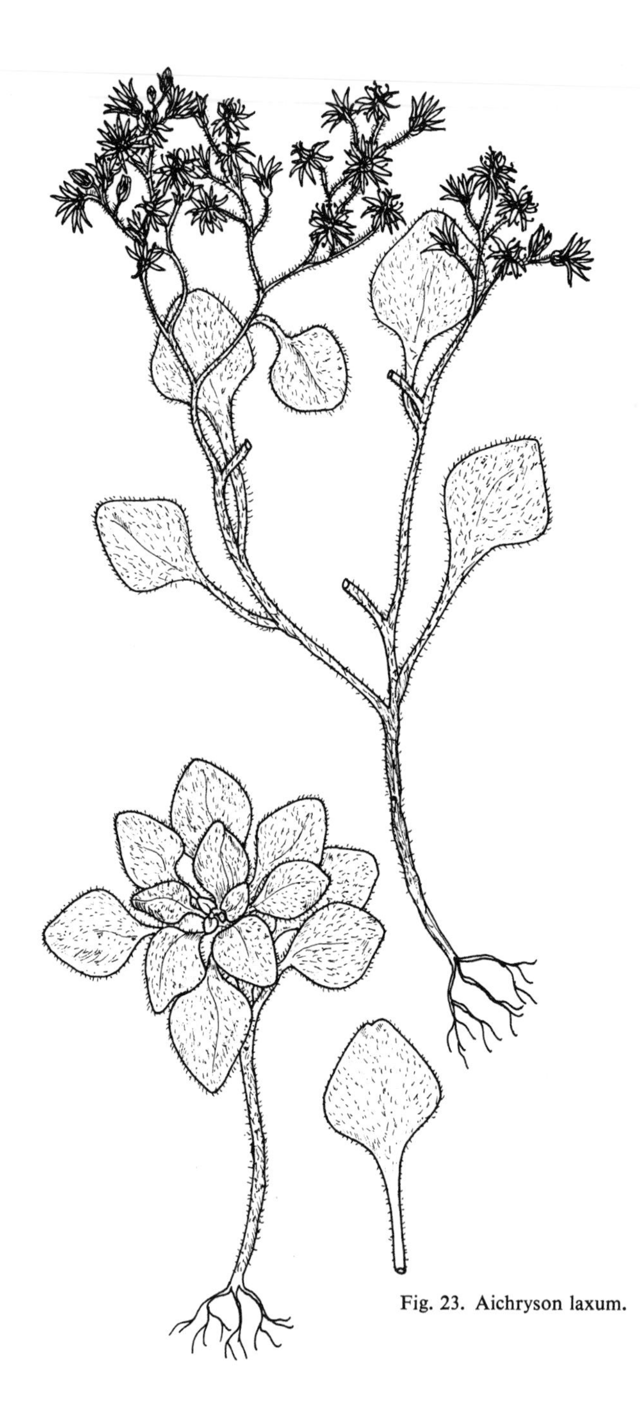

Fig. 23. Aichryson laxum.

Fig. 24. Monanthes muralis.

Fig. 25. Monanthes laxiflora.

Fig. 26. Bencomia brachystachya.

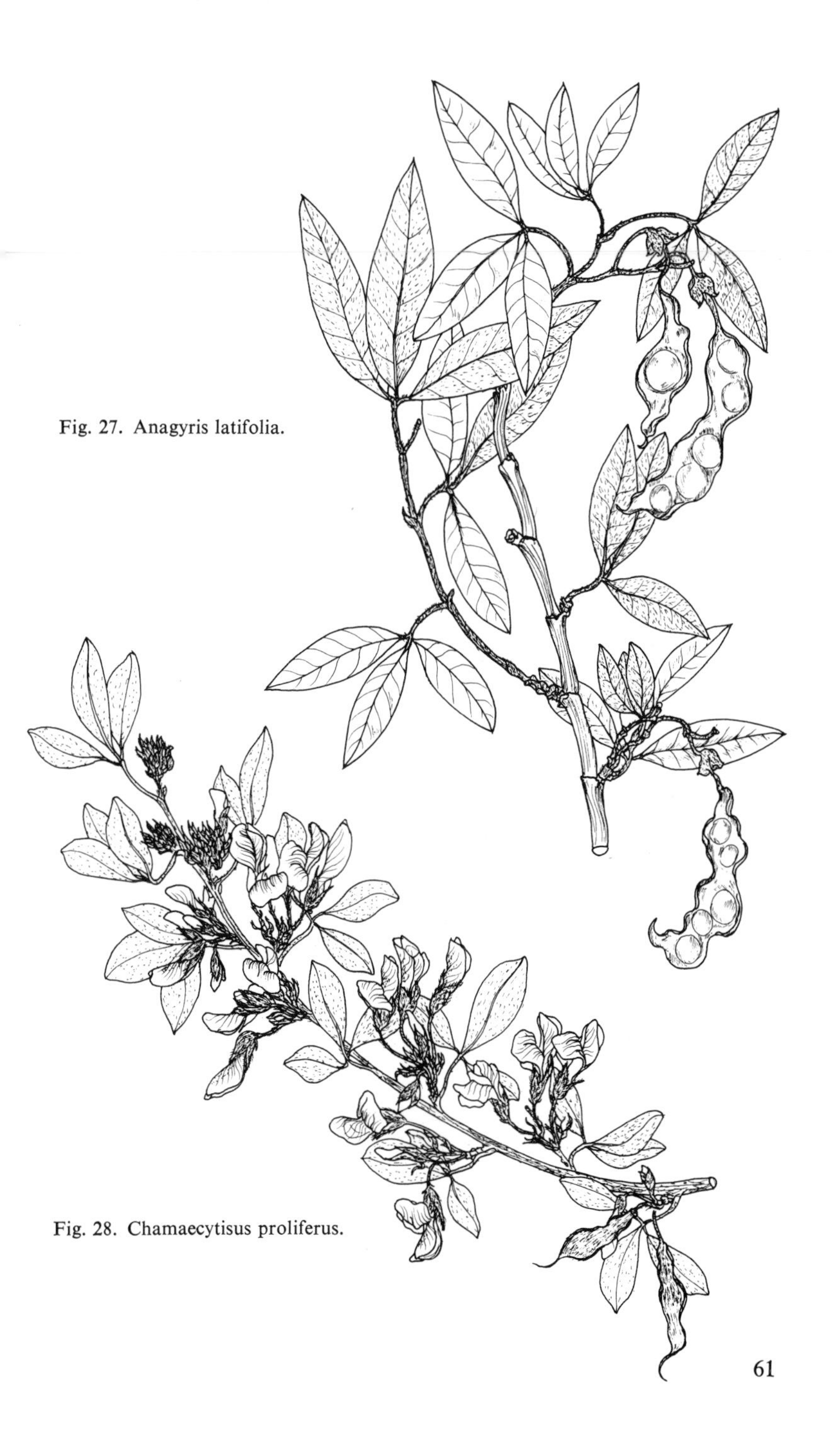

Fig. 27. Anagyris latifolia.

Fig. 28. Chamaecytisus proliferus.

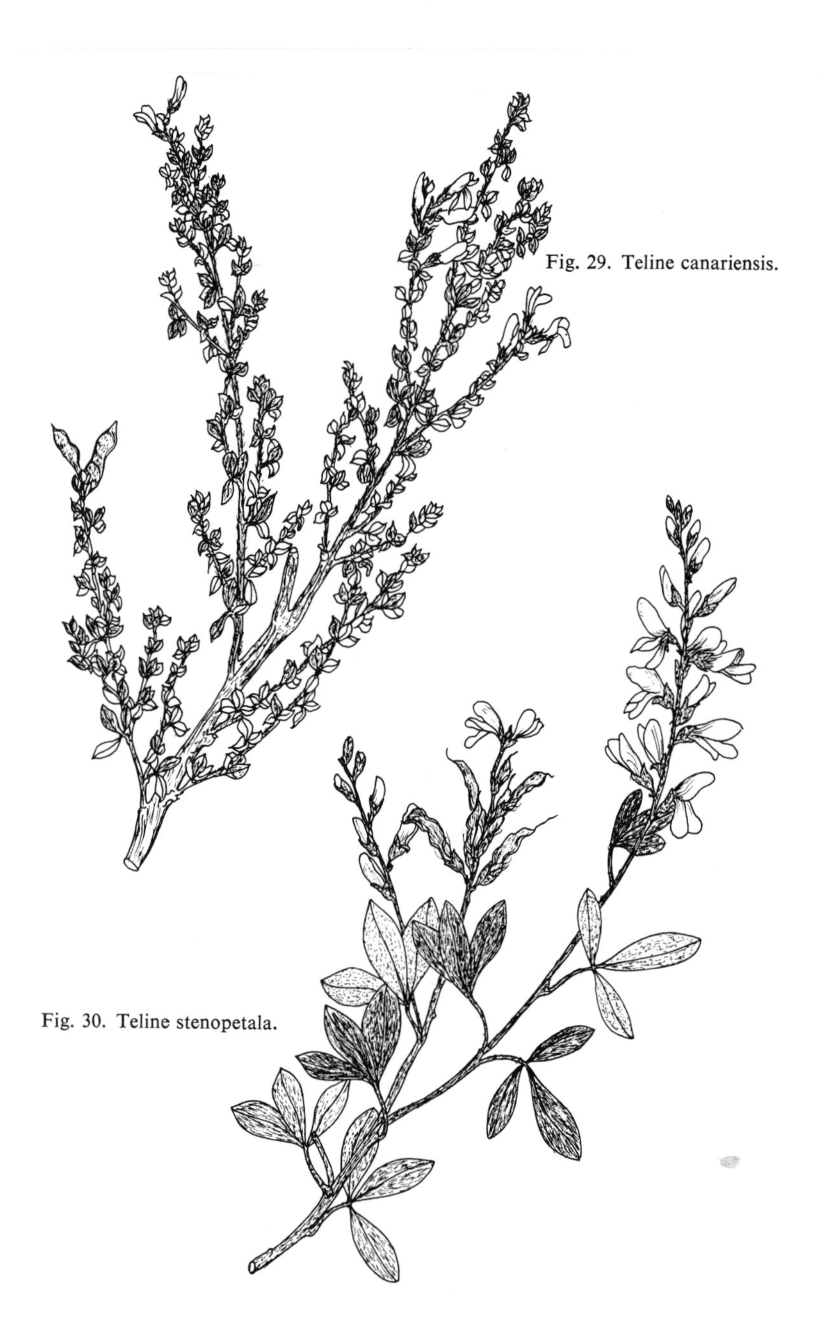

Fig. 29. Teline canariensis.

Fig. 30. Teline stenopetala.

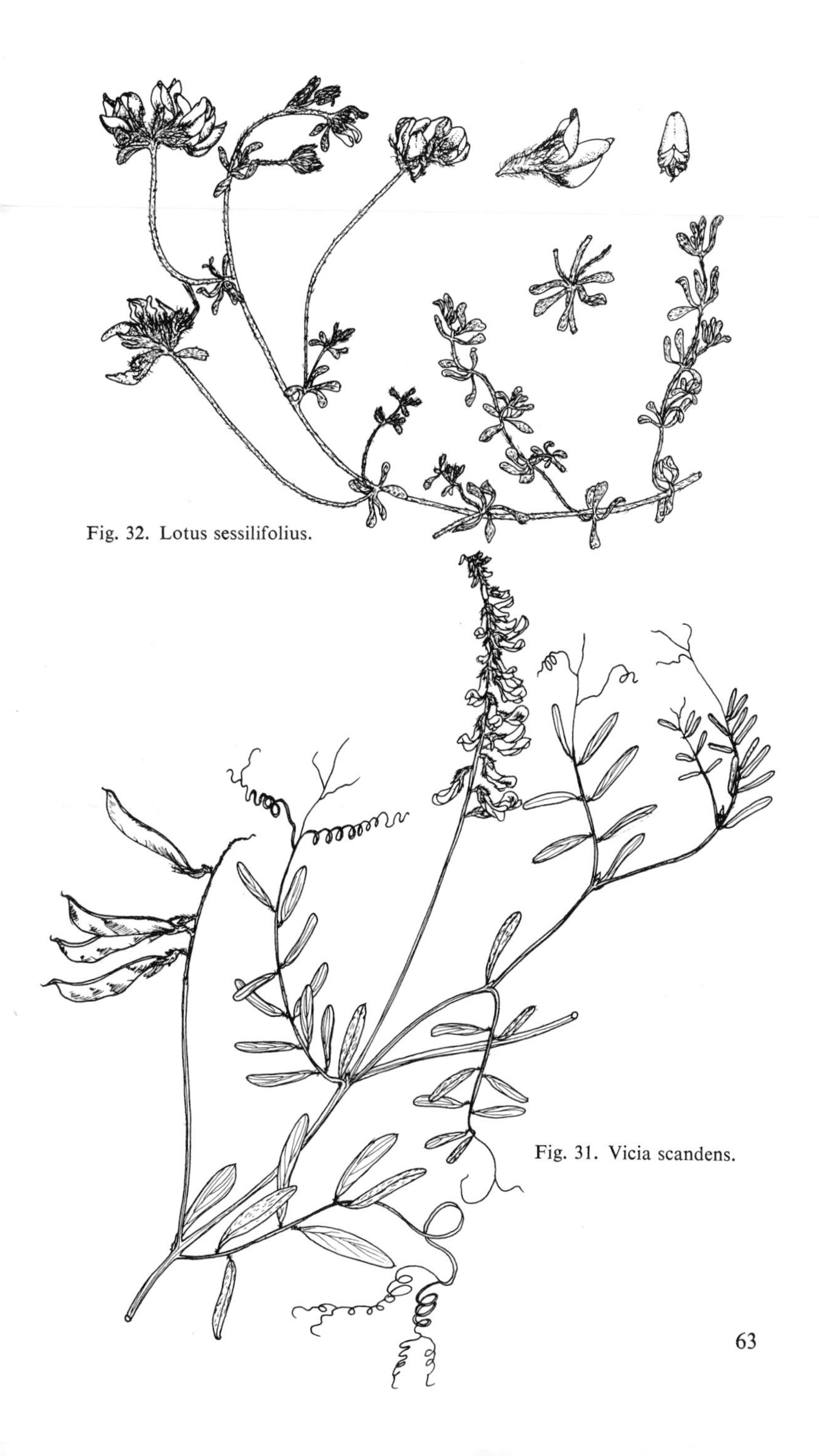

Fig. 32. Lotus sessilifolius.

Fig. 31. Vicia scandens.

Fig. 34. Lotus berthelotii.

Fig. 33. Lotus glaucus.

Fig. 35. Dorycnium broussonetii.

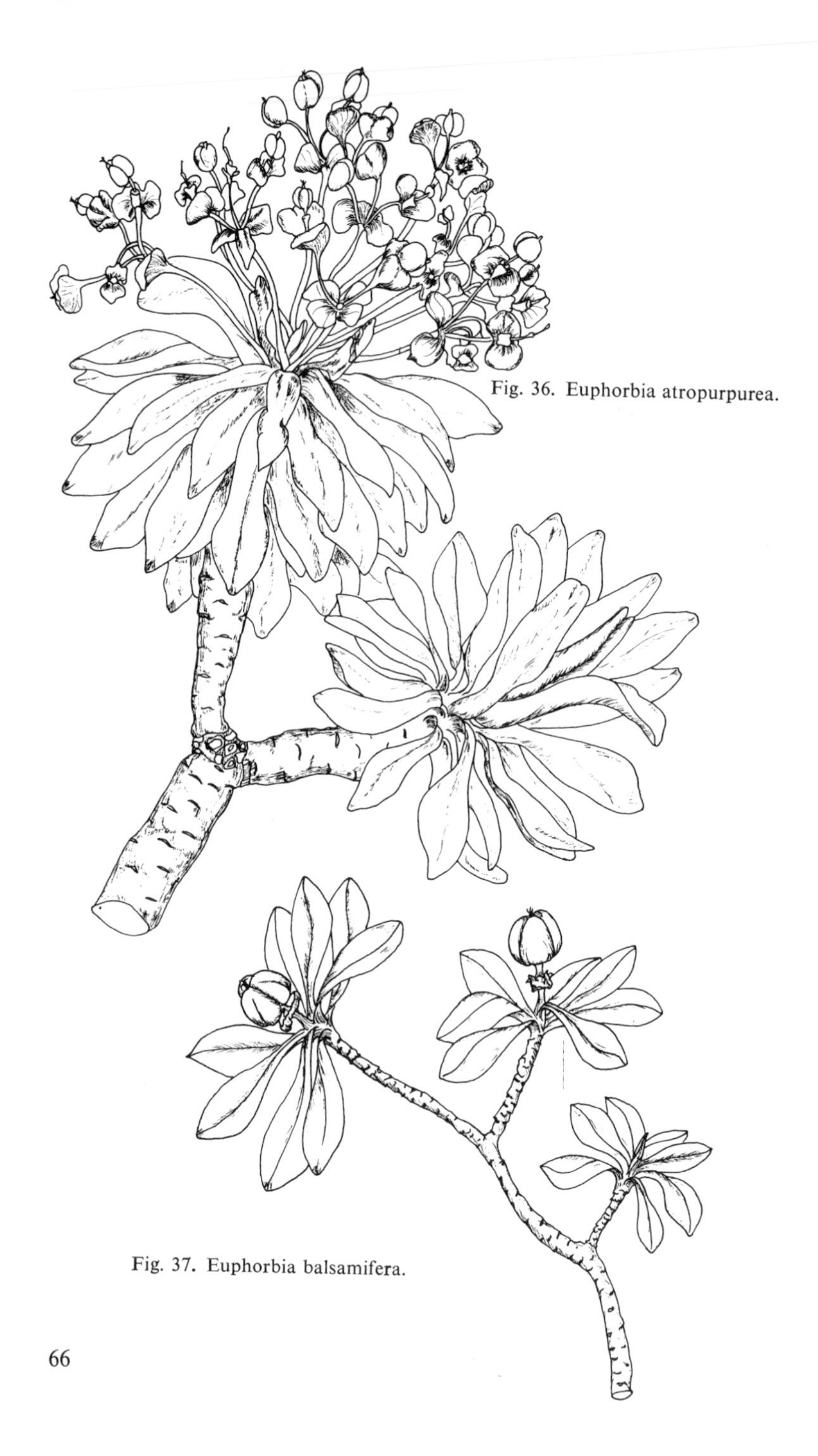

Fig. 36. Euphorbia atropurpurea.

Fig. 37. Euphorbia balsamifera.

Fig. 38. Cneorum pulverulentum.

Fig. 39. Rhamnus integrifolia.

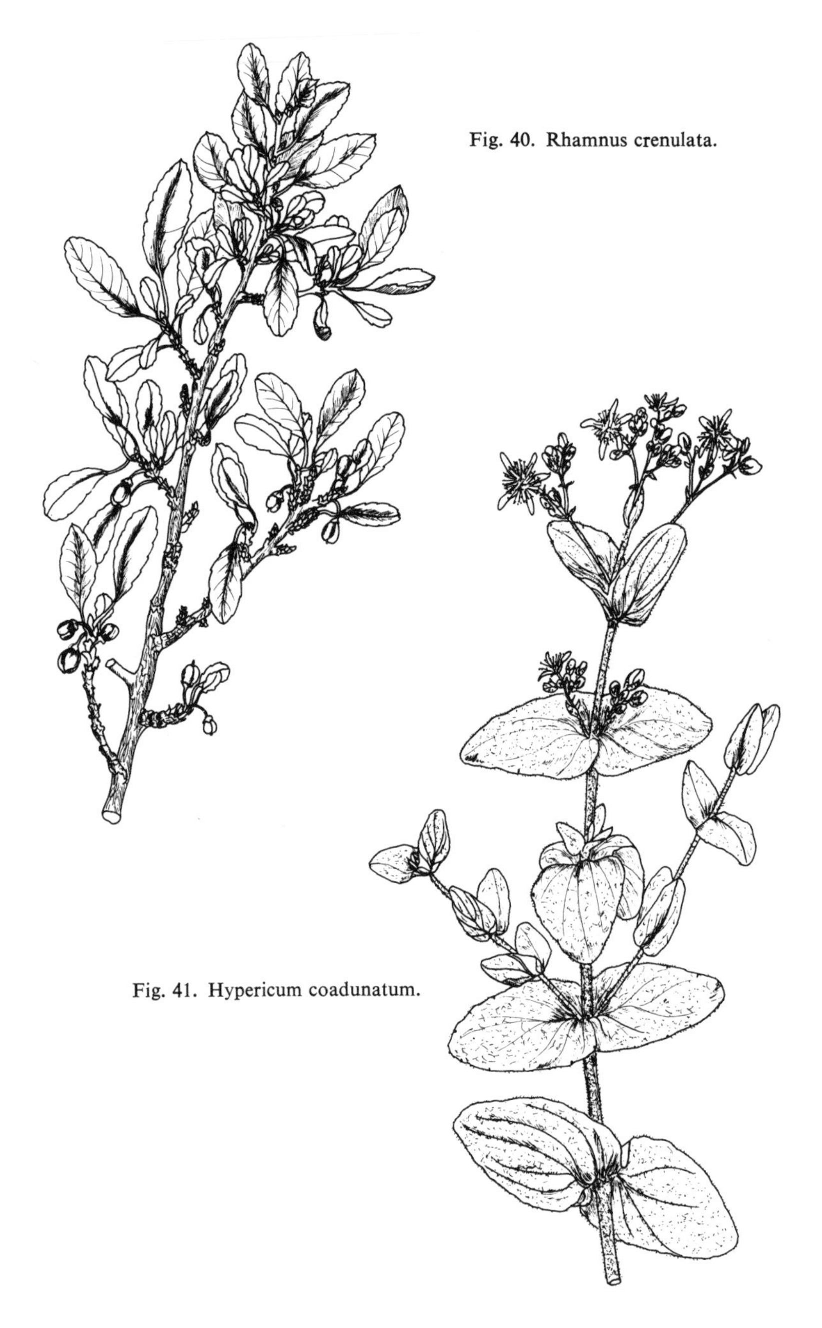

Fig. 40. Rhamnus crenulata.

Fig. 41. Hypericum coadunatum.

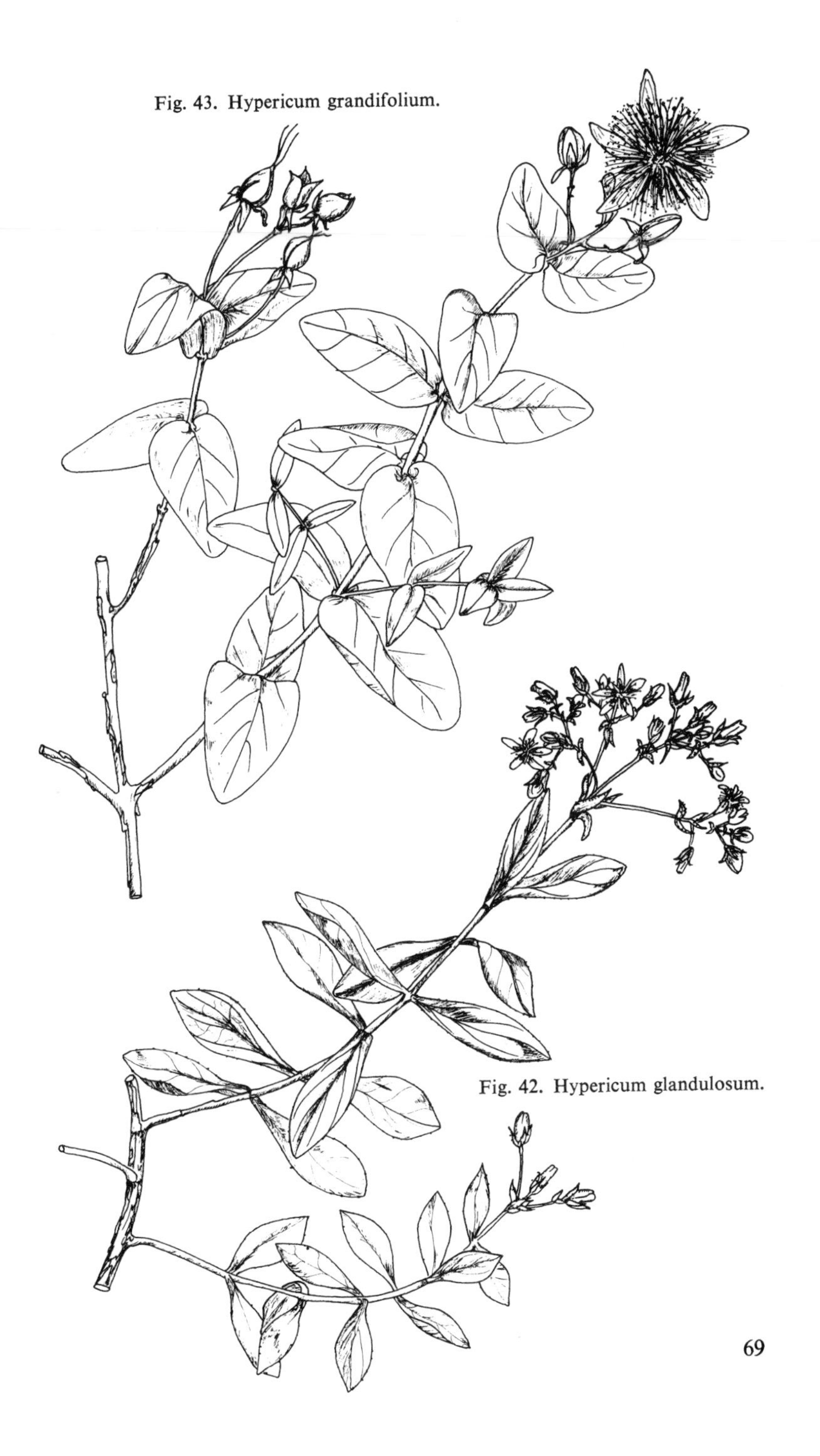

Fig. 43. Hypericum grandifolium.

Fig. 42. Hypericum glandulosum.

Fig. 47. Cistus osbeckifolius.

Fig. 44. Hypericum reflexum.

Fig. 46. Viola cheiranthifolia.

Fig. 45. Visnea mocanera.

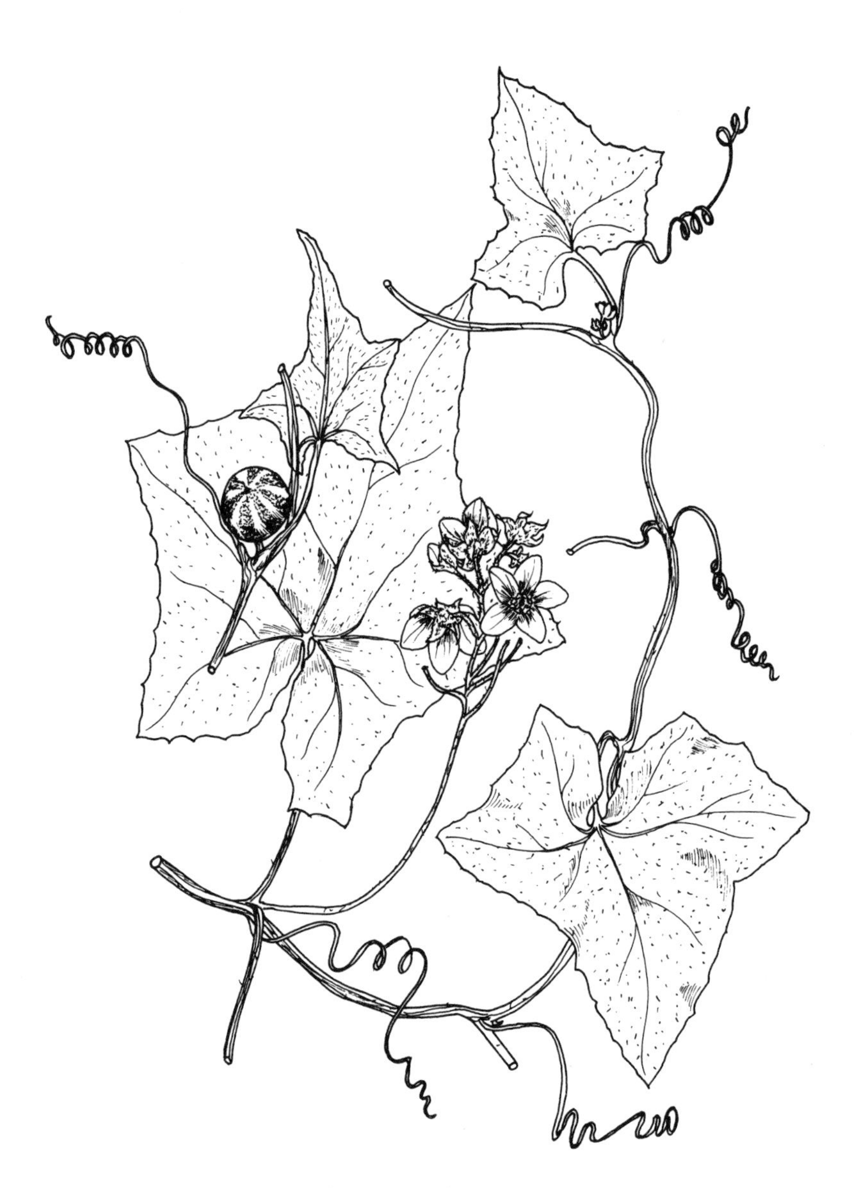

Fig. 48. Bryonia verrucosa.

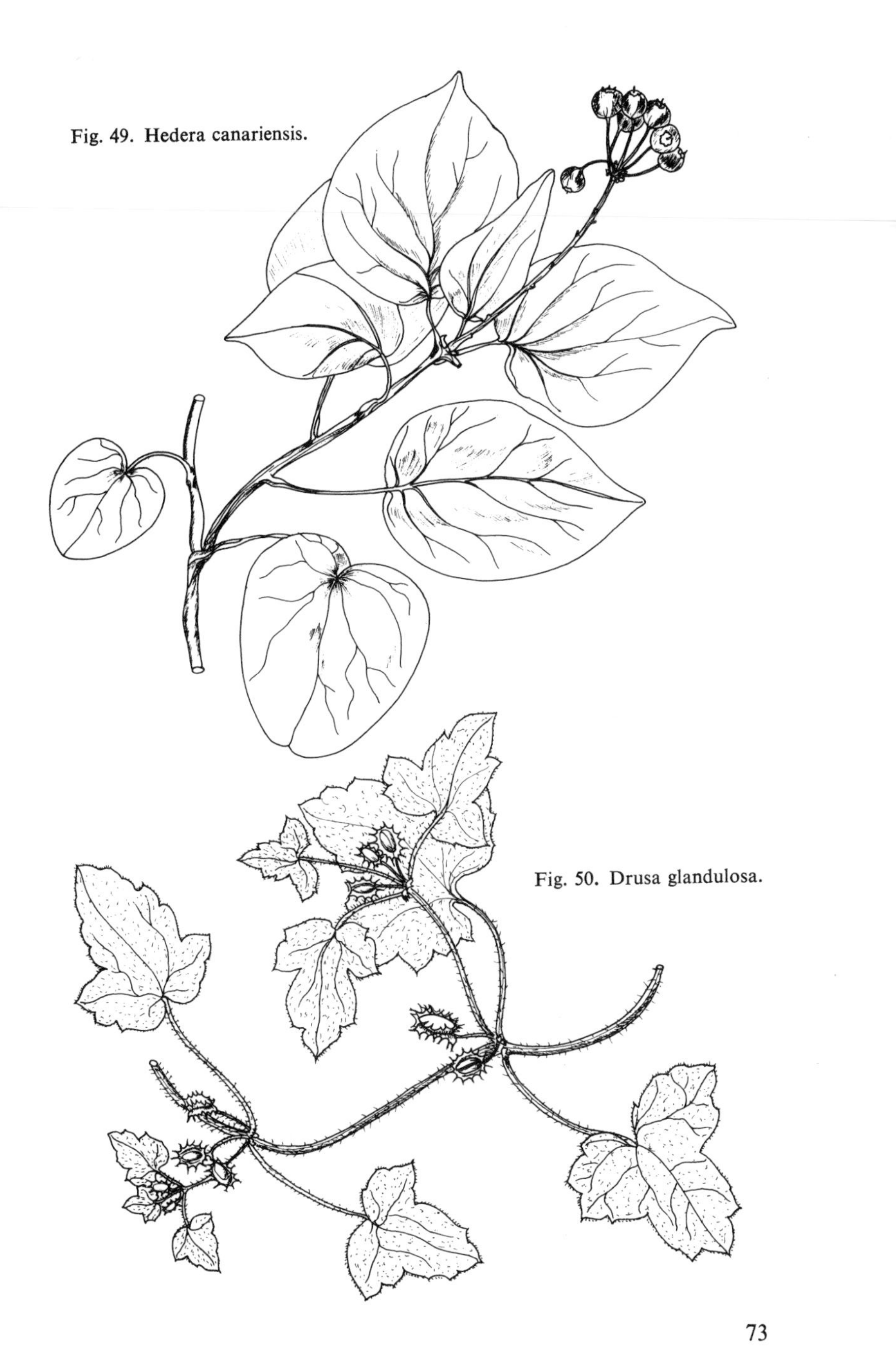

Fig. 49. Hedera canariensis.

Fig. 50. Drusa glandulosa.

Fig. 51. Pimpinella junionae.

Fig. 52. Tinguarra montana.

Fig. 53. Cryptotaenia elegans.

Fig. 54. Bupleurum salicifolium.

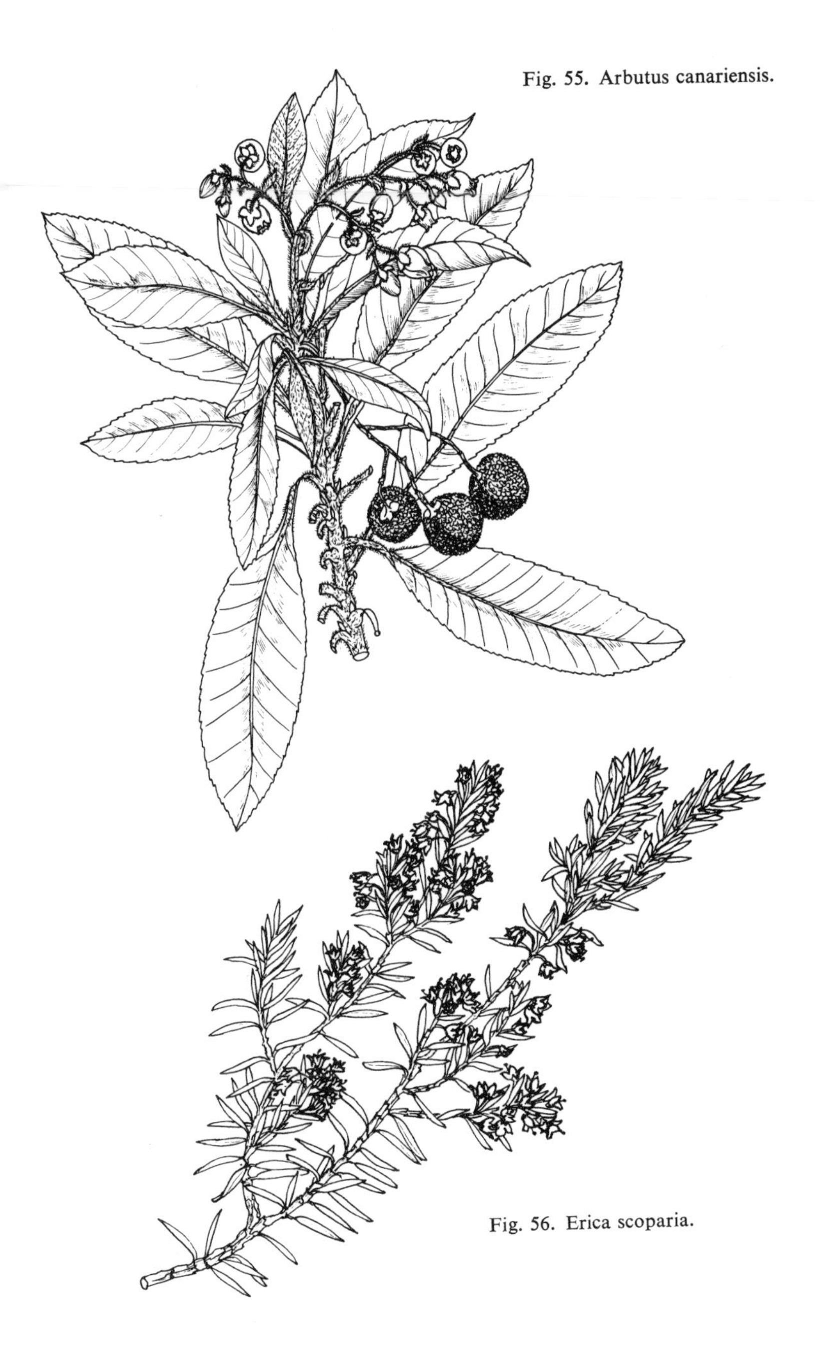

Fig. 55. Arbutus canariensis.

Fig. 56. Erica scoparia.

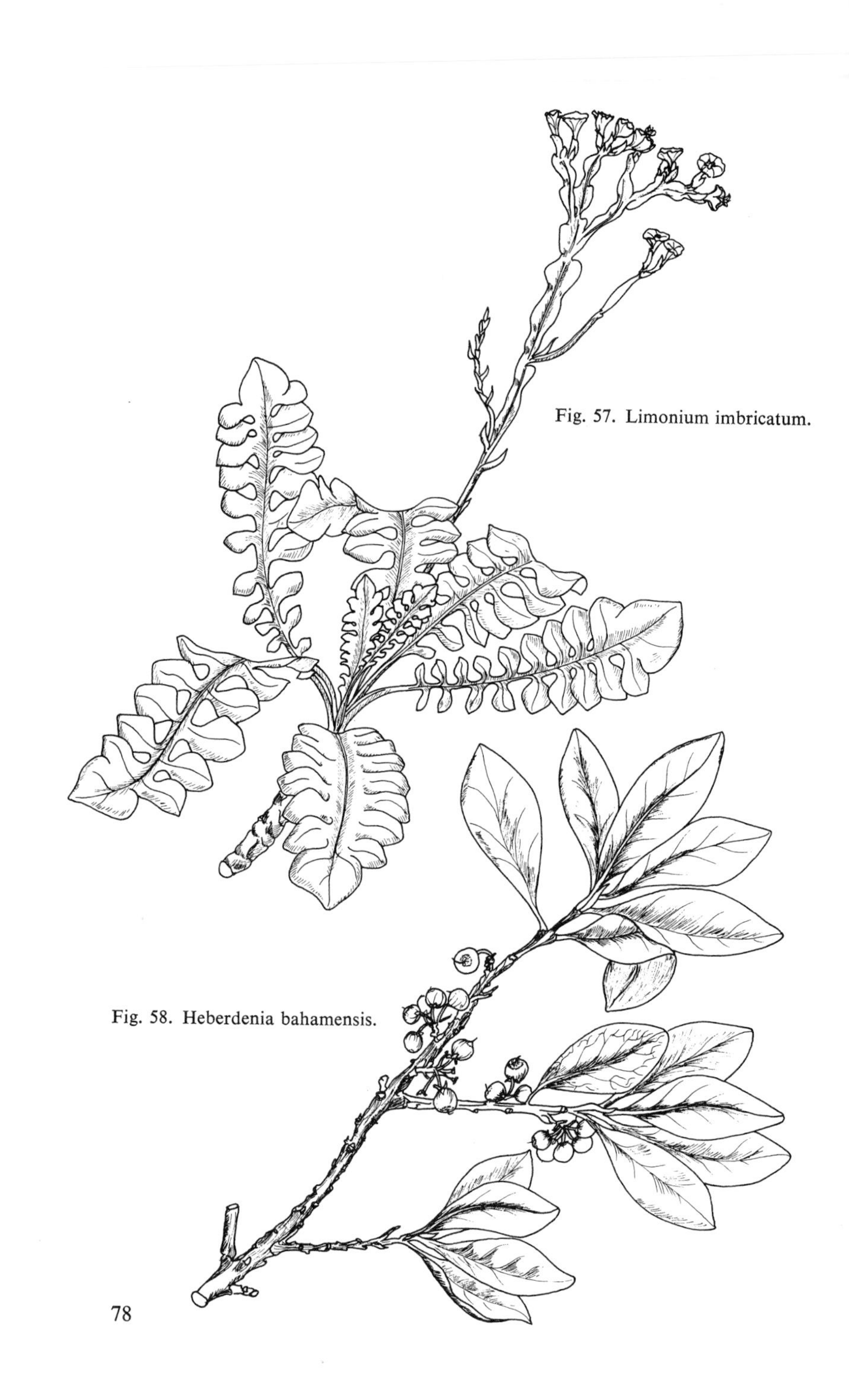

Fig. 57. Limonium imbricatum.

Fig. 58. Heberdenia bahamensis.

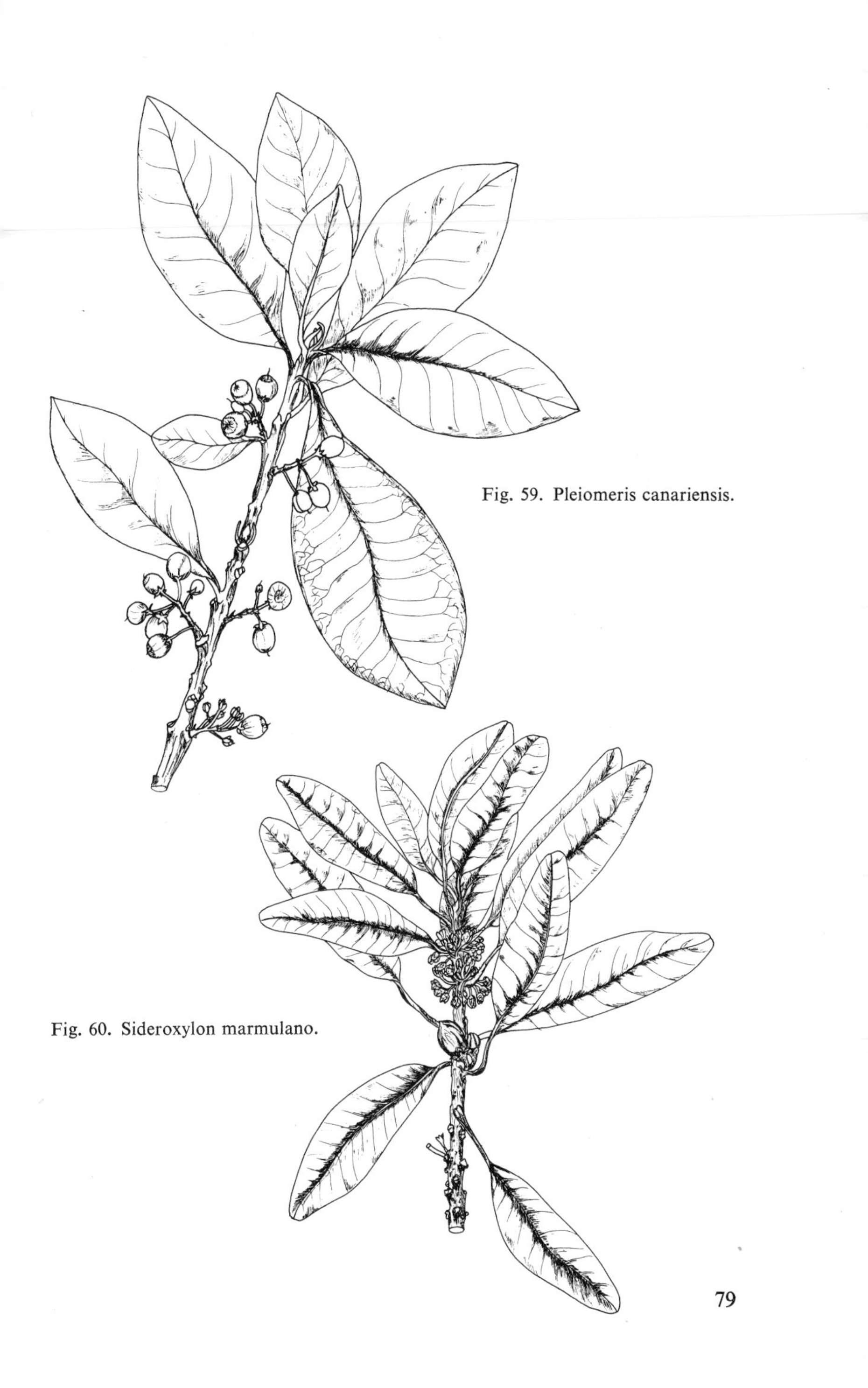

Fig. 59. Pleiomeris canariensis.

Fig. 60. Sideroxylon marmulano.

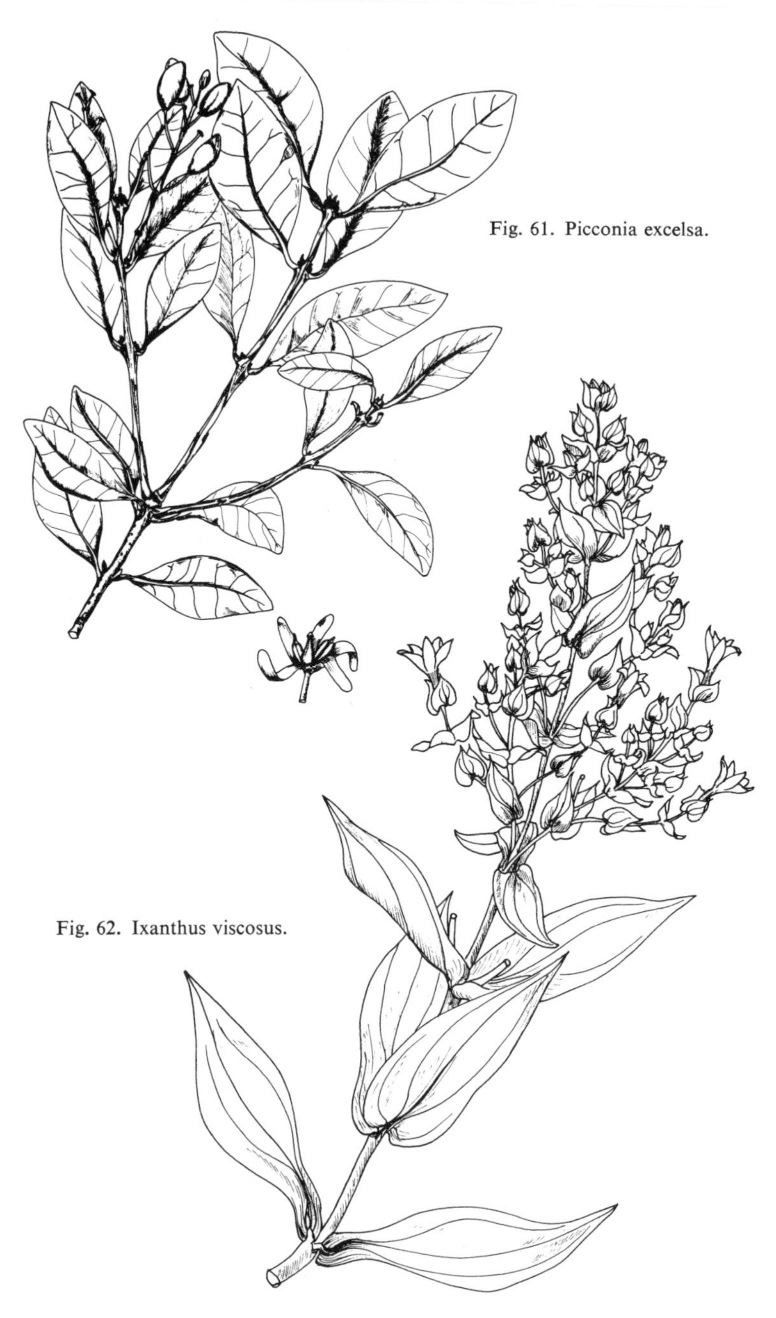

Fig. 61. Picconia excelsa.

Fig. 62. Ixanthus viscosus.

Fig. 64. Ceropegia dichotoma.

Fig. 63. Periploca laevigata.

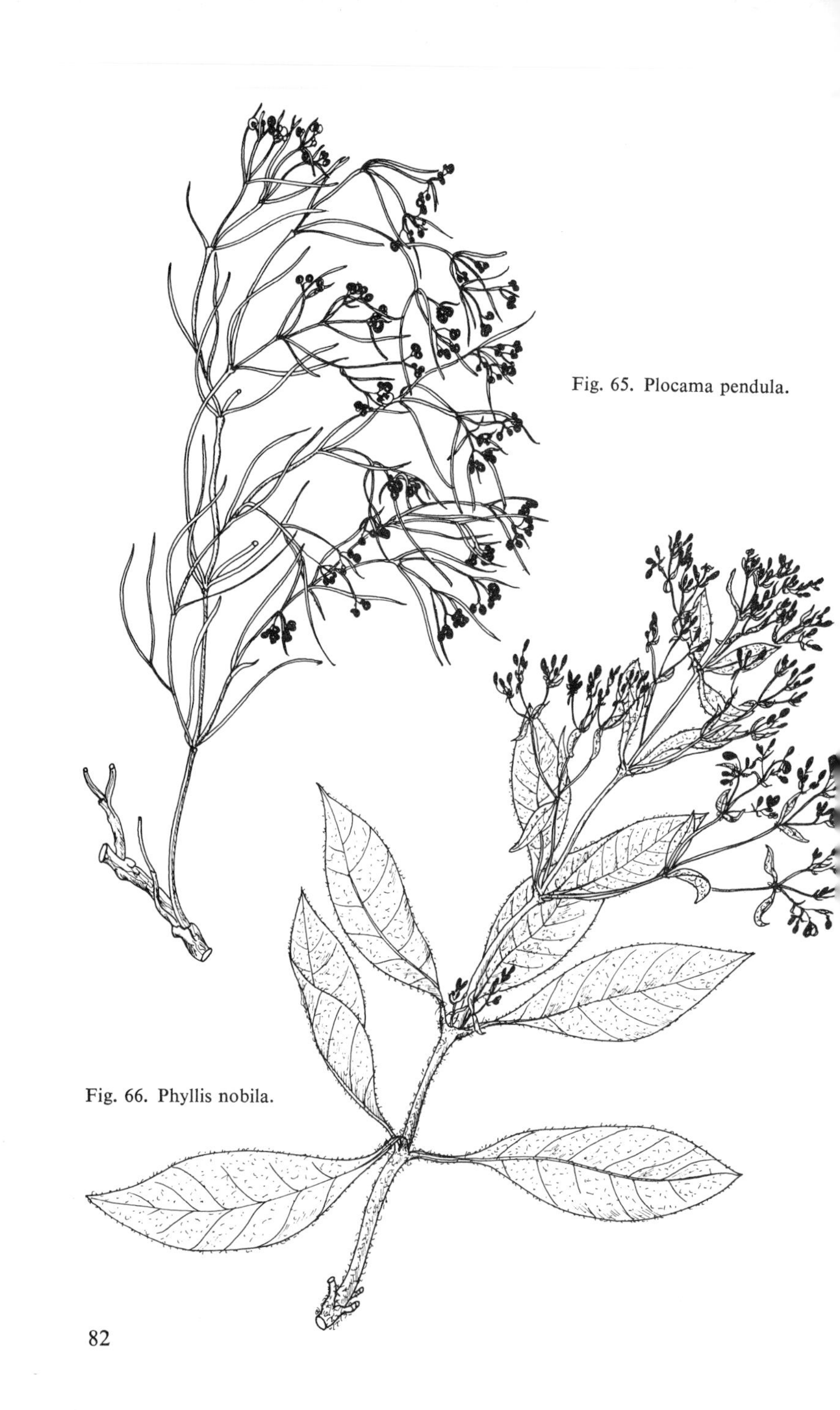

Fig. 65. Plocama pendula.

Fig. 66. Phyllis nobila.

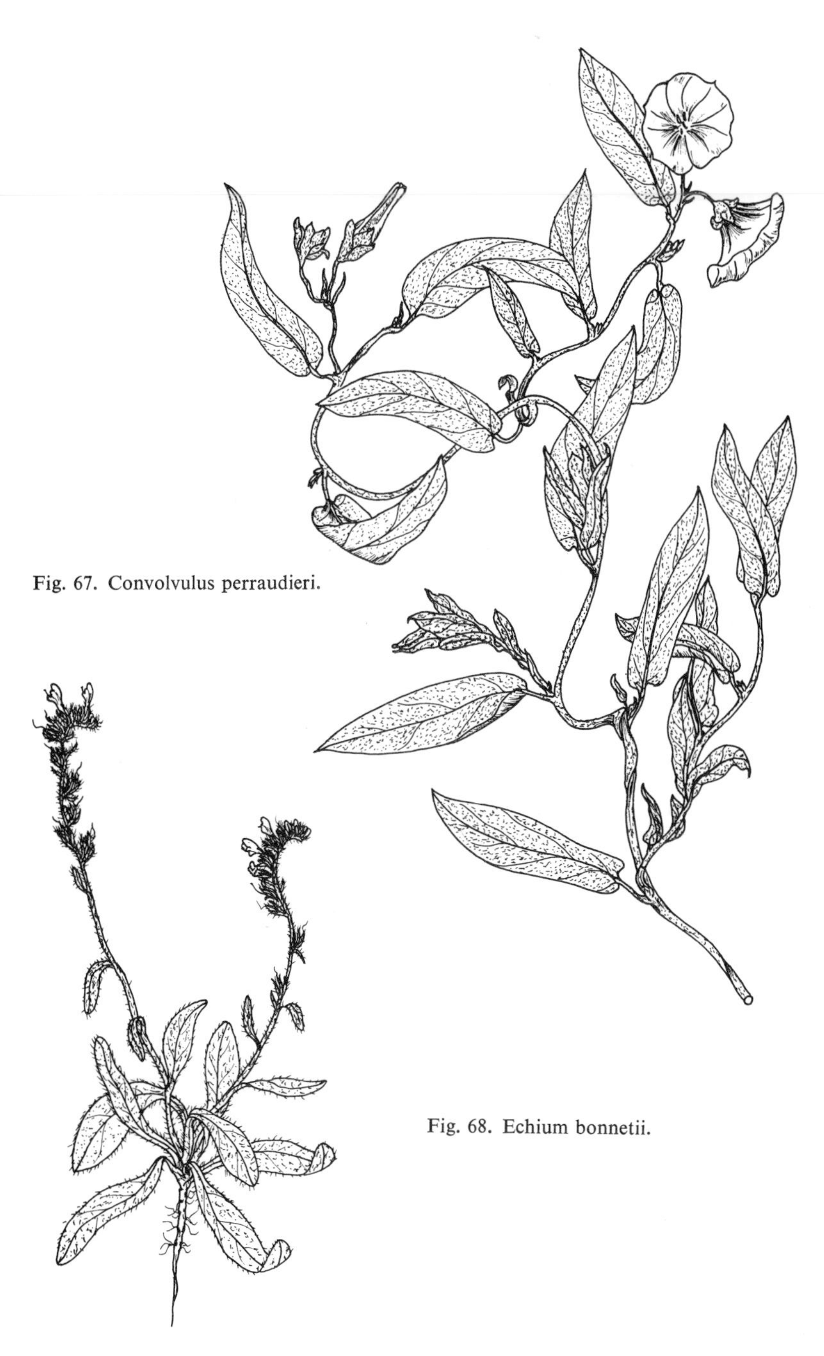

Fig. 67. Convolvulus perraudieri.

Fig. 68. Echium bonnetii.

Fig. 69. Echium handiense.

Fig. 70. Echium sventenii.

Fig. 71. Messerschmidia fruticosa.

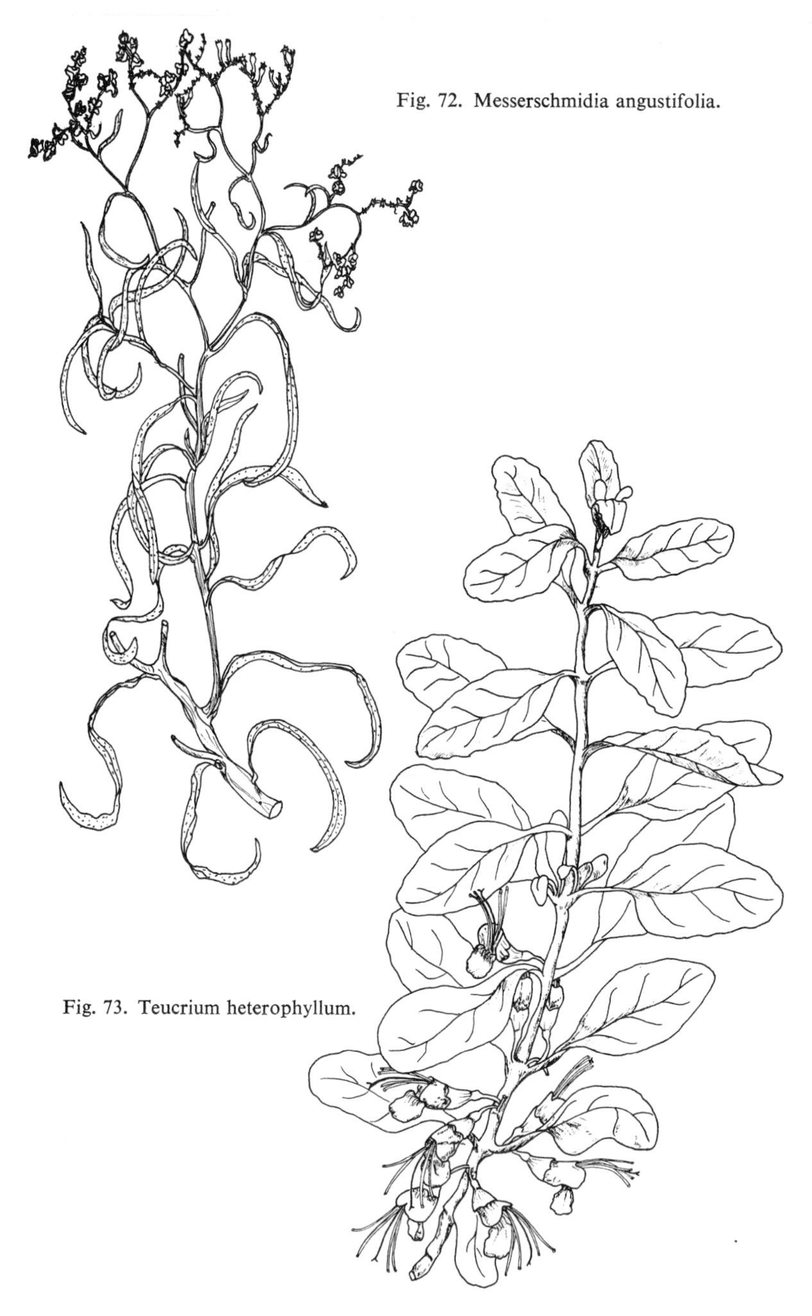

Fig. 72. Messerschmidia angustifolia.

Fig. 73. Teucrium heterophyllum.

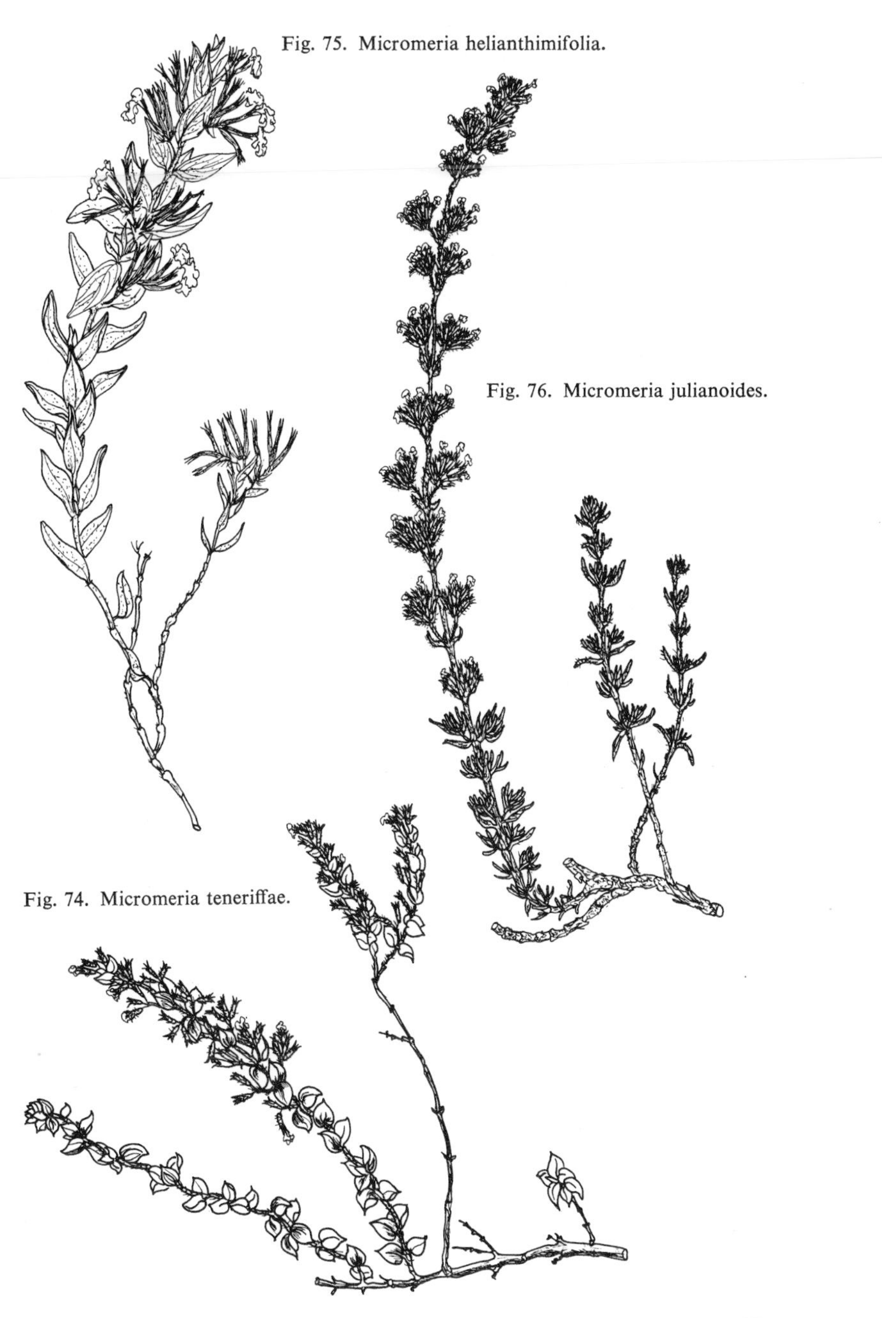

Fig. 75. Micromeria helianthimifolia.

Fig. 76. Micromeria julianoides.

Fig. 74. Micromeria teneriffae.

Fig. 77. Salvia broussonetii.

Fig. 79. Sideritis nervosa.

Fig. 78. Sideritis cystosiphon.

Fig. 81. Bystropogon canariense.

Fig. 80. Sideritis lotsyi.

Fig. 83. Scrophularia glabrata.

Fig. 82. Lavandula pinnata.

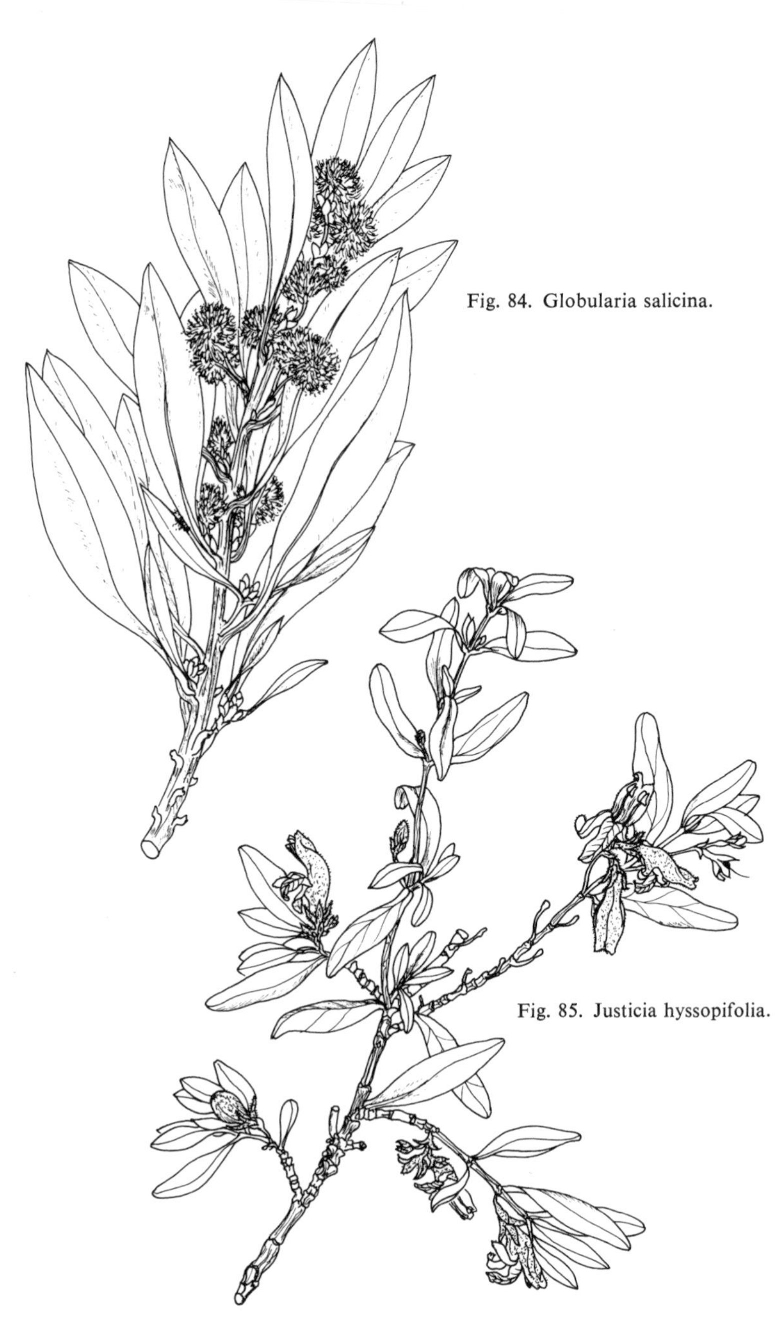

Fig. 84. Globularia salicina.

Fig. 85. Justicia hyssopifolia.

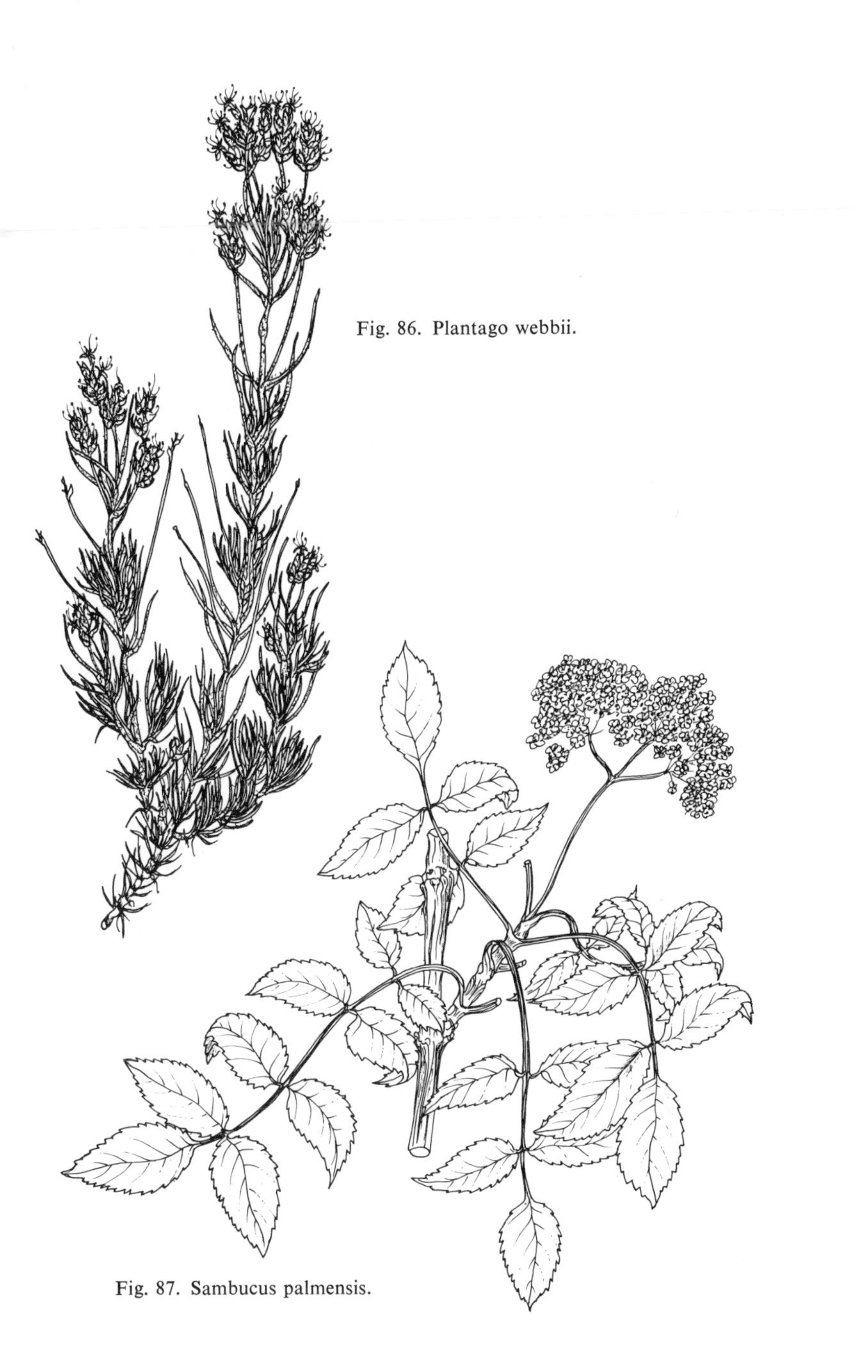

Fig. 86. Plantago webbii.

Fig. 87. Sambucus palmensis.

Fig. 89. Allagopappus dichotomus.

Fig. 88. Pterocephalus lasiospermus.

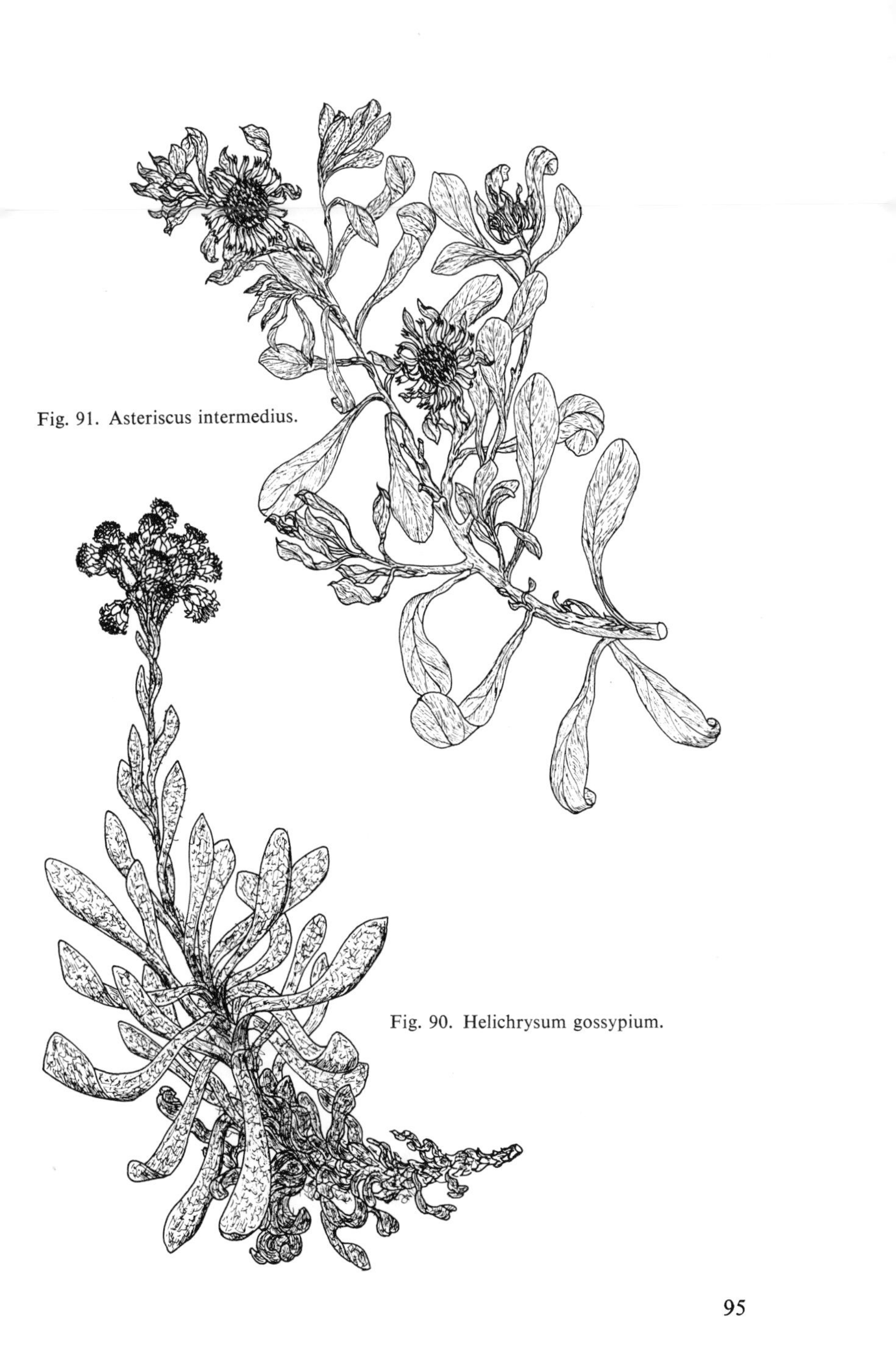

Fig. 91. Asteriscus intermedius.

Fig. 90. Helichrysum gossypium.

Fig. 92. Schizogyne sericea.

Fig. 93. Lugoa revoluta.

Fig. 94. Gonospermum gomeraeum.

Fig. 95. Argyranthemum frutescens.

Fig. 96. Argyranthemum gracile.

Fig. 97. Tanacetum ptarmaciflorum.

Fig. 99. Senecio appendiculatus.

Fig. 98. Senecio cruentus.

Fig. 101. Carlina xeranthemoides.

Fig. 100. Senecio hermosae.

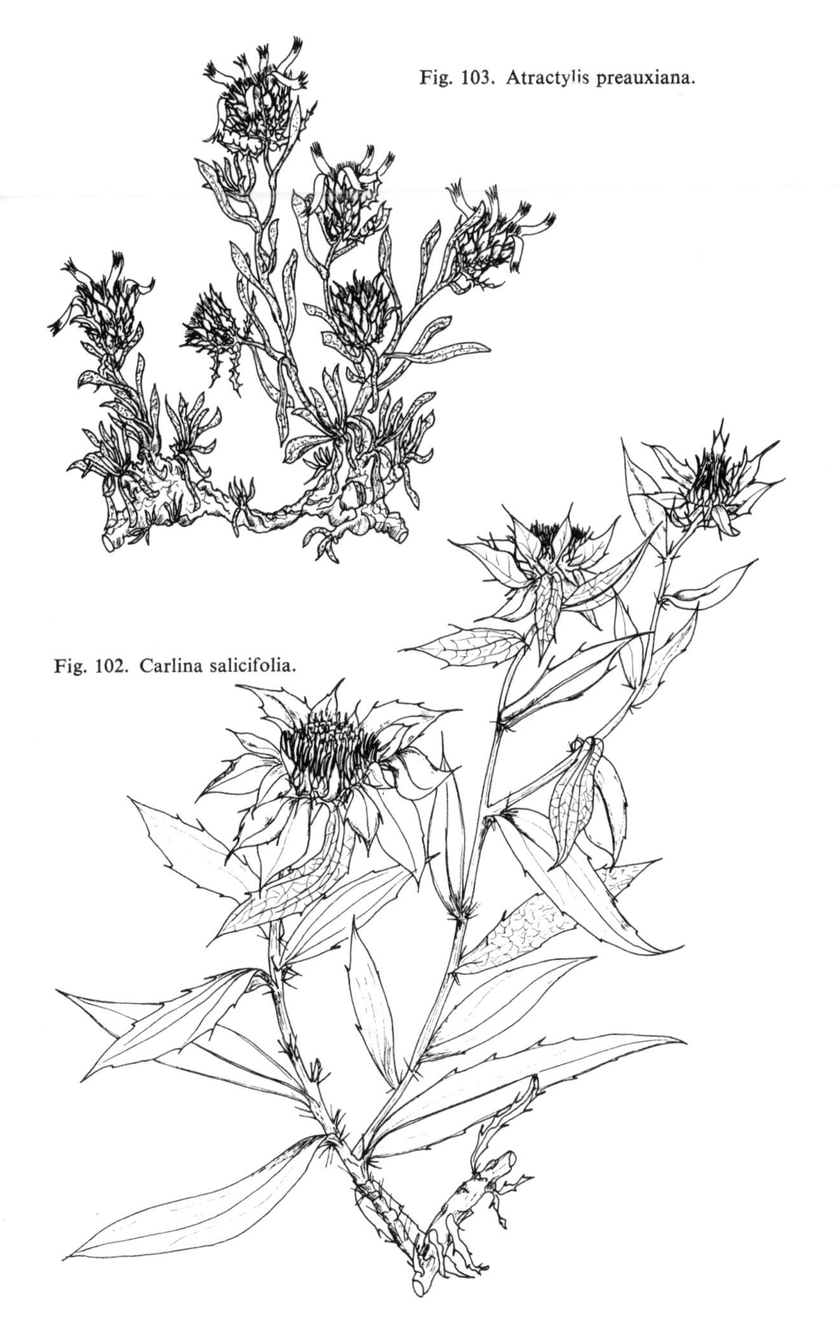

Fig. 103. Atractylis preauxiana.

Fig. 102. Carlina salicifolia.

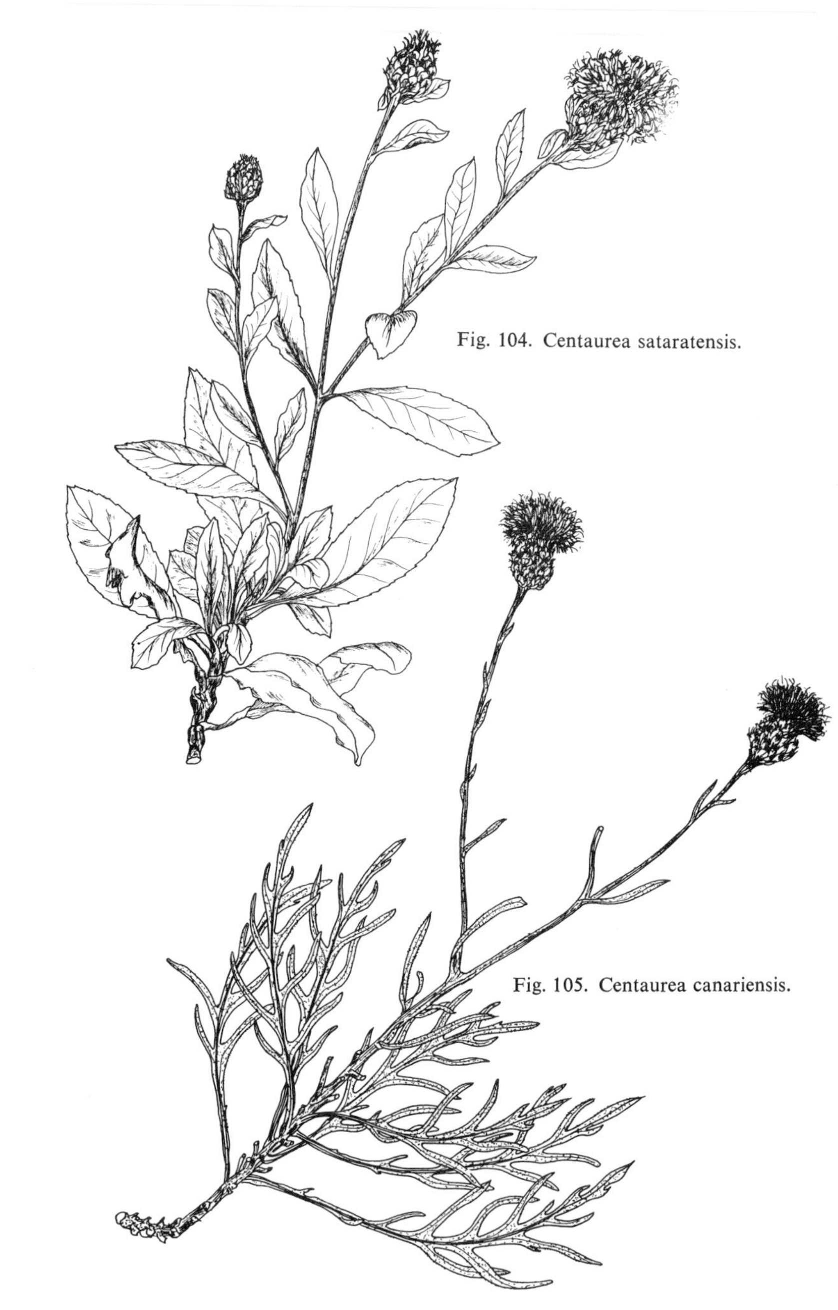

Fig. 104. Centaurea sataratensis.

Fig. 105. Centaurea canariensis.

Fig. 106. Tolpis crassiuscula.

Fig. 107. Sonchus capillaris.

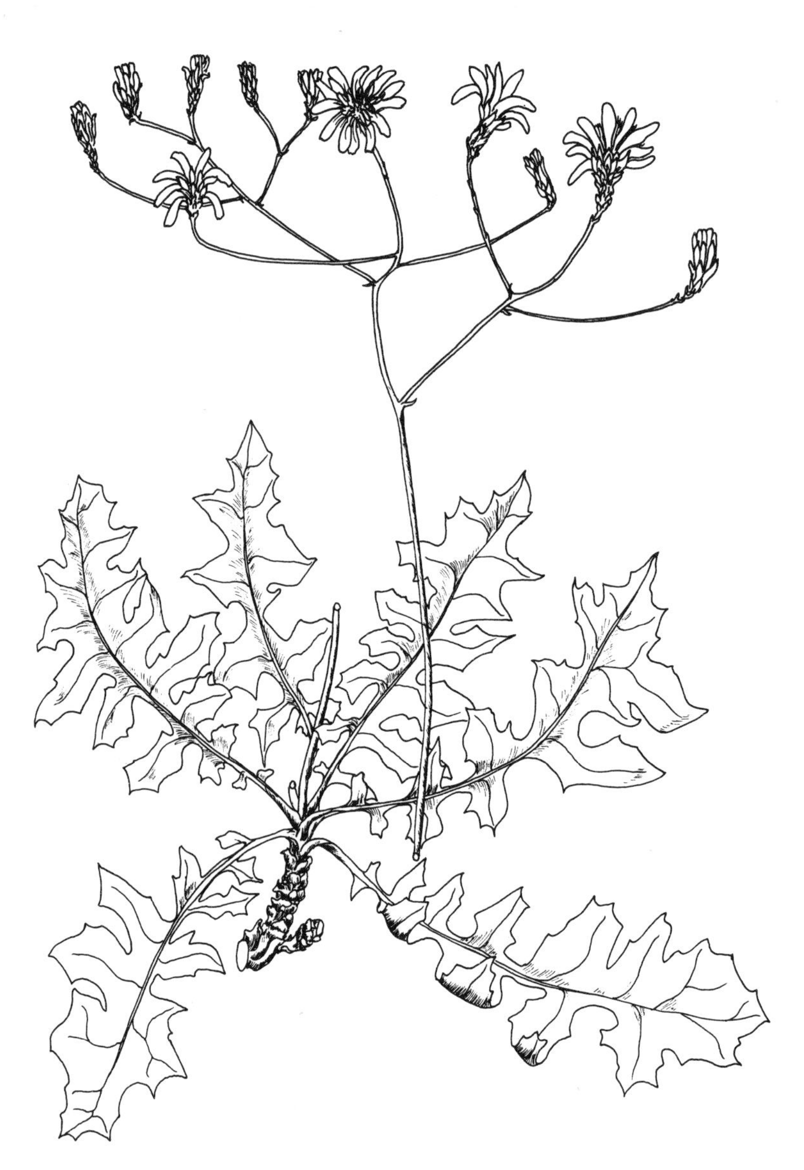

Fig. 108. Sonchus fauces-orci.

Fig. 109. Reichardia famarae.

Fig. 110. Asparagus arborescens.

Fig. 111. Pancratium canariense.

Fig. 112. Tamus edulis.

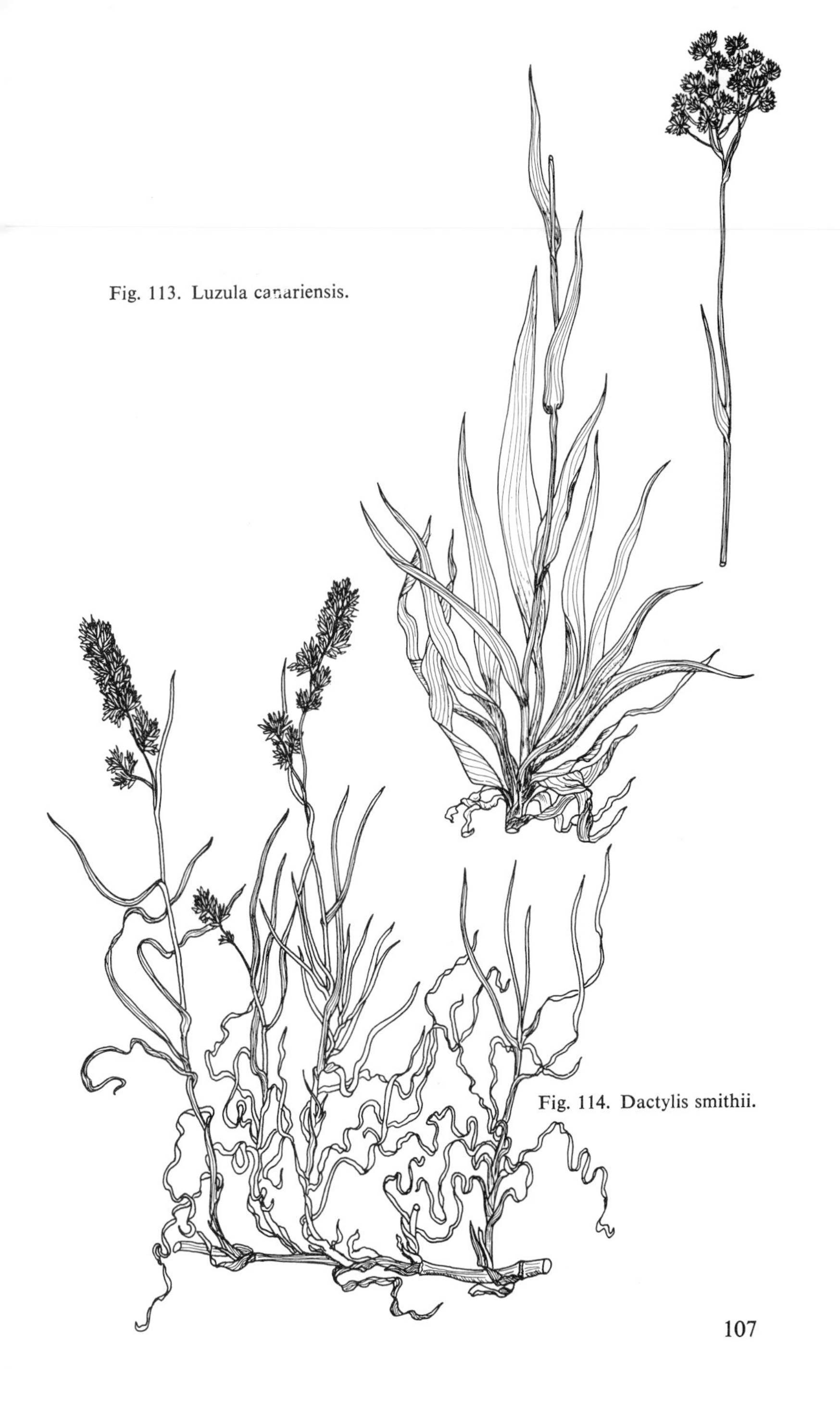

Fig. 113. Luzula canariensis.

Fig. 114. Dactylis smithii.

Fig. 116. Dracunculus canariensis.

Fig. 115. Brachypodium arbuscula.

Fig. 117. Gennaria diphylla.

Fig. 118. Habenaria tridactylites.

Colour Plates

General Views

XVIII. Las Cañadas del Teide, Tenerife, Subalpine zone dominated by the Pico de Teide 3718 m. The pumice and lava slopes support a rich endemic flora.

XIX. Las Cañadas del Teide, Tenerife Mid-Winter snow.

XX. High mountain pine forests of Tenerife under Winter snow. Fuente de Joco 1800 m.

XXI. Pinar de Tamadaba, Gran Canaria. **Pinus canariensis** woodland overlooking the montane zone of the island in the Roque Nublo region.

XXII. High Mountain pine forest below El Sombrerito on the south side of Tenerife above the town of Vilaflor.

XXIII. The montane zone of Gran Canaria in the botanically rich Caldera de Tirajana area below Paso de la Plata.

XXV. Barranco del Rio, La Palma. A track through the dense laurel woodland on the east side of the island.

XXIV. Sierra de Anaga, Tenerife. Central ridge with laurel forest along the crests.

XXVI. La Fortaleza de Chipude, Gomera. The small village of Chipude is dominated by a table-mountain with a rich cliff-flora.

XXVII. Teno mountains at the western end of Tenerife. Masca, a small hamlet nestles between steep cliffs with a host of local endemic plants.

XXVIII. Coastal vegetation at Punta de Teno. Tenerife.

XXIX. Cliff vegetation of Roque Cano at Vallehermoso, Gomera. The cliff faces abound with species such as **Sonchus ortunoi** and **Aeonium viscatum.**

XXX. Teno mountains, Tenerife. Cliff vegetation at Roque del Fraile on the north coast. **Ceropegia dichotoma** and **Euphorbia aphylla** combine with species such as **Aeonium tabuliforme** to make the area one of great interest to succulent enthusiasts.

XXXI. South coast of Tenerife from Montaña Roja to El Medano.

XXXII. Coastal slopes composed of recent lavas in El Golfo, Hierro. The rough, rocky slopes are covered with **Cneorum pulverulentum.**

XXXIII. Fuerteventura, dry dissected plains of the central region near Pajara.

XXXIV. Sandy beaches on the west coast of Lanzarote in the region just west of the Famara Massif.

Colour Plates

Species Illustrations

119. Davallia canariensis.

120. Adiantum reniforme.

121. Cheilanthes maderensis.

122. Pinus canariensis (female).

123. Pinus canariensis (male).

124. Juniperus cedrus.

125. Juniperus phoenicea.

126. Kunkeliella canariensis.

127. Rumex lunaria.

128. Polycarpaea smithii.

129. Polycarpaea latifolia.

131. Dicheranthus plocamoides.

130. Gymnocarpos salsoloides.

132. Apollonias barbusana.

133. Crambe laevigata.

134. Crambe sventenii.

135. Descurainia bourgaeana.

136. Parolinia intermedia.

137. Cheiranthus scoparius.

138. Cheiranthus scoparius var. cinereus.

139. Cheiranthus virescens.

140. Reseda crystallina.

141. Monanthes niphophila.

142. Monanthes praegeri.

143. Monanthes polyphylla.

144. Aichryson parlatorei.

145. Aichryson bollei.

146. Aichryson palmense.

147. Aeonium cuneatum.

148. Aeonium subplanum.

149. Aeonium undulatum.

150. Aeonium holochrysum.

151. Aeonium tabuliforme.

152. Aeonium manriqueorum.

153. Aeonium sedifolium.

154. Aeonium urbicum.

155. Aeonium lancerottense.

156. Aeonium valverdense.

157. Aeonium spathulatum.

158. Aeonium simsii.

159. Aeonium nobile.

160. Aeonium smithii.

161. Greenovia aurea.

162. Dendriopoterium menendezii.

163. Marcetella moquiniana.

164. Adenocarpus viscosus.

165. Adenocarpus foliolosus.

166. Adenocarpus foliolosus.

168. Teline canariensis.

167. Teline linifolia.

169. Teline microphylla.

170. Chamaecytisus proliferus.

171. Spartocytisus filipes.

172. Spartocytisus supranubius.

173. Retama monosperma.

174. Ononis angustissimus.

175. Lotus spartioides.

176. Lotus dumetorum.

177. Lotus hillebrandii.

178. Lotus maculatus.

179. Geranium canariense.

180. Euphorbia canariensis.

181. Euphorbia canariensis.

183. Euphorbia paralias.

182. Euphorbia handiensis.

185. Euphorbia aphylla.

184. Euphorbia balsamifera.

186. Euphorbia obtusifolia.

187. Euphorbia regis-jubae.

188. Euphorbia bourgaeana.

189. Euphorbia mellifera.

190. Euphorbia bravoana.

191. Cneorum pulverulentum.

192. Ruta pinnata.

193. Ruta oreojasme.

194. Ilex canariensis.

195. Maytenus canariensis.

196. Lavatera phoenicea.

197. Lavatera acerifolia.

198. Hypericum canariense.

199. Hypericum glandulosum.

200A. Cistus symphytfolius.

200. Cistus monspeliensis.

201. Ferula linkii.

202. Astydamia latifolia.

203. Erica arborea.

204. Limonium spectabile.

205. Limonium fruticans.

208. Limonium arborescens.

207. Limonium bourgaei.

206. Limonium rumicifolium.

209. Limonium pectinatum.

210. Ceropegia fusca.

211. Ceropegia dichotoma.

212. Ceropegia hians.

213. Caralluma burchardii.

214. Caralluma burchardii.

215. Plocama pendula.

217. Convolvulus floridus.

216. Rubia fruticosa.

218. Convolvulus fruticulosus.

219. Convolvulus glandulosus.

220. Convolvulus caput-medusae.

221. Convolvulus scoparius.

222. Echium giganteum.

223. Echium brevirame.

224. Echium aculeatum.

225. Echium hierrense.

226. Echium strictum.

227. Echium wildpretii.

228. Echium simplex.

229. Echium pininana.

230. Echium onosmifolium.

231. Echium onosmifolium.

232. Echium auberianum.

235. Echium webbii.

233. Echium leucophaeum.

234. Echium decaisnei.

236. Myosotis latifolia.

237. Teucrium heterophyllum.

238. Thymus origanoides.

239. Salvia canariensis.

240. Nepeta teydea.

242. Bystropogon plumosus.

243. Lavandula minutolii

241. Cedronella canariensis.

244. Sideritis macrostachys.

247. Sideritis dendrochahorra.

246. Sideritis cystosiphon.

245. Sideritis infernalis.

249. Sideritis dasygnaphala.

248. Sideritis argosphacelus.

252. Sideritis nutans.

251. Sideritis gomeraea.

250. Sideritis candicans.

253. Micromeria pineolens.

254. Micromeria varia.

255. Micromeria herphyllimorpha.

256. Withania aristata.

257. Solanum lidii.

259. Isoplexis isabelliana.

258. Solanum vespertilio.

260. Isoplexis canariensis.

261. Isoplexis canariensis.

262. Scrophularia langeana.

263. Scrophularia smithii.

264. Scrophularia calliantha.

265. Campylanthus salsoloides.

266. Lyperia canariensis.

267. Kickxia scoparia.

268. Kickxia heterophylla.

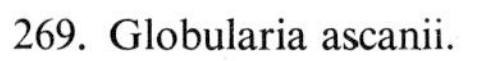

269. Globularia ascanii.

270A. Plantago famarae.

270. Plantago asphodeloides.

271. Viburnum rigidum.

272. Pterocephalus dumetorum.

273. Canarina canariensis.

274. Asteriscus sericeus.

275. Asteriscus stenophyllus.

276. Helichrysum monogynum.

277. Vieraea laevigata.

278. Gonospermum canariense.

279. Argyranthemum gracile.

281. Argyranthemum foeniculaceum.

280. Argyranthemum broussonetii.

282. Argyranthemum haouarytheum.

283. Argyranthemum escarrei.

284. Argyranthemum teneriffae.

285. Argyranthemum ochroleucum.

286. Argyranthemum lidii.

287. Argyranthemum winteri.

289. Argyranthemum frutescens (Gomera).

288. Argyranthemum frutescens (Tenerife).

290. Tanacetum ptarmaciflorum.

291. Tanacetum ferulaceum.

292. Senecio papyraceus.

293. Senecio murrayi.

294. Senecio webbii.

295. Senecio heritieri.

296. Senecio kleinia.

297. Centaurea junoniana.

298. Centaurea arbutifolia.

299. Sonchus palmensis.

300. Sonchus arboreus.

301. Sonchus leptocephalus.

302. Sonchus canariensis.

305. Sonchus acaulis.

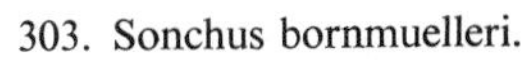

303. Sonchus bornmuelleri.

304. Sonchus congestus.

306. Sonchus platylepis.

307. Sonchus ortunoi.

308. Sonchus brachylobus.

309. Sonchus brachylobus var. canariae.

310. Sonchus tuberifer.

311. Sonchus hierrensis.

313. Sonchus tectifolius.

312. Sonchus radicatus.

314. Tolpis laciniata.

315. Prenanthes pendula.

316. Pulícaria canariensis.

317. Sventenia bulpleuroides.

318. Launaea arborescens.

319. Andryala cheiranthifolia.

320. Phoenix canariensis.

321. Phoenix canariensis.

322. Dracaena draco.

323. Asparagus pastorianus.

325. Melica teneriffae.

324. Semele androgyna
var. gayae.

Species Descriptions

Key to Families

How to use the keys to plants described in this book

The keys provided in the following section are for quick identification of plant species. Each key is composed of pairs of sets of contrasting characters. The first of each pair of sets of characters is preceded by a number **1, 2, 3, 4** etc. and the second, contrasting, set of characters has the same number as the first. Each plant specimen should be examined and compared with the sets of characters starting at **1.** to see with which it best agrees, for example if **1.** states *Trees or shrubs with cones* and the specimen is woody with cones then the user proceeds to **2.** and considers whether the plant is best fitted to the first dichotomy of **2.**, for example *Leaves opposite* or the second *Leaves alternate* and so on.

If the plant does not have the structure described in the first dichotomy of **1.** then the user should seek out the second dichotomy of **1.** which may be on the same page or up to several pages on. The user then proceeds as before comparing the plant with the next sets of characters until, by process of elimination, the Family (eg. Scrophulariaceae) Genus (eg. *Scrophularia*) and finally the species (eg. *Scrophularia arguta*) is reached. Identification can then be confirmed by using the description and distribution information provided coupled with, in many cases, the illustrations. The numbers following some species descriptions refer to the numbered illustration(s) of that species.

1. Trees and shrubs with cones, seeds not enclosed in an ovary.
  2. Leaves opposite or whorled, scale or needle-like, cones more or less globular, fleshy .. .. **Cupressaceae**
  2. Leaves alternate or fasciculate on short shoots, female cones with thick, woody scales .. .. **Pinaceae**
1. Trees shrubs or herbs bearing flowers composed of perianth, stamens and ovary, seeds enclosed in an ovary.

3. Leaves usually parallel-veined, parts of the flower usually in threes or in multiples of 3 .. .. .. (p. 228) **Monocotyledones**
3. Leaves usually net-veined, parts of the flower usually in fours or fives or some multiple of 4 or 5 or a large indefinite number .. .. **(Dicotyledones)**
4. Perianth of 2 distinct whorls (petals and sepals).
5. Petals free.
6. Ovary superior.
7. Carpels free.
8. Stamens numerous, at least 3 times as many as petals.
9. Shrubs or scrambling plants with thorny stems .. **Rosaceae**
9. Herbs with pubescent stems .. .. .. **Ranunculaceae**
8. Stamens twice as many as petals or fewer .. **Crassulaceae**
7. Carpels and styles united or ovary of 1 carpel.
10. Flowers regular.
11. Stamens more than twice as many as petals or stamens and petals numerous.
12. Stamens all united below into a tube round the style .. **Malvaceae**
12. Stamens free or in bundles.
13. Carpel 1, trees or shrubs with white flowers .. .. **Rosaceae**
13. Carpels 2 or more or if 1 then plants not woody.
14. Styles several, free or connate only at the base, stamens united into bundles below.
15. Leaves opposite, herbs or shrubs .. .. .. **Guttiferae**
15. Leaves alternate, trees .. .. **Ternstroemiaceae**
14. Style 1, stigma simple, stamens free .. .. **Cistaceae**
11. Stamens twice as many as petals or fewer.
16. Trees or shrubs.
17. Leaves small, scale-like, flowers numerous in dense spikes .. .. **Tamaricaceae**
17. Leaves not scale-like, not particularly small.
18. Seeds with a white fluffy aril .. .. **Celastraceae**
18. Seeds not arillate.
19. Leaves with tubular, sheathing stipules (ochrae) .. **Polygonaceae**
19. Leaves without tubular sheathing stipules.
20. Fruit a siliqua or silicula .. .. **Cruciferae**
20. Fruit a capsule or berry.
21. Leaves dotted with translucent glands .. .. **Rutaceae**
21. Leaves not gland-dotted.
22. Shrubs with yellow flowers borne on the petioles **Cneoraceae**
22. Trees or shrubs with small inconspicuous flowers in cymes, not borne on petioles .. **Rhamnaceae**
16. Herbs.
23. Leaves compound or lobed.
24. Leaves succulent .. .. **Zygophyllaceae**
24. Leaves not succulent .. .. **Geraniaceae**
23. Leaves entire.
25. Stipules present.
26. Leaves opposite, stipules small, scarious .. .. **Caryophyllaceae**
26. Leaves alternate, stipules tubular, sheathing .. **Polygonaceae**
25. Stipules absent.
27. Leaves opposite, fruit a capsule.

28. Style 1 .. .. **Frankeniaceae**
28. Styles 2–5 .. .. **Caryophyllaceae**
27. Leaves spirally arranged, fruit a siliqua or silicula .. **Cruciferae**
10. Flowers zygomorphic.
29. Flowers saccate or spurred at the base.
30. Leaves much divided, stamens 2 .. .. .. **Papaveraceae**
30. Leaves entire, stamens 5 .. .. **Violaceae**
29. Flowers not saccate or spurred.
31. Petals entire, stamens 10, united into a tube or all but 1 united .. .. **Leguminosae**
31. Petals fimbriate or lobed, stamens free .. .. **Resedaceae**
6. Ovary inferior or partly so.
32. Woody climbers, fruit a fleshy drupe or berry .. **Araliaceae**
32. Herbs or small shrubs, fruit composed of 2 dry, indehiscent carpels, usually separating when ripe .. **Umbelliferae**
5. Petals united at least at the base.
33. Ovary superior.
34. Stamens united into a tube, flowers with a large upper petal .. .. **Leguminosae**
34. Stamens not united into a tube, flowers not with a large upper petal.
35. Stamens twice as many as corolla-lobes.
36. Shrubs, leaves not peltate, carpels united .. .. **Ericaceae**
36. Succulent herbs, leaves peltate, carpels free .. **(Umbilicus)**
35. Stamens as many as or fewer than corolla-lobes.
37. Ovary deeply 4-lobed with 1 ovule in each locule.
38. Leaves alternate .. .. **Boraginaceae**
38. Leaves opposite .. .. **Labiatae**
37. Ovary not deeply 4-lobed.
39. Flowers zygomorphic.
40. Flowers in dense heads surrounded by an involucre .. .. **Globulariaceae**
40. Flowers not in dense heads, involucre absent.
41. Fruit a pair of long follicles, seeds with a long pappus .. .. **Asclepiadaceae**
41. Fruits not long follicles, seeds never with a pappus.
42. Stamens 4 or 2.
43. Fruit an explosive capsule, inflorescences axillary .. .. **Acanthaceae**
43. Fruit a dehiscent capsule, inflorescences usually terminal or if axillary then stamens more than 2 .. .. **Scrophulariaceae**
42. Stamens 5, flowers only slightly zygomorphic **Solanaceae**
39. Flowers regular.
44. Trees with drupaceous or small, berry-like fruits.
45. Juice milky .. .. **Sapotaceae**
45. Juice not milky.
46. Stamens fewer than corolla-lobes .. .. **Oleaceae**
46. Stamens equalling corolla-lobes.
47. Fruits red, stamens alternate with corolla-lobes, leaves usually with spiny margins or tip **Aquifoliaceae**

47. Fruits brownish, stamens opposite the corolla-lobes, leaf-margins never spiny .. .. **Myrsinaceae**
44. Shrubs or herbs, fruits usually capsulate or berry-like.
48. Stamens fewer than corolla-lobes.
49. Fruit a capsule .. .. **Scrophulariaceae**
49. Fruit a berry or small berry-like drupe .. **Oleaceae**
48. Stamens equalling corolla lobes.
50. Calyx large, brightly coloured, stamens opposite corolla-lobes .. .. **Plumbaginaceae**
50. Calyx not large or brightly coloured, stamens alternate with corolla-lobes.
51. Leaves opposite .. .. **Gentianaceae**
51. Leaves alternate or radical.
52. Petals, sepals and stamens 4, flowers in dense, oval heads or long, dense spikes .. **Plantaginaceae**
52. Petals, sepals and stamens usually 5, flowersusually in lax cymes, racemes or panicles.
53. Fruit a capsule .. .. **Convolvulaceae**
53. Fruit a berry.. .. .. **Solanaceae**
33. Ovary inferior.
54. Flowers in heads surrounded by an involucre.
55. Anthers fused into a tube round the style .. .. **Compositae**
55. Anthers free .. .. **Dipsacaceae**
54. Flowers not in heads with an involucre.
56. Leaves in whorls of 4 or more, flowers regular, petals 4 **Rubiaceae**
56. Leaves not in whorls, flowers zygomorphic or if regular then petals 5.
57. Herbs, climbing with tendrils .. .. **Cucurbitaceae**
57. Shrubs or herbs, tendrils absent.
58. Shrubs, inflorescences many-flowered .. .. **Caprifoliaceae**
58. Herbs, inflorescences few-flowered .. .. .. **Campanulaceae**
4. Perianth segments all petaloid, all sepaloid or absent.
59. Perianth segments all petaloid.
60. Stipules sheathing, scarious .. .. **Polygonaceae**
60. Stipules not sheathing, not scarious, often absent.
61. Leaves in whorls of 4 or more .. .. .. **Rubiaceae**
61. Leaves not in whorls.
62. Flowers in heads surrounded by an involucre.
63. Stamens free, flowers hermaphrodite .. .. **Dipsacaceae**
63. Anthers fused into a tube round the style, flowers hermaphrodite or unisexual .. .. .. **Compositae**
62. Flowers not in dense heads though usually in umbels **Umbelliferae**
59. Perianth segments sepaloid or absent.
64. Trees or shrubs.
65. At least the male flowers in catkins.
66. Fruit a 2-valved capsule .. .. **Salicaceae**
66. Fruit a drupe, samara, berry or nutlet but not capsulate.
67. Leaves entire, fruit a somewhat fleshy drupe .. **Myricaceae**
67. Leaves pinnately lobed, fruit a berry, samara or nutlet .. .. **Rosaceae**

65. Flowers not in catkins.
68. Ovary inferior .. .. **Santalaceae**
68. Ovary superior.
69. Plants with milky latex, capsules 3-lobed .. .. **Euphorbiaceae**
69. Plants without milky latex, fruit not a 3-lobed capsule.
70. Trees or robust shrubs.
71. Ovary with 2– to 4–locules, fruit a drupe .. **Rhamnaceae**
71. Ovary with 1 cell, fruit an ellipsoidal berry .. **Lauraceae**
70. Shrubs, fruit dry or a very small berry.
72. Perianth segments enlarging in fruit, leaves fleshy .. .. **Chenopodiaceae**
72. Perianth segments not enlarging in fruit, leaves not fleshy.
73. Flowers hermaphrodite, fruit a small berry (*Bosia*) **Amaranthaceae**
73. Flowers unisexual, fruit a small, dry achene .. **Urticaceae**
64. Herbs.
74. Leaves opposite.
75. Fruit a capsule .. .. **Caryophyllaceae**
75. Fruit an achene enclosed in the perianth .. .. **Urticaceae**
74. Leaves alternate.
76. Fruit a trigonous nut .. .. **Polygonaceae**
76. Fruit an achene enclosed in the perianth .. .. **Urticaceae**

PINACEAE —————— Pine Family

Trees. Leaves simple, needle-like, grooved, in clusters on lateral shoots. Flowers monoecious, in cones. Male cones small, the female large and woody in fruit.

PINUS

Evergreen trees with needle-like leaves borne on short shoots. Male flowers in short, catkin-like clusters. Female flowers forming ovoid to conical cones.

P. canariensis Chr. Sm. ex DC. Up to 30 m. Bark thick, reddish-brown. Leaves up to 30 cm, in threes, slender, acute, densely crowded. Cone about 10–20 cm. ***Pino Canario*. (122, 123)**

TENERIFE, HIERRO, GRAN CANARIA, LA PALMA: Very common in mountain regions above 1000 m. Locally dominant in many places. Particularly well developed pine forests occur at Tamadaba on GRAN CANARIA, Esperanza, Vilaflor and Agua Manza on TENERIFE, the central region of HIERRO and the Caldera, Cumbrecita and Fuencaliente regions of LA PALMA.

P. canariensis is not native on GOMERA where a few trees have, however, been planted in the Alto Garajony region.

CUPRESSACEAE

Resiniferous, monoecious or dioecious trees or shrubs with opposite or whorled leaves. Fruit a cone or berry-like.

JUNIPERUS

Monoecious or dioecious shrubs or small trees. Leaves opposite or in whorls of 3, needle-like or scale-like. Cones axillary or terminal, ovoid to globose, indehiscent, berry-like. Seeds ovoid.

1. Adult leaves needle-like, fruit axillary .. .. .. **J. cedrus**
1. Adult leaves scale-like, fruits usually terminal .. .. **J. phoenicea**

J. cedrus Webb & Berth. Shrub or small tree up to 15 m. Branches more or less pendulous. Leaves needle-like, flat, in whorls of 3. Cone globose, axillary, reddish-brown when ripe. ***Cedro.*** **(124)**

TENERIFE: Agua Mansa, 1400 m, Las Cañadas, La Fortaleza, Degollada de los Cedros, Montaña Blanca, etc, 1900–2200 m, rare on cliffs; LA PALMA: High mountains of the outer rim of the Caldera de Tabouriente, Pico del Cedro, etc, 1600–2000 m, rare; GOMERA: Roque de Agando.

J. phoenicea L. Shrub or small tree up to 8 m. Adult leaves scale-like, more or less triangular, closely appressed to the twigs. Cones globose, terminal, black at first, becoming dark red when ripe. ***Sabina.*** **(125)**

TENERIFE: Sierra Anaga, Taganana, Roques de Anaga, etc, S. region, Arico, Guimar, Masca, N. coast, Los Silos, Las Cañadas, 200–1900 m; LA PALMA: Caldera de Tabouriente; GOMERA: N. and N.W. regions, locally very frequent, Agulo, Vallehermoso, Epina, Arure, 300–750 m; HIERRO: Locally common in the C. area of the island, Pico de Tenerife, El Golfo, Frontera, Sabinosa, etc; GRAN CANARIA: Tenteniguada.

DICOTYLEDONES

SALICACEAE —————— Willow Family

Dioecious trees or shrubs. Leaves alternate, simple. Flowers in catkins. Perianth absent. Male flowers with numerous stamens. Female flowers with an ovary of 2 carpels. Styles 2. Stigmas usually bifid. Fruit a 2-valved capsule.

SALIX

Buds with one outer scale. Flowers in catkins. Each flower subtended by a bract. Fruit a 2-valved capsule.

S. canariensis Chr. Sm. Tall shrub or small tree up to 10 m. Leaves oblong to lanceolate, entire or crenate, pubescent beneath. Catkins up to 6 cm long. ***Sauce.*** (2)

TENERIFE: Laurel woods, stream-sides, in damp places, Las Mercedes, Agua Garcia, Barranco Ruiz, S. valleys, Barranco del Infierno, Guia de Isora; GRAN CANARIA: Los Tiles de Moya, Guia, Barranco de la Angostura 300–1200 m; LA PALMA: Barranco de las Angustias, N. & N.E. regions, Los Tiles etc; GOMERA: Forest zone, Agando, El Cedro, Hermigua valley; HIERRO: Forests of El Golfo, Fuente de Tinco etc.

MYRICACEAE

Dioecious trees or shrubs. Flowers in catkins in the axils of bracts. Stamens 2-many. Female flowers with 2 or more bracteoles. Ovary superior, 1-celled. Style short.

MYRICA

Leaves alternate, simple, rather aromatic. Fruit a drupe.

M. faya Aiton. Evergreen shrub or small tree up to 10 m. Twigs with peltate hairs. Leaves oblanceolate, 4–12 cm; base cuneate; margins somewhat revolute. Catkins usually branched, borne amongst the leaves of the current year's growth. Fruit a reddish to black drupe with a rather fleshy, rough, waxy surface. ***Faya.*** **(3)**

ALL ISLANDS: Locally very common in forests and degraded forest scrub on the W. ISLANDS and GRAN CANARIA, 400–1500 m, often in association with *Erica arborea* ('Faya/Brezal'), very rare on the E. ISLANDS, LANZAROTE: Peñitas de Chache.

URTICACEAE ——————— Nettle Family

Shrubs or herbs. Leaves simple, stipulate, hairy, the hairs sometimes stinging. Flowers small, unisexual. Ovary free, superior. Styles 1 or 2. Fruit small, 1-seeded, dry.

1. Tall shrubs, inflorescences terminal .. .. .. **Gesnouinia**
1. Perennial herbs, sometimes woody at the base or annuals, inflorescences axillary.
   2. Leaves alternate, spiny .. .. **Forsskahlea**
   2. Leaves opposite, not spiny.
      3. Leaf-margins dentate, inflorescences stalked .. .. **Urtica**
      3. Leaf-margins entire, inflorescences sessile .. .. **Parietaria**

PARIETARIA

Perennial. Leaves opposite. Flowers polygamous with bisexual and male flowers.

P. filamentosa Webb & Berth. Woody-based shrublet with long, procumbent branches. Leaves ovate, acuminate. Inflorescences sessile, axillary.

TENERIFE: Cliffs of the S.W. region and on the N. coast, Ladera de Guimar, Barranco de Tamadaya, Tamaimo, Masca, Teno 200–450 m, often frequent, Orotava valley to Icod, occasional; LA PALMA: Mazo, Barranco de Santa Lucia.

URTICA

Leaves opposite, dentate, with stinging hairs; stipules present. Flowers monoecious or dioecious in small, axillary, spicate inflorescences.

1. Perennial, inflorescences shorter than leaves .. .. **U. morifolia**
1. Annual, inflorescences longer than leaves .. .. **U. stachyoides**

U. morifolia Poir. Woody perennial. Leaves ovate, cordate at base, coarsely dentate. Inflorescences small, long-pedunculate, shorter than the leaves. ***Ortigon.***

TENERIFE: Laurel forests, Las Mercedes, Taganana, Agua Garcia, Agua Mansa, Los Silos, Barranco de Añavigo etc, 500–1400 m; LA PALMA: N. coast, Garafia, Barranco Franceses, Los Tiles, La Galga; GOMERA: El Cedro, Las Hayas, Epina, Arure; HIERRO: El Golfo, Fuente de Tinco, Frontera, Jinamar; GRAN CANARIA: Barranco de Moya, San Mateo 800–1200 m, locally frequent.

U. stachyoides Webb. Annual herb with the inflorescences longer than the leaves.

TENERIFE: Guimar, Barranco de Badajoz, Orotava; GRAN CANARIA: La Atalaya, Caldera de Bandama; HIERRO: Valverde, central region, El Golfo, lower zone, occasional, 50–500 m.

GESNOUINIA

Shrubs. Flowers monoecious, 1 female and 2 male in each involucre. Involucres clustered into a panicle. Male flowers; calyx 4-partite, stamens 4; female flowers: ovary included, styles short.

G. arborea (L.) Gaudich. Small tree or shrub. Leaves entire, acuminate, 3-nerved, pubescent; stipules absent. Inflorescence a dense panicle. Seeds (achenes) enclosed by the calyx. ***Ortegon de los Montes.*** **(4)**

TENERIFE: Laurel forest zone, Sierra Anaga, Las Mercedes, Vueltas de Taganana 600–800 m, Icod el Alto, Barranco del Agua near Los Silos, rare; LA PALMA: Laurel woods at Los Tiles, Cubo de la Galga, Cumbre Nueva, Barlovento etc; GOMERA: El Cedro forest, Roque de Agando, Chorros de Epina, 600–1000 m; HIERRO: Forests of El Golfo; GRAN CANARIA: Los Tiles de Moya, Barranco de la Virgen, rare.

FORSSKAHLEA

Small shrubs or perennial herbs. Leaves alternate, dentate, spiny; stipules present. Flowers monoecious. Male flower with 1 stamen.

F. angustifolia Retz. Leaves densely lanate beneath, margins prickly. Inflorescences axillary, small, pinkish. ***Ratonera.***

ALL ISLANDS: Common in dry areas of the lower zone (N. Africa).

SANTALACEAE

Shrubs or herbs. Leaves alternate, simple, sometimes scale-like. Flowers small, hermaphrodite or unisexual. Perianth 5-lobed. Ovary inferior. Style 1. Fruit a small drupe. Hemiparasites.

1. Plants dioecious, leaves lanceolate, fleshy .. .. **Osyris**
1. Plants hermaphrodite, leaves small, scale-like .. .. **Kunkeliella**

OSYRIS

Dioecious, evergreen shrub up to 3 m. Leaves lanceolate. Perianth 3- to 4-parted.

O. quadripartita Salzm. ex Decne. Robust. Leaves lanceolate, leathery. Flowers small, in axillary clusters of 2–3. Fruit a fleshy, red drupe.

TENERIFE: Cuevas Negras de Los Silos, 300 m. Locally frequent on N. slopes (S. Spain, N. Africa).

KUNKELIELLA

Hemiparasitic *Ephedra*-like shrubs. Leaves small, scale-like. Inflorescences axillary, 1-flowered. Flowers hermaphrodite, 5-parted. Fruit a small, succulent drupe.

1. Branchlets glabrous (Gran Canaria) .. .. .. **K. canariensis**
1. Branchlets setulose (Tenerife) .. .. **K. psilotoclada**

K. canariensis Stearn. Shrub up to 80 cm. Lower branches brownish, woody; upper green, flexuous, glabrous. Flowers small, cream-coloured. Perigon lobes acute. Ripe fruits globose, white, about 6 cm across, 1-seeded. **(126)**

GRAN CANARIA: Riscos de Guayadeque, 700 m, in *Teline/Euphorbia* scrub, rare.

K. psilotoclada (Svent.) Stearn. Like *K. canariensis* but the branchlets setulose and the flowers larger with the perigon lobes obtuse.

TENERIFE: Barranco de Masca, dry rocky slopes, 800 m, v. rare.

POLYGONACEAE —————— Dock Family

Herbs or shrubs. Leaves alternate, simple, with a membranous sheath (ochreae) in place of stipules encircling the nodes of the stem. Flowers hermaphrodite. Perianth of 3–6 segments, enlarging and becoming membranous in fruit. Ovary superior. Fruit a 1-seeded nut, flattened, winged or 3-angled.

RUMEX

Shrubs or perennial herbs. Flowers unisexual. Perianth in 2 whorls of 3, the inner enlarging and becoming hard to form the valves in fruit. Stamens 6. Fruit a small nut.

1. Shrub, leaves ovate, truncate at base, inflorescence branched .. .. **R. lunaria**
1. Herb, leaves ovate-deltoid, hastate at base, inflorescence usually simple .. .. **R. maderensis**

R. lunaria L. Shrub. Leaves often wider than long, truncate at base, rounded at apex. Inflorescence a compound panicle. Valves reniform or orbicular. ***Vinagrera.*** **(127)**

ALL ISLANDS: Very common shrub in the lower zone, in *Euphorbia* communities; GRAN CANARIA: Cuesta de Silva, Moya, Tafira, Fataga etc; TENERIFE: Orotava valley, Teno, Sierra Anaga.

R. maderensis Lowe. Perennial herb, leaves ovate-deltoid. Panicle dense, valves membranaceous, suborbicular.

TENERIFE: Agua Mansa, El Tanque, Garachico 600–1200 m, local; LA PALMA: S. region, Sta. Cruz, Mazo, rare; GOMERA: Cumbre de Vallehermoso, Epina, 600 m; HIERRO: Risco de Jinamar, Fuente de Tinco, Frontera.

R. vesicarius a small herb with large, reddish inflated valves and entire, lanceolate leaves is frequent on the S. slopes of TENERIFE and some of the other islands, the minute *R. bucephalophorus* is a very common weed and *R. acetosella* is frequent in forest clearings on GOMERA.

CHENOPODIACEAE

Herbs or shrubs, often succulent. Leaves alternate or opposite, exstipulate. Flowers hermaphrodite or unisexual, solitary or in small cymes. Perianth 1- to 5-merous, often enlarging in fruit. Ovary superior or semi-inferior. Fruit usually an achene.

1. Leaves more or less flat, ovary connate with the receptacle in fruit .. .. **Beta**
1. Leaves more or less cylindrical, ovary not connate with the receptacle in fruit.
   2. Flowers subtended by two bracteoles, leaves green .. **Traganum**
   2. Flowers not subtended by 2 bracteoles, leaves white-lanate .. .. **Chenolea**

Mediterranean genera with non-endemic species in coastal regions include *Suaeda*, *Salsola* and *Atriplex*.

BETA

Glabrous herbs. Leaves flat. Flowers hermaphrodite, in few-flowered cymes or solitary in more or less spicate inflorescences, 5-parted. Ovary connate with the receptacle in fruit.

1. Annual, leaves cordate .. .. **B. patellaris**

1. Perennial, leaves hastate or sagittate.
  2. Leaves ovate or deltoid .. .. **B. procumbens**
  2. Leaves more or less linear.. .. .. **B. webbiana**

B. patellaris Moq. Procumbent annual, up to 60 cm. Leaves triangular-ovate, usually cordate. Inflorescence lax with leafy bracts. Cymes 1- to 3-flowered. Stigmas 2. Fruits usually single. **(6)**

ALL ISLANDS: Common in coastal regions and in dry rocky areas to 250 m.

B. procumbens Sm. Very variable, procumbent or scrambling perennial. Stems herbaceous. Leaves long-petiolate, hastate or sagittate, ovate to deltoid, the margins remotely sinuate-lobed. Flowers solitary or in groups of 2–3. **(5)**

TENERIFE: N. coast region, Taganana to Teno, locally very common, S. coast, El Medano, Los Cristianos; GRAN CANARIA: N. coast, Las Palmas to Galdar, Playa de Jinamar, Telde, up to 300 m; HIERRO: LA PALMA, GOMERA: Local, in coastal areas; FUERTEVENTURA: S. region, Gran Tarajal to Jandia; LANZAROTE: N. coast region, Playa de Famara, Arrecife.

B. webbiana Moq. Like *B. procumbens* but the stems woody at the base, and the leaves linear-sagittate, sometimes lobed.
GRAN CANARIA: N. coast, Cuesta de Silva, Guia, etc, Agaete to San Nicolas; TENERIFE: Puerto de la Cruz; FUERTEVENTURA: C. region, Betancuria, Pajara, Puerto Rosario.

Two widespread Mediterranean species, *B. macrocarpa* Guss. and *B. vulgaris* L., are also common in coastal regions of the Canaries.

CHENOLEA

Pubescent herbs or subshrubs. Leaves linear to linear-lanceolate, entire, cylindrical. Perianth 5-parted, the segments developing a spine on the back in fruit.

C. tomentosa (Lowe) Maire (*C. canariensis* Moq.) Erect perennial up to 50 cm. Stem and leaves white-villous, later becoming glabrescent. Flowers solitary or in fascicles of 2–3 in the axils of the upper leaves.

TENERIFE: Coastal regions, Punta de Teno, S. coast, El Medano, Los Cristianos, etc; GRAN CANARIA: E. coast, Melanara, Arinaga, S. region, Maspalomas, Mogan; FUERTEVENTURA: Frequent in coastal areas, dunes etc, Gran Tarajal, Jandia, Cofete, Lobos, Corralejos, Puerto Rosario; LANZAROTE: Arrecife, Playa de Famara, Graciosa; HIERRO: Cuesta de Sabinosa. (N. Africa).

TRAGANUM

Small branched shrubs up to 1 m. Branches villous. Leaves entire, cylindrical. Flowers axillary, solitary, subtended by 2 bracteoles. Seeds small, dry.

T. moquinii Webb. Robust. Stems striate. Leaves green-yellowish, narrowly oblong to oval, acute. Flowers densely lanate.

FUERTEVENTURA: Locally common, plant of summits of sand-dunes, Jandia, Matas Blancas, Cofete, Corralejos, Puerto Rosario, etc; LANZAROTE: Playa de la Caleta, Graciosa; GRAN CANARIA: Jinamar, Maspalomas; TENERIFE: S. coast, El Medano, Los Cristianos. (N. Africa).

AMARANTHACEAE

Herbs or shrubs. Leaves opposite or alternate, entire. Flowers hermaphrodite or unisexual. Perianth of 3–5 segments, free or connate at the base, dry and scarious. Stamens 3–5. Ovary unilocular. Fruit a berry or dry and membranous.

BOSIA

Dioecious shrubs. Flowers 5-parted; male flowers with rudimentary ovary; female flowers with 5 stamenodes. Stigmas 3, small, fleshy. Fruit a small berry.

B. yervamora L. Shrub up to 3 m. Branches greenish, slender. Leaves up to 7 cm long, ovate-lanceolate, alternate, petiolate, glabrous. Flowers in short terminal or axillary, racemose inflorescences, greenish with two scarious bracts at the base of the pedicel. Fruits greenish-black, about the size of a small pea. ***Hierbamora.***

TENERIFE: Locally common particularly along the N. coast, Barranco Hondo, Santa Ursula, La Rambla, Los Silos; GRAN CANARIA: N. part of the island, Tafira, Moya, Agaete, Bandama etc; LA PALMA: W. coast region, Tazacorte, Santa Cruz; GOMERA: Barranco de la Villa, Agulo, Vallehermoso, dry slopes in the lower zone.

CARYOPHYLLACEAE ——— Carnation Family

Herbs or small shrubs. Leaves usually opposite or whorled, entire. Scarious stipules sometimes present. Flowers regular. Sepals 4–5, free or fused. Petals 4–5, free, sometimes absent. Stamens 8–10. Ovary superior. Stigmas 1 or 2–5. Fruit a capsule dehiscing with as many or twice as many teeth as styles.

1. Shrubs up to 50 cm, in coastal regions, stigma simple .. **Gymnocarpos**
1. Small shrublets or herbs usually less than 20 cm, stigmas lobed.
  2. Stipules present, papery, sometimes caducous.
    3. Stigmas 1 or 2.
      4. Inflorescence bracts usually very conspicuous, sometimes longer than the flowers, leaves aristate .. **Paronychia**
      4. Bracts inconspicuous, always much shorter than the flowers, leaves not aristate.
        5. Leaves linear, succulent, blue-glaucous .. .. **Dicheranthus**
        5. Leaves lanceolate-elliptical, flat, greenish-yellow .. **Herniaria**
    3. Stigmas 3 or 5 .. .. **Polycarpaea**
  2. Stipules absent.
    6. Sepals free.
      7. Styles 2, capsule with 2 teeth, leaves setaceous .. **Bufonia**
      7. Styles 3–5, capsules with at least as many teeth as styles, leaves not setaceous.
        8. Capsules with as many teeth as styles, petals pinkish **Minuartia**
        8. Capsules with twice as many teeth as styles, petals white .. .. **Cerastium**
    6. Sepals joined to form a calyx tube .. .. .. **Silene**

MINUARTIA

Annual to perennial decumbent herbs without stipules. Capsule teeth as many as styles, wide and obtuse.

1. Perennial with broad leaves, petals shorter than sepals .. **M. platyphylla**
1. Annual with linear leaves, petals equalling or longer than sepals .. .. **M. webbii**

M. platyphylla (Christ) McNeill. Glandular, decumbent, perennial, often somewhat woody at base. Stems up to 40 cm. Leaves ovate to suborbicular, up to 1·5 cm long, apiculate. Inflorescences 1- to few-flowered. Calyx segments lanceolate, longer than the pink petals.

FUERTEVENTURA: Jandia region, especially on north coast cliffs up to 600 m, Cofete. LANZAROTE: Famara massif, locally frequent up to 700 m. (Riscos de Famara).

The narrower-leaved Mediterranean species *M. geniculata* (Poiret) Thell. also occurs in the same area of both islands but is distinguishable by its linear to lanceolate, grey to reddish tinged leaves.

M. webbii McNeill & Bramwell (*Rhodalsine gayana* Webb, non *Minuartia gayana* Maire). Annual to perennial glandular herb. Stems divaricate. Leaves linear to linear-lanceolate. Calyx teeth elliptical; margins scarious. Petals pink, longer than or equal to the calyx.

FUERTEVENTURA: Coastal sands, Gran Tarajal, N. region dunes between Toston and Corralejos, locally common.

CERASTIUM

Perennial herbs. Flowers in cymose inflorescences. Sepals free. Petals white, bifid. Nectaries present. Styles 5. Fruit a cylindrical capsule with twice as many teeth as styles.

C. sventenii Jalas. Stems up to 50 cm, glandular-hairy, somewhat woody at base. Leaves lanceolate, acute, up to 3 cm long, pubescent, the midrib prominent beneath. Inflorescences 6- to 8-flowered. Sepals erect, glandular, the margins scarious. Petals slightly longer than calyx.

LA PALMA: High mountains of the rim of the Caldera de Tabouriente, Los Roques, Siete Fuentes 1900–2000 m, Cumbre Vieja, rare.

BUFONIA

Erect, perennial herbs, branching at the base. Leaves linear, setaceous. Inflorescences cymose-paniculate. Sepals 4. Petals 4, shorter than or equalling the sepals, white. Styles 2. Capsule with 2 teeth. Seeds few, compressed.

B. teneriffae Christ. Much branched. Perennial up to 35 cm. Petals almost equalling the shorter pair of outer sepals. Capsule 2–3 mm. Seeds flattish with tuberculate margin and rugose faces.

TENERIFE: Las Cañadas, locally common in parts of the subalpine zone, 2000 m. Los Azulejos, Los Roques etc. (Records from GRAN CANARIA probably refer to the southern European *B. paniculata* which resembles this species in some facies but is annual in habit).

SILENE

Perennial herbs or small shrubs. Calyx tube with up to 30 veins and 5 small teeth. Petals clawed. Stamens 10. Styles 3–5. Fruit capsular with 6 (8 or 10) teeth. Carpophore present.

1. Calyx with prominent dark brown to purple veins.
 2. Leaves narrowly lanceolate, flowers suberect at anthesis (subalpine zone of Tenerife) .. .. **S. nocteolens**
 2. Leaves lanceolate to spathulate, flowers inclined at anthesis (forest cliffs of La Palma) .. .. .. **S. pogonocalyx**
1. Calyx with pale brown or green veins.
 3. Leaves glabrous.
 4. Flowers reddish, leaves spathulate (Hierro) .. .. **S. sabinosae**

4. Flowers white, leaves obovate to oblanceolate (Tenerife) **S. lagunensis**

3. Leaves pubescent at least beneath.
   5. Leaves ovate-spathulate, calyx veins brownish, flowers usually inclined at anthesis (Tenerife) .. .. .. **S. berthelotiana**
   5. Leaves lanceolate, calyx veins pale green, flowers erect at anthesis (Gomera) .. .. **S. bourgaeii**

S. lagunensis Chr. Sm. Perennial cliff plant, the base and stems woody. Lower leaves petiolate, obovate to oblanceolate, glabrous, acuminate, the margins minutely ciliate to erose; cauline leaves lanceolate, sessile. Inflorescence simple with 2- to 5-flowered, more or less sessile dichasia. Flowers inclined at anthesis, on very short pedicels. Calyx densely glandular-hairy, narrowly oblong, enlarging somewhat in fruit, the veins greenish or light brown; teeth 3 mm, acute. Petals white, the limb about 6 mm. Capsules ovoid, longer than the carpophore.

TENERIFE: Forest and xerophytic cliffs in the Anaga region and along the N. coast. Punta del Hidalgo, Taganana, Pico Inglés, 200–700 m, rare.

S. bourgaeii Webb ex Christ. Like *S. lagunensis* but leaves lanceolate, finely pubescent especially beneath. Flowers pedicellate, erect at anthesis. Calyx veins pale green.

GOMERA: North coast cliffs up to 500 m. Agulo, Hermigua, rare.

S. sabinosae Pitard. Like *S. lagunensis* but very robust. Leaves spathulate, mucronate. Flowers reddish. Capsule oblong-conical.

HIERRO: El Golfo region from E. of Frontera to Riscos de Sabinosa, forest and lower zone cliffs, 200–600 m, very local.

S. berthelotiana Webb ex Christ. Small shrub. Leaves ovate, long petiolate, pubescent, the margins ciliate. Inflorescence laxly branched with long-pedunculate dichasia. Flowers inclined to suberect. Calyx veins brown; teeth acute. Petals pale pink.

TENERIFE: Cliffs in the deep barrancos of the S. region of the island between Guia de Isora and Guimar, Barranco de Añavigo, Barranco del Fraile, 500–1000 m.

S. nocteolens Webb & Berth. Subshrub up to 1 m, the base woody. Leaves tufted at base of flowering stems, narrowly lanceolate, acute. Inflorescences long; dichasia 2- to 6-flowered. Pedicels longer than calyx. Flowers suberect at anthesis. Calyx pubescent; veins dark brown, prominent; teeth small, less than 3 mm, acute. Petals pinkish-white. Capsules ovate, longer than carpophore. **(11)**

TENERIFE: Subalpine zone of Las Cañadas. Locally very common, La Fortaleza, Ucanca etc, ca. 2000 m.

S. pogonocalyx (Svent.) Bramwell. Lax perennial with woody base. Stems up to 80 cm. Leaves lanceolate to spathulate, acute or obtuse, shortly and sparsely pubescent on margins and lower midrib. Inflorescence long, lax; dichasia 4- to 5-flowered, branched. Pedicels subtended by small linear bracts. Flowers inclined at anthesis. Calyx shorter than pedicels, pubescent; veins deep purple-brown; teeth 2 mm, acute. Petals pink or white. Capsule ovoid, twice as long as carpophore.

LA PALMA: Locally common on cliffs in laurel and pine forests 600–1800 m. La Galga, Los Tilos, La Cumbrecita.

POLYCARPAEA

Small shrublets with woody stock or perennial herbs. Leaves in verticillasters.

Stripules scarious. Flowers 5-parted. Petals (staminodes) present. Style trifid at tip. Capsule 3-valved.

1. Leaves densely to sparsely pubescent.
  2. Leaves lanceolate to ovate, very succulent; apex obtuse or acute (Coastal rocks and dunes) .. .. .. **P. nivea**
  2. Leaves linear to narrowly spathulate, not or only slightly succulent; apex aristate.
    3. Densely branched shrublet with silvery leaves, cymes dense .. .. **P. tenuis**
    3. Sparsely branched, leaves grey-brown, cymes lax .. **P. aristata**
1. Leaves more or less glabrous.
  4. Leaves more or less orbicular.
    5. Leaves glabrous, glaucous, very succulent, apex obtuse (plant of cliff-faces) .. .. **P. carnosa**
    5. Leaves subglabrous, not glaucous, fleshy but more or less flat, apex usually aristate (plant of roadsides and forest tracks) .. .. **P. latifolia**
  4. Leaves linear to lanceolate or bradly spathulate.
    6. Leaves lanceolate to broadly spathulate, inflorescence very dense (common on roadsides, coastal regions and in forests) .. .. **P. divaricata**
    6. Leaves long, linear, inflorescence diffuse, dichotomously branched (forest cliffs on La Palma) .. .. **P. smithii**

P. tenuis Webb ex Christ. Dense, much branched, woody at base. Leaves linear to subspathulate, white-silvery, pubescent, aristate. Cymes axillary and terminal, dense. **(8)**

TENERIFE: Subalpine zone of Las Cañadas, 1800–2200 m, Boca de Tauce, La Fortaleza, Topo de la Grieta, Montaña de Deigo Hernandez, El Portillo, Los Roques etc; common.

P. aristata Chr. Sm. Lax, dichotomously branched, leaves narrow, spathulate, grey-green, sparsely pubescent, aristate. Cymes axillary and terminal, small, lax.

TENERIFE: Lower zone and pine forests 200–1200 m. Agua Manza, Teno, Barranco del Fraile near Guia de Isora, El Medano, sporadic. Plants from the Tamadaya Pine forest of GRAN CANARIA seem to belong to this species. The leaves are, however, silvery and the plants rather small and densely branched.

P. latifolia Poiret. Procumbent, base woody. Leaves orbicular, succulent, subglabrous, apex filiform, acute. Axillary inflorescences sessile, small, terminal large. **(129)**

TENERIFE: Sierra de Anaga, Roque de los Pasos, Las Mercedes, Taganana, Los Silos, roadsides and tracks, wet places in the forest zone; GOMERA: N.W. region, Chorros de Epina, Lomo de Carretón, Bosque de La Haya, 600–1000 m; locally common.

P. carnosa Chr. Sm. Very woody, pendulous cliff plant. Leaves orbicular, very fleshy, glabrous, glaucous, apex obtuse. Inflorescences usually terminal, subglobular. **(9)**

TENERIFE: Coastal zones of the mountains of Teno and Anaga, El Fraile W. of Buenavista, Masca, Tamaimo, Risco Blanco, Ingueste de San Andrés, chasmophyte, usually on basalt cliffs; GOMERA: N.E. region in similar habitats, Barranco de la Villa, La Laja, 50–500 m; locally abundant.

P. divaricata Aiton (*P. teneriffae* Lam.) Small procumbent shrublet with or without woody base. Leaves lanceolate to broadly spathulate, glabrous or sparsely pubescent, the apex acute or obtuse. Inflorescence large, terminal, much branched with long fruiting branches. **(10)**

TENERIFE: Locally very common, N. coast, Playa de Castro, La Rambla, Teno, Guimar etc, 5–1200 m; GOMERA: Locally abundant in most parts of the island; LA PALMA: Fuencaliente, Barranco de las Angustias, N. coast etc; GRAN CANARIA; HIERRO; LANZAROTE; FUERTEVENTURA.

P. nivea Aiton. Prostrate plant with very woody rootstock. Branches short, ascending. Leaves lanceolate to ovate, very succulent, very densely silvery pubescent, obtuse. Inflorescences small, silvery, globular to lax and divaricately branched. **(7)**

TENERIFE: Playa Las Americas, El Medano, Los Cristianos etc., dunes and coastal rocks; GRAN CANARIA: Punta de Arinaga, Melenara, S. coast; FUERTEVENTURA: Matas Blancas de Jandia, Cotillo, Cofete etc; LANZAROTE: Playa de Famara, Graciosa; locally frequent, 0·50 m, always near the sea.

P. smithii Link. Pendulous cliff plant with long hanging stems. Leaves long, linear, succulent, glabrous, obtuse. Inflorescence large, dichotomously branched, diffuse. **(128)**

LA PALMA: Forest cliffs 500–1000 m, Cubo de la Galga, Tunel de la Galga, Barranco de los Tiles, Barranco del Agua, La Caldera etc; locally frequent.

GYMNOCARPOS

Shrubs. Leaves opposite or in groups at nodes, entire. Flowers in terminal cymes. Sepals 5, shortly aristate. Corolla absent. Stamens 5. Stigma simple.

G. salsoloides Webb ex Christ. Shrub up to 50 cm. Stems white-greyish brown. Leaves linear to lanceolate, succulent, up to 2 cm long, entire. Inflorescences greyish-green, about 1 cm across. **(130)**

TENERIFE: S. slopes and Punta de Teno, near sea-level; locally abundant.

Plants from the E. islands and GRAN CANARIA have shorter leaves and pale, reddish-brown inflorescences and seem to be referable to the North African *G. decander* Forsk.

DICHERANTHUS

Peduncles 3-flowered, the lateral flowers functionally male. Flowers 5-parted. Corolla absent. Stamens 2-3. Style 1, bifid or trifid.

D. plocamoides Webb. Dwarf shrub up to 30 cm. Stems erect or ascending. Leaves linear, blue-glaucous, succulent. Inflorescences terminal, cymose, flattish. Calyx lobes greyish-pink. **(131)**

TENERIFE: Teno region, locally frequent up to 500 m. Masca, Teno, Valle de Santiago, Los Silos. GOMERA: N. coast from San Sebastian to Vallegranrey, up to 400 m; locally very common at Vallehermoso.

PARONYCHIA

Flowers small, in dense cymes. Bracts silvery. Sepals 5. Petals absent or rudimentary. Stamens 2–5. Stigmas 2. Capsules with 5 valves or indehiscent.

1. Leaves paired, opposite, lanceolate .. .. .. **P. canariensis**
1. Leaves in groups of 3–5 at nodes, linear .. .. .. **P. gomerensis**

P. canariensis Juss. Perennial with woody bases to the stems, procumbent. Leaves opposite, lanceolate acute or acuminate; stipules papery. Inflorescenses in dense cymes at ends of branches. ***Nevadilla*** **(12)**

TENERIFE, GRAN CANARIA, LA PALMA, GOMERA, HIERRO: Locally very common in the xerophytic zone up to 800 m.

P. gomerensis (Burchard) Svent. & Bramwell. Like *P. canariensis* but with 3–5 linear leaves at each node. Inflorescences smaller and less dense.

GOMERA: Local in the S. and S.W. of the island, up to 800 m, Chipude, Vallegranrey; rare.

HERNIARIA

Flowers in axillary or terminal glomerules. Calyx of 4–5 sepals. Petals 4, rudimentary or absent. Stamens 2–4. Stigmas 2. Capsule indehiscent, shorter than calyx.

H. canariensis Chaudr. Prostrate woody-based perennial. Leaves more or less flat, lanceolate-elliptical, clustered. Flowers very small, in clusters along the younger branches.

TENERIFE: S. coast, locally frequent in sandy and rocky places up to 400 m El Medano, Adeje; GRAN CANARIA: S.E. region on the coast; rather rare, Arinaga.

RANUNCULACEAE

Herbs. Leaves alternate, simple or compound. Flowers hermaphrodite, regular. Sepals and petals 5. Stamens many. Ovary of many 1-seeded, free carpels.

RANUNCULUS

Perennial herbs. Flowers in cymose panicles. Petals yellow, glossy. Achenes numerous with a persistent style.

R. cortusifolius Willd. Robust, hairy plant with tubers. Basal leaves up to 30 cm across, orbicular-cordate, shallowly lobed. Inflorescence subcorymbose. Flowers up to 5 cm across. Achenes smooth, black when ripe. ***Morgallana.*** **(13)**

TENERIFE: Very common in the laurel forest zone and on damp rocks in the lower zone, 200–1500 m, Teno, Masca, Sierra Anaga, Adeje, Icod etc; GRAN CANARIA: Very frequent in the mountain zone below Cruz de Tejeda, 800–1500 m, San Mateo, Sta. Brigida, Andén Verde, Moya etc; LA PALMA: Cubo de la Galga, Barlovento; GOMERA: Forests of the central region, El Cedro, Agando, Epina to Arure; HIERRO: forests of El Golfo; LANZAROTE: Famara Massif, Haria; FUERTEVENTURA: Jandia, Pico de la Zarza.

Several non-endemic *Ranunculus* species commonly occur as weeds, these include *R. aquatilis* (in irrigation channels), *R. muricatus*, *R. arvensis* and *R. ophioglossifolius*.

LAURACEAE ———— Bay Family

Trees or tall shrubs. Leaves evergreen, entire, alternate, sometimes with glands. Dioecious, hermaphrodite or polygamous. Flowers regular, small, greenish, 4- to 6-parted. Ovary superior. Fruit an elliptical berry.

1. Leaves with glands.
  2. Leaves with small glands in the axils of midrib and side veins of lower leaf surface .. .. **Laurus**
  2. Leaves with a single pair of large, hairy glands near the base of the midrib .. .. **Ocotea**
1. Leaves without glands.
  3. Leaves lanceolate, acute, light green, turning reddish when old .. .. **Persea**

3. Leaves broadly lanceolate to ovate, obtuse, dark green, glossy .. .. **Apollonias**

LAURUS

Leaves glandular. Flowers dioecious. Anthers opening by two valves.

L. azorica (Seub.) Franco. Up to 20 m. Young twigs tomentose. Leaves lanceolate, the young ones tomentose beneath. Fruit ovoid 1–1·5 cm, black when ripe. ***Loro.*** **(15)**

TENERIFE; LA PALMA, GOMERA, HIERRO: Very common in the cloud forest zone of all the western islands, often dominant, Las Mercedes, Agua Garcia, El Cedro etc, 500–1500 m; GRAN CANARIA: N. coast region, Moya, Arucas, Barranco de la Virgen etc. Becoming rare due to excessive exploitation of the forest zone of the island.

OCOTEA

Leaves with two large glands towards the base of the midrib. Flowers dioecious or polygamous. Anthers with 4 valves.

O. foetens (Aiton) Benth. Tall trees up to 40 m. Leaves broadly lanceolate to ovate. Fruits acorn-like in a basal cup. ***Til.*** **(17)**

TENERIFE: Deeper parts of laurel-forest valleys, Vueltas de Taganana, Agua Garcia, Las Mercedes, Icod, 500–1000 m, rather rare; LA PALMA: Frequent only in the N.E. part of the island, Los Tilos nr. Los Sauces, Barranco del Rio, Barlovento, Cubo de la Galga; GOMERA: El Cedro, Barranquillos de Vallehermoso, locally frequent; HIERRO: In the laurel woods of El Golfo above Frontera, 500–800 m; GRAN CANARIA: A few examples remain at Los Tiles de Moya.

PERSEA

Leaves without glands, pale green. Flowers usually hermaphrodite. Anthers with 4 valves.

P. indica (L.) Spreng. Tree up to 20 m. Twigs finely sericeous when young. Leaves lanceolate, acute or obtuse, pale green, becoming reddish when old. Fruit about 2 cm, ellipsoid, bluish-black when ripe. ***Viñatigo.*** **(16)**

TENERIFE: Locally frequent in the forest zones along the N. coast of the island from Anaga to Los Silos, sporadic in the S, Las Mercedes, Vueltas de Taganana, Agua Garcia, Agua Mansa, Orotava valley, Guimar, Adeje 500–1500 m; LA PALMA: N.E. and N. coast forest zones, 400–1500 m, locally common, Garafia, Gallegos, Barranco Franceses, Barlovento to La Galga and Barranco del Rio, Cumbre Nueva; GOMERA: Very common in the forests of El Cedro, Agando, Arure, Vallehermoso; HIERRO: Laurel woods of the El Golfo region, Jinamar, Frontera, Fuente de Tinco; GRAN CANARIA: Very sporadic in the N., Moya, Barranco de la Virgen.

APOLLONIAS

Leaves without glands, dark green, glossy. Flowers hermaphrodite. Anthers with 2 valves.

A. barbujana (Cav.) Bornm. Tree or tall shrub up to 25 m. Leaves broadly lanceolate to ovate, dark green, glossy, the margins somewhat revolute. Fruits ovoid, 1–1·5 cm, with a short cupule. ***Barbusano.*** **(14) (132)**

TENERIFE: Forest zone and on cliffs in the lower zone, locally frequent, Sierra Anaga, Agua Garcia, Barranco Ruiz, Cuevas Negras, Teno, 300–1200 m; LA PALMA: N.E. region and sporadically in the S. part of the island, Las Breñas, Cumbre Nueva, Los Tiles, Barlovento; GOMERA: Central region, El Cedro,

Benchijigua, Arure, 600–1000 m; HIERRO: Frequent in the forests of El Golfo; GRAN CANARIA: Very sporadic in the N., Los Tiles de Moya, Barranco de la Virgen, Teror.

subsp. ***ceballosi*** Svent. Like the typical species but with broadly ovate leaves.

GOMERA: Cliffs near Epina in the region above Alojera 600–1000 m, rare.

PAPAVERACEAE —— Poppy, Fumitory Family

Herbs. Leaves entire or several-times divided. Flowers solitary or in racemes or spikes, hermaphrodite, regular or zygomorphic. (The following characters are those of the fumitory group. Several species of poppy have been introduced to the islands.) Sepals 2, small, caducous. Petals 4 in two dissimilar whorls, the outer pair with a short spur, the inner narrower and often more or less joined. Stamens 2 each with 3 branches. Ovary superior. Fruit a small nutlet.

FUMARIA

Small herbs. Leaves compound. Flowers 4-petalled, 2-lipped with a spurred, hooded upper petal, 2 narrow laterals and the lower keel-like. Sepals 2. Stamens 2. Fruit 1-seeded.

1. Flowers red, corolla very narrow .. .. **F. coccinea**
1. Flowers pink with blackish tips, corolla broad .. .. **F. praetermissa**

F. coccinea Lowe. Spreading annual. Leaves with oval segments. Tendrils absent Inflorescences long-stalked, axillary. Flowers red. Corolla very narrow. Fruit smooth.

TENERIFE: Teno, Santa Cruz, S. slopes, Tamaimo, Guia de Isora, up to 400 m; LANZAROTE, FUERTEVENTURA, HIERRO. Probably also on GRAN CANARIA, GOMERA and LA PALMA but the genus in the Canaries needs detailed study. Records of *F. montana* Schmidt from the Canaries probably refer to this species.

F. praetermissa Pugsl. Leaf-segments flat, broad. Corolla with wing-petals turned outwards, pink with purple-blackish tips. Fruit obscurely keeled.

LANZAROTE: Yaiza, Haria, Arrecife; FUERTEVENTURA: Jandia, Mañas Blancas, Betancouria; TENERIFE: Buenavista, La Guancha, Guia de Isora, Hoya de Malpais, 400–600 m; GRAN CANARIA: Roque Nublo, Juncalillo. Probably widespread in the islands and many records of *F. muralis* appear to refer to this species.

CRUCIFERAE ———— Mustard Family

Shrubs or herbs. Leaves alternate. Flowers usually hermaphrodite, usually regular. Sepals 4. Petals 4, free, clawed. Stamens usually 6. Ovary superior. Stigma capitate or bilobed. Fruit a dehiscent capsule opening by two valves from below; when at least 3 times longer than wide known as a siliqua, when less than 3 times longer than wide a silicula which can sometimes be indehiscent.

1. Fruit a silicula, petals white.
 2. Fruits 4-ribbed or winged, leaves with dentate margins or pinnatifid .. .. **Crambe**
 2. Fruits flattish, not ribbed or winged, leaves entire with entire margins .. .. **Lobularia**
1. Fruit a siliqua, petals pink, mauve or yellow.
 3. Leaves 1- to 3-pinnatifid, petals yellow.
  4. Perennial shrubs .. .. **Descurainia**
  4. Annual or biennial herbs .. .. **Erucastrum**

3. Leaves entire, petals pink or mauve.
  5. Siliquas with a pair of long, bilobed horns .. .. **Parolinia**
  5. Siliquas without horns .. .. **Cheiranthus**

CRAMBE

Small shrubs or perennial cliff plants. Glabrous to strigose. Hairs stiff, unbranched. Sepals erecto-patent. Petals white with a short claw at base. Fruit a silicula with the lower segment sterile and forming a stalk; upper segment ovoid to globose, indehiscent, 4-ribbed or winged. Stigma sessile.

1. Leaves lyrate-pinnatifid to pinnatifid with linear lobes.
  2. Leaves glaucous, smooth (Fuerteventura) .. .. **C. sventenii**
  2. Leaves rough, scabrous, green (S. Tenerife) .. .. **C. arborea**
1. Leaves more or less entire with dentate to laciniate margins.
  3. Leaves linear with a few weak teeth (W. Gran Canaria) **C. scoparia**
  3. Leaves lanceolate to ovate, coarsely toothed or laciniate.
    4. Leaves glabrous (Tenerife: Masca) .. .. .. **C. laevigata**
    4. Leaves scabrous to strigose, usually rough to the touch.
      5. Leaves more or less sessile or with a short, narrowly winged petiole.
        6. Panicle branches more or less patent, leaves sessile (N.W.–S.W. Tenerife) .. .. **C. scaberrima**
        6. Panicle branches suberect or erect, leaves with a short, winged petiole (Gomera) .. .. .. **C. gomeraea**
      5. Leaves with a distinct, unwinged petiole which sometimes has a pair of distinct auricles.
        7. Margins deeply laciniate (S. Tenerife) .. .. **C. arborea**
        7. Margins dentate.
          8. Stems very rough, spiny, leaf-margins with pungent teeth (Gran Canaria) .. .. **C. pritzelii**
          8. Stems more or less smooth, leaf-margins with blunt teeth.
            9. Leaves very large, up to 50 cm, membranous, not very rough to the touch, pedicels very slender (La Palma) .. .. **C. gigantea**
            9. Leaves less than 20 cm, more or less thick and rough, pedicels stiff .. .. **C. strigosa**

C. arborea Webb ex Christ. Shrub up to 2 m. Stem ridged. Leaves ovate, coarsely laciniate to pinnatifid with more or less linear lobes, rough. Inflorescence paniculate, the branches slender. Petals twice as long as calyx. Fruits 4-ribbed. **(18)**

TENERIFE: S. region, on basalt and phonolite cliffs at Ladera de Guimar, 400–600 m; local and rather rare.

C. strigosa L'Hér. Small shrub up to 1·5 m. Leaves ovate, rough; margins irregularly crenulate-dentate; petioles of adult leaves usually with a pair of subopposite auricles. Inflorescence paniculate, intricately branched. Petals much longer than calyx. Fruit weakly 4-ribbed.

TENERIFE: Locally common on cliffs and banks in laurel forests and occasionally in the lower zone, Cruz de Taganana, El Bailadero, Bajamar, Agua Mansa, Icod de los Vinos, Los Silos, Barranco del Agua etc, 400–1000 m; GOMERA: Central region near Roque de Agando, 1000 m, forest cliffs, Chorros de Epina; HIERRO: Forest cliffs between Jinamar and Frontera, Fuente de Tinco, 600–800 m; LA PALMA.

C. gigantea (Ceb. & Ort.) Bramwell. Like *C. strigosa* but up to 4 m. Leaves very large, up to 50 cm long, thin, not rough. Inflorescence very large, intricately dichotomously branched with filiform pedicels. Fruits with 2 well-developed ribs.

LA PALMA: Laurel forest ravines in the N.E. region, Los Tiles, Barranco del Agua above Los Sauces, Cumbre Nueva, 500–700 m, sporadic.

C. scaberrima Webb ex Bramwell. Similar to *C. strigosa* in habit but with broadly ovate, very rough, glaucous leaves narrowing abruptly to a very short, winged petiole or more usually sessile. Inflorescence robust, the branches more or less patent. Flowers large. Petals 3 times as long as sepals.

TENERIFE: Mountains of the Teno region from Buenavista to Masca, sea-level to 600 m, Valle de Santiago del Teide, Tamaimo, Bandas del Sur, Adeje, Guimar, usually on basalt cliffs; abundant at Teno, rare elsewhere.

C. pritzelii Bolle. Like *C. strigosa* but the stems very rough, spiny. Leaves lanceolate to elliptical, very rough especially below; margins dentate, the teeth pungent. Inflorescence branches patent.

GRAN CANARIA: N. coast from Lentiscal to Agaete valley, Los Tiles de Moya; locally frequent in forest and lower zones, 400–800 m.

C. gomeraea Webb ex Christ. Small shrub. Leaves lanceolate to ovate, rough, petiolate, usually somewhat sinuately lobed towards the base. Petiole usually narrowly winged. Panicle branches erect or suberect. Flowers tending to be clustered at tips of branches. Fruit 4-ribbed.

GOMERA: E. region from Barranco de la Villa to Tagamiche and Roque Agando 600–1000 m, W. region, Lomo de Carretón, Vallegranrey; generally rare.

C. laevigata DC. ex Christ. Small shrubs with ovate or ovate-lanceolate, very coarsely toothed, glabrous, glaucous leaves. Panicle small, lax with few patent branches. ***Col del Risco.*** **(133)**

C. sventenii Bramwell & Sunding. Leaves lyrate-pinnatifid, glaucous, smooth. Inflorescences paniculate with suberect branches. Fruit strongly 2-winged. **(19)** **(134)**

FUERTEVENTURA: S. region near Gran Tarajal, mountain cliffs of Montaña Vigán and Montaña Cardones, 400 m; rare.

C. scoparia Svent. Slender cliff plant, often very woody and robust at the base. Leaves caducous, linear to narrowly lanceolate, the margins weakly toothed. Inflorescence with a few long, pendulous branches. Flowers small. Fruit 4-ribbed.

GRAN CANARIA: High mountain cliffs of the W. and S.W. region, Barranco de Tejeda, San Nicolas, Guayedra etc, 1000–1600 m; very rare; TENERIFE: Valle de Masca, rocks and walls below the village of Masca, 200–1200 m; locally frequent.

LOBULARIA

Perennial herbs or small shrublets with medifixed hairs. Sepals patent. Petals white, entire. Fruits small, flat, round, with 1–2 (–3) seeds per locule.

1. Sepals pinkish or reddish (La Palma) .. .. .. **L. palmensis**
1. Sepals green, sometimes with dark tip .. .. .. **L. intermedia**

L. intermedia Webb & Berth. Greyish-white, pubescent perennial often with woody base. Leaves entire, linear-lanceolate, obtuse. Sepals green, occasionally with dark tip. Fruit small suborbicular to obovate, flat. **(20)**

ALL ISLANDS: Locally very common on dry rocks in the lower zone up to 1000 m, occasionally on forest cliffs to 1250 m (Agua Mansa).

L. palmensis Webb. Small, very woody shrublet with erect-ascending stems. Leaves narrowly lanceolate, acute or obtuse. Sepals pinkish or reddish, petals broad.

LA PALMA: Locally frequent in and below the forest zones, on cliffs, 500–1600 m La Cumbrecita, La Galga, Los Tiles, Tijarafe.

L. lybica. A N. African species with large fruits occurs on Lanzarote and Fuerteventura in coastal regions.

DESCURAINIA

Small shrubs. Lower leaves densely fasciculate, 1- to 3-pinnate. Flowers in dense to lax racemes, yellow. Petals clawed. Siliquas more or less 4-angled; valves keeled. Seeds usually slightly winged.

1. Sepals less than 2·5 mm; claw ± equalling petal.
  2. Leaves ± sessile; pedicels of siliquas patent .. .. **D. bourgaeana**
  2. Leaves shortly petiolate; pedicels of siliquas ascending .. **D. gilva**
1. Sepals 3–4 mm; claw much shorter than petal.
  3. Lower leaves 1-pinnate; lobes linear, acute.
    4. Plants branched at base, canescent, 1–1·5 m (Tenerife) **D. gonzalezi**
    4. Plants branched above, green, 60–80 cm (Gran Canaria) .. .. **D. preauxiana**
  3. Lower leaves 2- to 3-pinnatisect; lobes linear-lanceolate to ovate, obtuse or acute.
    5. Lower leaves petiolate, usually 3-pinnatisect .. .. **D. millefolia**
    5. Lower leaves subsessile to sessile, 2-pinnatisect.
      6. Leaf-lobes elliptic-ovate or spathulate; siliquas 20-seeded (Gran Canaria).. .. .. **D. artemisoides**
      6. Leaf-lobes linear-lanceolate; siliquas 28–32-seeded (Tenerife) .. .. **D. lemsii**

D. gonzalezi Svent. Shrub 100–150 cm, branches at base, woody, hairy, densely leafy. Leaves pinnate to more or less entire; lobes linear, acute. Inflorescences usually unbranched. Siliquas glabrous, erect or arcuate, 2 cm long. Seeds reddish brown.

TENERIFE: Las Cañadas del Teide, very local near Los Roques and Arenas Negras on dry cinder slopes, 2000 m.

D. bourgaeana Webb ex O. E. Schulz. Like *D. gonzalezi* but with 2-pinnatisect leaves with linear-lanceolate segments which are usually toothed at the tip. Fruiting pedicels patent; siliquas ascending, 16-seeded. Seeds brown. ***Hierba Pajonera.*** **(135)**

TENERIFE: Las Cañadas de Teide, common in the subalpine zone in rocky areas, locally dominant, El Portillo to Montaña Blanca, Parador Nacional etc, 1800–2200 m.

D. lemsii Bramwell. Small shrub differing from *D. bourgaeana* by its silvery-grey leaves with narrow, acute lobes and its erect fruiting pedicels with many-seeded (28–32) siliquas.

TENERIFE: Upper pine forest zone on N. side of Pico de Teide, Cumbres de Pedro Gil, Fuente de Joco, Montaña Ayesa, 1600–2000 m; locally frequent.

D. millefolia (Jacq) Webb & Berth. Very variable small shrub up to 1 m. Leaves often rosetted towards stem apex, usually 3-pinnatisect; lobes often toothed, ovate to rounded-ovate. Racemes dense, congested. Fruiting pedicels patent to ascending, often curved. Siliquas 10–20-seeded. Seeds reddish-brown.

TENERIFE: Common from Anaga to Teno in xerophytic scrub and on cliffs

200–1000 m. Taganana, Icod el Alto, Teno, Masca etc; LA PALMA: Frequent in the central region on dry rocky slopes and cliffs and in pine woodland, 600–1500 m. El Time, El Paso, Barranco del Rio etc.; GOMERA: Coastal regions of the N. side, San Sebastian, Lomo Fragoso to Vallegranrey, 200–600 m, frequent.

D. gilva Svent. Small shrub up to 40 cm. Leaves sessile, more or less erect, 2-pinnatisect, greyish-felted. Inflorescences simple or branched. Fruiting pedicels erect or diverging. Siliquas 16–24 seeded. Seeds chestnut-brown.

LA PALMA: High mountains of the Caldera region above Tijarafe, 1700–2000 m. Open sunny places in upper pine forest zone; locally abundant.

D. preauxiana Webb ex O. E. Schulz. Small branched shrub, 60–80 cm. Leaves usually more or less rosulate at tips of stems, pinnatisect, with up to 6 pairs of linear to filiform lobes, sparsely glandular-hairy. Siliqua pedicels suberect. Siliquas 18–27-seeded.

GRAN CANARIA: Central and S. regions 400–1600 m. Tejeda, Roque Nublo area, Tirajana valley, Fataga, Tazartico; locally frequent on dry exposed slopes.

D. artemisoides Svent. Small shrub about 50 cm. Leaves 2-pinnatisect with 6–10 pairs of primary segments; lobes broad, elliptic-spathulate, obtuse. Fruits about 20-seeded. Seeds brown.

GRAN CANARIA: Rare in the Agaete, Guayedra area in the lower zone on moist, shady cliffs and in ravines, Los Berrezales, 400–600 m.

PAROLINIA

Erect shrubs. Stems and leaves covered with stellate hairs. Leaves linear to narrowly lanceolate, entire. Petals clawed, pink or white. Siliqua with 2 pronounced bifid appendages.

1. Siliqua appendages shortly bifid or trifid; siliqua 2- to 3-septate (Gomera) .. .. **P. schizogynoides**
1. Siliqua appendages deeply bifid; siliqua 4- to 8-septate.
  2. Siliqua less than 2 cm, 4-septate (Tenerife) .. .. **P. intermedia**
  2. Siliqua more than 2 cm, 5- to 8-(9-)septate (Gran Canaria) .. .. **P. ornata**

P. ornata Webb, 1–1·5 m. Leaves linear-lanceolate, up to 10 cm long. Inflorescences 10–30 flowered. Sepals erect, 5–6 mm. Petals spathulate, pink. Siliqua 2–2·5 cm long, each valve 5- to 8-seeded; appendages deeply bifid. ***Dama.***

GRAN CANARIA: S. region from Guayadeque to Mogan, on dry, exposed slopes 400–600 m. Locally very common, also sporadically on the W. side of the island, San Nicolas, Agaete, 400 m.

P. intermedia Svent. & Bramwell. Like *P. ornata* but more compact, with shorter leaves and the siliquas less than 2 cm with each valve 4-(5-)seeded. (**136**)

TENERIFE: Punta de Teno, 50–200 m, locally frequent in xerophytic scrub, below Chio on dry S.W. slopes, 300 m.

P. schizogynoides Svent. Dwarf, compact shrub up to 80 cm. Leaves linear, 3 cm long. Racemes up to 20-flowered. Petals narrow. Siliqua about 1 cm, each valve 2- to 3-(4-)seeded; appendages shallowly bifid or trifid at tip.

GOMERA: S.W. region, Valle de Argaga, 200–350 m, locally frequent; coast near Alojera, rare.

Matthiola bolleana Webb ex Christ, an annual species with mauve flowers and distinctive, often curled fruits, is abundant in the S. part of FUERTEVENTURA.

CHEIRANTHUS

Small shrubs with branched or medifixed hairs. Sepals erect, the inner saccate.

Petals mauve-violet, long-clawed. Stigma 2-lobed. Fruit a siliqua; valves with distinct median vein. Seeds in 1 or 2 rows.

1. Leaves with unbranched medifixed hairs .. .. .. **C. scoparius**
1. Leaves with forked, 3-armed hairs .. .. .. **C. virescens**

C. scoparius Brouss. Shrub with erect or ascending stems. Leaves linear to linear lanceolate, pubescent, all hairs medifixed; margins entire or toothed. Flowers pale to deep mauve. Siliquas more or less erect. ***Alheli del Teide.*** **(137) (138)**

TENERIFE: Subalpine zone of Las Cañadas, very common, 1800–2200 m, Valle de Santiago del Teide (var. *cinereus*), 400–500 m, locally very common; GRAN CANARIA: (subsp. *lindleyi*) locally common in the high mountains of the C. region, Cruz de Tejeda, Paso de la Plata, Pinos de Galdar etc, 1500–1800 m.

C. virescens Webb ex Christ. Like *C. scoparius* but the leaves linear to lanceolate, margins usually at least remotely toothed. Hairs 3-armed. Inflorescences lax. ***Alheli Montuño.*** **(139)**

TENERIFE, LA PALMA, HIERRO, GOMERA, GRAN CANARIA: Locally frequent in the lower and forest zones up to about 1000 m.

ERUCASTRUM

Annual to biennial herbs with lyrate-pinnatifid leaves. Flowers in lax terminal racemes. Sepals erecto-patent. Petals yellow, clawed. Fruit a linear, tortulose siliqua. Valves strongly 1-veined. Beak short, more or less conical.

1. Upper cauline leaves petiolate, petals about 6 mm .. **E. cardaminoides**
1. Upper cauline leaves sessile, petals about 8–9 mm .. **E. canariense**

E. canariense Webb & Berth. Lower leaves lyrate, densely hispid; upper leaves sessile, entire or dentate. Inflorescence with several branches. Petals about 8–9 mm.

LANZAROTE: N. region, Haria, Famara Massif, La Caleta etc; FUERTEVENTURA: Matas Blancas, La Oliva, Betancouria, locally abundant; GRAN CANARIA.

E. cardaminoides (Webb ex Christ) O. E. Schulz. Lower leaves segmented almost to the midrib, pubescent; upper leaves petiolate, more or less glabrous. Inflorescence with one or few branches. Petals about 6 mm. **(22)**

TENERIFE: S. region, Guimar, Barranco de Bufadero, Valle Seco, Punta de Teno; GRAN CANARIA: N. and W. regions, Bahia de Confital, Anden Verde, coastal zone up to 200 m; HIERRO: Tamaduste, El Golfo; GOMERA: N. coast, Agulo, Puerto de Vallehermoso; FUERTEVENTURA.

RESEDACEAE ——————— Mignonette Family

Herbs or small shrubs. Leaves simple or pinnate. Inflorescence spicate or racemose. Flowers hermaphrodite, zygomorphic calyx. Petals entire or lobed. Stamens free, inserted on a nectar-secreting disc. Carpels united below into a superior ovary. Fruit a capsule, open at the top.

RESEDA

Annual to perennial herbs or shrublets. Leaves entire or lobed. Petals and sepals 4–8. Stamens 10–25, their bases forming a disc. Fruit an elongate capsule opening at the top.

1. Leaves linear, entire, flowers white .. .. .. **R. scoparia**
1. Leaves with a single pair of lobes, flowers yellow .. **R. crystallina**

Also 2 weedy yellow-flowered European species *R. lutea*—leaves pinnate, capsules about 1·5 cm long, *R. luteola*—leaves simple, capsules about 0·5 cm.

The N. African *Oligomeris subulata* Del. with inconspicuous petals also occurs in dry coastal regions.

R. scoparia Brouss. Small erect or prostrate shrub. Leaves linear, fleshy, obtuse. Flowers small. Petals white. **(21)**

TENERIFE: Locally abundant in the S. Adeje, El Medano, 0–300 m, Punta de Teno, 25 m; GOMERA: Coastal regions of the N. & W., Puerto de Vallehermoso, Argaga; GRAN CANARIA: Rather rare, Melenara, Punta Arinaga, Isleta etc, coastal regions.

R. crystallina Webb & Berth. Perennial or annual herb. Leaves usually with a single pair of linear to oblanceolate lobes. Flowers yellow. Capsules somewhat warty. **(140)**

GRAN CANARIA: W. & S.W., near coasts, Andén Verde, San Nicolas, Mogan 10–100 m; LANZAROTE: Famara Massif, Haria, Playa de Famara; FUERTEVENTURA: Jandia region, Morro Jable, Gran Tarajal, Corralejo.

CRASSULACEAE ———— Houseleek Family

Herbs or shrubs. Leaves succulent, entire. Flowers in racemes or cymes, 5- to 32-parted. Sepals usually united at base. Petals free (united in *Umbilicus*). Stamens twice as many as petals. Carpels as many as petals, free, usually with a hypogynous scale at the base, forming distinct, dry capsules in fruit.

1. Basal leaves orbicular, peltate, inflorescence a long raceme .. .. **Umbilicus**
1. Basal leaves not peltate, inflorescence more or less cymose.
  2. Flowers 5-parted .. .. **Sedum**
  2. Flowers 6- to 32-parted.
    3. Hypogynous scales very large, conspicuous, fan-shaped **Monanthes**
    3. Hypogynous scales small, inconspicuous, 2-horned to quadrate, sometimes absent.
      4. Hypogynous scales absent, flowers 18- to 32-parted.. **Greenovia**
      4. Hypogynous scales present, flowers 6- to 12-parted.
        5. Annual to perennial herbs, hypogynous scales 2-horned or digitate .. .. **Aichryson**
        5. Robust perennials, woody at least at the base, hypogynous scales usually quadrate .. .. .. **Aeoniun**

UMBILICUS

Perennials with tuberous or rhizomatous stock. Basal leaves suborbicular, more or less peltate; cauline leaves usually linear-lanceolate. Flowers 5-parted, in a long raceme. Calyx small. Corolla tubular or campanulate, the lobes erect. Stamens 10.

U. heylandianus Webb & Berth. Up to 1 m. Basal leaves peltate; cauline leaves more or less lanceolate, becoming linear in the inflorescence. Flowers pedicellate, horizontal or drooping. Corolla bright yellow.

LA PALMA: High mountains of the Caldera region, Cumbres de Garafia, Topo Alto de los Corralejos, Roque de los Muchachos etc, above 1600 m; GRAN CANARIA: Mountains of the C. region, Cruz de Tejeda to Artenara, very local 1300–1500 m. (Iberian Peninsula).

The European-Mediterranean species ***U. horizontalis*** (Guss.) DC. is very common on rocks and walls in the Canary Islands. It is a smaller plant with more or less sessile flowers and a short, yellowish-brown corolla.

MONANTHES

Small, smooth, papillose or pubescent, herbaceous or subshrubby perennials (1 annual). Leaves opposite or alternate and densely rosulate, succulent. Flowers with filiform pedicels, 6- to 9-parted. Petals linear. Hypogynous scales very large, fan-shaped. Stamens twice as many as petals. Carpels 6–8.

1. Small annual up to 5 cm .. .. **M. icterica**
1. Perennials.
 2. Leaves pubescent or ciliate.
  3. Rosettes very small, less than 1 cm, calyx arachnoid-hairy .. .. **M. minima**
  3. Rosettes at least 1·5 cm, calyx not arachnoid-hairy.
   4. Rosettes about 3 cm, leaves spathulate, glandular-pubescent (Tenerife, Guimar) .. .. .. **M. adenoscepes**
   4. Rosettes about 2 cm, leaves oblanceolate, pubescent (Tenerife, S. coast of Anaga) .. .. **M. dasyphylla**
 2. Leaves glabrous or papillate.
  5. Leaves more or less opposite, not rosulate.
   6. Leaves very thick, fleshy, ovate, usually silvery .. **M. laxiflora**
   6. Leaves slender, linear-lanceolate, usually reddish-tinged (Tenerife, Anaga) .. .. **M. anagensis**
  5. Leaves in a dense rosette.
   7. Stem short, thick, unbranched or sometimes stoloniferous, usually bulboid.
    8. Rosettes very small, about 1 cm, flowers with a reddish tinge (Tenerife, Anaga) .. .. .. **M. praegeri**
    8. Rosettes at least 1·5 cm, flowers usually not reddish-tinged.
     9. Rosettes lax, leaves usually with purplish streaks or blotches.
      10. Leaves angular in cross-section, stem very robust, up to 1·5 cm (Tenerife, subalpine zone) .. .. **M. niphophila**
      10. Leaves more or less rounded in cross-section, stem usually less than 1 cm .. .. .. **M. brachycaulon**
     9. Rosettes very dense, leaves pale green.
      11. Stems unbranched, rosettes usually flat or concave .. .. **M. pallens**
      11. Stems usually branched or plants growing in dense clusters, rosettes convex .. .. .. **M. silensis**
   7. Stems branched, creeping, usually slender, not bulboid.
    12. Small shrublet with slender branches, leaves purplish at least beneath .. .. **M. muralis**
    12. Dense cushion or mat-forming plants, leaves not purplish.
     13. Rosettes more or less globose, flowering stems and calyx with long, whitish hairs .. .. .. **M. polyphylla**
     13. Rosettes elongate, narrow beneath, flowering stems and calyx finely pubescent.
      14. Leaves long, 3–6 mm wide, rosettes up to 1·5 cm (Gomera) .. .. **M. amydros**

14. Leaves short, about 2–2·5 mm wide, rosettes usually less than 1 cm (Tenerife, La Palma) .. **M. subcrassicaulis**

M. icterica (Webb ex Bolle) Praeger. Small annual up to 5 cm. Branches dark, divaricate. Leaves alternate, sessile, fleshy, blunt. Inflorescences 4-flowered with minute bracts. Flowers 6-parted. Scales reddish.

TENERIFE: W. and S.W. regions, Los Silos, Teno, Guia de Isora etc, locally abundant; GOMERA: N. coast from San Sebastian to Vallehermoso, 100–1000 m, rocks and cliffs in the Winter and Spring months.

M. minima Bolle. Small, stoloniferous rosette plant with ciliate, blunt leaves. Flowering stems erect, about 2 cm. Flowers 3–6, very small. Calyx arachnoid-hairy.

TENERIFE: S. slopes of Anaga, Barranco Seco, Barranco de Bufadero, rare.

M. adenoscepes Svent. Rosette unbranched, 3–4 cm. Leaves many, densely crowded, glandular-pubescent. Inflorescences several, axillary, more or less erect. Flowers 2–5, about 6 mm across. Calyx purplish. Hypogynous scales pale.

TENERIFE: Cliff-faces, crevices and overhangs, Ladera de Guimar, 200–600 m, locally frequent.

M. dasyphylla Svent. Rosette lax, about 2 cm. Leaves few, pubescent. Inflorescences axillary, suberect. Flowers about 5 mm, reddish.

TENERIFE: S. coast of Sierra Anaga on coastal rocks between San Andrés and Igueste, rare.

M. laxiflora (DC.) Bolle. Diffuse, tortuous or hanging plant. Leaves opposite, sessile, egg-shaped, usually silvery-grey or sometimes dark-greenish, very variable. Inflorescences lax, few-flowered. Flowers yellowish to purplish. Hypogynous scales small. (**25**)

TENERIFE, GOMERA, GRAN CANARIA, LANZAROTE, FUERTEVENTURA: Locally abundant on rocks and cliffs in the lower and forest zones, up to 1200 m, found only in the Jandia region of Fuerteventura and confined to the N. part of Lanzarote.

M. anagensis Praeger. Like *M. laxiflora* but the leaves subopposite, slender, linear-lanceolate, usually reddish-tinged and the calyx segments much narrower.

TENERIFE: Sierra de Anaga, frequent along the central ridge, Pico Inglés, Cruz de Afur, El Bailadero etc, Taganana, 100–700 m.

M. praegeri Bramwell. Stem thick, bulb-like. Rosette very small, usually less than 6 mm. Leaves club-shaped. Flowers few, reddish. (**142**)

TENERIFE: A few places along the coast on the N. side of Sierra Anaga, Bajamar, Punta del Hidalgo etc. 80–300 m, locally common.

M. brachycaulon (Webb & Berth.) Lowe. Stem thick, bulb-like. Rosette solitary or with stolons, at least 1·5 cm and up to 5 cm across. Leaves oblanceolate-spathulate, usually rounded in cross-section or the upper surface flat or concave; midrib and base purplish; apex obtuse. Inflorescences axillary. Flowers 5 to 10, up to 1 cm across, purplish-green.

TENERIFE: Locally frequent on the N. side of the island, Agua Mansa, La Guancha, San José, Los Silos etc, up to 1500 m; GRAN CANARIA: Common in the N. and C. regions, Barranco de Guiniguada, Tenteniguada, Santa Brigida, Teror, Agaete, Barranco de Tejeda, Juncalillo etc, 200–1600 m.

M. niphophila Svent. Like *M. brachycaulon* but never stoloniferous and the leaves angular in cross-section with a long, attenuate apex. (**141**)

TENERIFE: Subalpine zone of Las Cañadas, Montaña de Diego Hernandez, La

Fortaleza, S. walls of Las Cañadas near Topo de la Grieta, 2000–2300 m, rare.

M. pallens (Webb ex Christ) Christ. Unbranched rosette plant. Rosette dense, flat or concave with numerous axillary inflorescences. Leaves pale green, densely papillose. Inflorescences leafy. Flowers small.

TENERIFE: W. region, Masca, Tamaimo, Valle de Santiago del Teide, locally common on dry rocks, 300–600 m, Sierra Anaga; HIERRO: Sabinosa; GOMERA: Fuente Blanca, Barranco de Vallehermoso.

M. silensis (Praeger) Svent. Like *M. pallens* but the rosettes much smaller, often branched or growing in dense clumps, convex. Inflorescences usually leafy only at the base.

TENERIFE: N.W. region, Los Silos, Cuevas Negras, Buenavista, Roque del Fraile, Punta de Teno, shady coastal rocks up to 450 m, locally common.

M. muralis (Webb ex Bolle) Christ. Small branched shrublet. Rosettes dense, terminal. Leaves with deep purple markings. Inflorescences terminal, 3- to 7-flowered. Calyx and petals with small red dots. **(24)**

HIERRO: Frequent on rocks and walls in the Valverde region, El Golfo, Frontera, Sabinosa etc, 300–700 m; LA PALMA: S. region, Fuencaliente, Barranco de las Angustias etc; GOMERA: N. and N.W. regions, Vallehermoso, Chorros de Epina, Lomo de Carretón, Vallegranrey, Fortaleza de Chipude etc, up to 1000 m, locally common.

M. polyphylla Haw. Dense mat or cushion-forming rosette plant. Rosettes many, more or less globose, pale green to pinkish. Inflorescences terminal, few-flowered. Pedicels and calyx densely white-hairy. **(143)**

GRAN CANARIA, TENERIFE, LA PALMA: Locally common on cliffs and earthy banks, usually in shade, coasts to the forest zones, 50–700 m, very common on the N. coast of TENERIFE.

M. subcrassicaulis (O. Ktze.) Praeger. Like *M. polyphylla* but usually not mat-forming, the rosettes elongate, tapering below, the flowering stems and calyx finely pubescent.

TENERIFE: W. region, Teno, Masca, Los Silos etc, S. coast, Guimar, Adeje; LA PALMA: E. coast, La Galga, Barlovento, locally abundant.

M. amydros Svent. Like *M. subcrassicaulis* but the leaves broader, up to 6 mm wide, the rosettes much larger, up to 1·5 cm and the flowers larger, ca. 1 cm.

GOMERA: Locally very common on the N. side of the island, Barranco de la Villa, Majona, Agulo, Vallehermoso, Epina etc, 20–800 m.

SEDUM

Succulent herbs. Leaves alternate. Inflorescence cymose. Flowers 5-parted. Stamens twice as many as petals. Carpels equal in number to petals.

S. lancerottense Murray. Small creeping herb. Leaves light green, fleshy, glabrous. Flowers yellow, about 0·5 cm.

LANZAROTE: N. region, Riscos de Famara, Haria, El Rio etc, 400–700 m, locally common on rocks and cliffs.

S. rubens L., a small reddish annual species with pinkish or white flowers is common in dry areas of the W. islands.

AICHRYSON

Small annual to perennial herbs. Branching often pseudodichotomous. Leaves alternate, entire, somewhat fleshy. Calyx 6- to 12-parted, green, fleshy. Petals yellow, as many as calyx segments. Stamens twice as many as petals. Hypogynal scales 2-horned or dentate. Carpels as many as petals.

1. Perennial shrublets (Lanzarote and Fuerteventura).
  2. Leaves sessile, viscid-hairy (Lanzarote) .. .. .. **A. tortuosum**
  2. Leaves shortly stalked, hairy but with few glandular hairs (Fuerteventura) .. .. **A. bethencourtianum**
1. Annual to triennial herbs.
  3. Plants hairy, often densely so.
    4. Leaf-blade broadest near base, stems with ascending to more or less erect branches.
      5. Hairs gland-tipped (La Palma).. .. .. .. **A. palmense**
      5. Hairs not gland-tipped.
        6. Leaves with black glands on the margins (La Palma) **A. bollei**
        6. Leaves without black glands on margins .. .. **A. laxum**
    4. Leaf-blade broadest at or above the middle, stems simple or with patent-divaricate branches.
      7. Plants about 15–20 cm, purplish red throughout (Gran Canaria) .. .. **A. porphyrogennetos**
      7. Plants less than 12 cm, only the stems sometimes reddish.
        8. Petals shorter than sepals (La Palma) .. .. **A. brevipetalum**
        8. Petals longer than sepals .. .. **A. parlatorei**
  3. Plants glabrous or subglabrous.
    9. Leaf-margins with purple crenulations .. .. .. **A. punctatum**
    9. Leaf-margins without purple crenulations .. .. **A. pachycaulon**

A. tortuosum (Aiton) Webb & Berth. Small dense shrublet up to 10 cm. Stems woody at base. Leaves crowded towards tips of stems, sessile, more or less spathulate or obovate, fleshy, viscid-hairy, the apex rounded. Inflorescences short, leafy. Flowers 8-parted, about 1 cm across. Petals golden-yellow, ciliate. Floral scales more or less quadrate.

LANZAROTE; Haria, Famara cliffs, Mirador del Rio, up to 700 m, locally frequent; FUERTEVENTURA.

A. bethencourtianum Bolle. Like *A. tortuosum* but more or less herbaceous, dense. Leaves spathulate to suborbicular, shortly petiolate, fleshy, densely hairy. Floral scales cuneate to subpalmate. ***Pelotilla.***

FUERTEVENTURA: Jandia region, on cliffs of the N. side of the Peninsula, N. & C. regions, La Oliva, Vallebron etc.

A. laxum (Haw.) Bramwell. (*A. dichotomum* (DC.) Webb & Berth.) Erect, softly hairy annual or biennial. Stems up to 30 cm, dichotomously branched, the branches usually more or less erect. Leaves subrosulate, petiolate, subrhomboidal to suborbicular, usually broadest towards the base, somewhat fleshy, densely hairy. Inflorescences flattish, diffuse. Flowers yellow, 9- to 12-parted. Floral scales irregularly lobed. **(23)**

TENERIFE, GRAN CANARIA, LA PALMA, HIERRO, GOMERA: Locally very frequent in forest regions, on rocks, cliffs, banks, walls etc., 400–1200 m, also in shady places in the lower zone.

A. porphyrogennetos Bolle. Like *A. laxum* but with widely divaricate branches, the stems and leaves purplish, leaves broadest near the middle and the inflorescence lax.

GRAN CANARIA: E. and N.E. parts of the island in shady ravines, 600–1000 m, Tenteniguada, Barranco de Moya, local.

A. palmense Webb ex Bolle. Like *A. laxum* but very viscid with dense gland-tipped hairs, the young leaves often rather inrolled and the petals pale yellow. **(146)**

La Palma: Frequent on dry, shady rocks particularly on the E. side of the island, Fuencaliente, El Charco, Barranco del Rio, Barlovento, La Galga, Barranco de Los Tilos etc, 300–800 m.

A. bollei Webb ex Bolle. Small annual or biennial with dense, appressed hairs. Leaves densely hairy, the margins with black, crenate glands. Flowers pale yellow. **(145)**

La Palma: Very local but sometimes abundant, La Cumbrecita, N. coast region, Cueva de la Zarza, Barranco Franceses, Barranco del Rio, Tijarafe, pine or laurel woods up to 1600 m.

A. parlatorei Bolle. Small annual up to 10 cm. Leaves obrhomboidal, densely hairy, often reddish, the margins with a few small, black crenations. Inflorescence diffuse. Flowers 8- to 9-parted, golden yellow. **(144)**

Tenerife, Gran Canaria, La Palma, Gomera, Hierro: Locally very common on dry rocks and walls from 30–1000 m.

A. brevipetalum Praeger. Similar to *A. parlatorei* but the flowers very small, 6- to 7-parted with very short, erect or suberect petals.

La Palma: Known only from a few places in the C. and E. parts of the island, Barranco del Rio, La Galga, Cumbre Nueva, La Caldera.

A. punctatum (Chr. Sm.) Webb & Berth. Annual to biennial, glabrous or subglabrous herb. Leaves fleshy, ovate-rhomboidal, the margins with dark crenulations. Inflorescences leafy, lax. Flowers golden-yellow, 6- to 8-parted.

Tenerife, Gran Canaria, La Palma, Hierro, Gomera: Cliffs, rocks, slopes, ravines usually in damp or wet soils in forests or N. coasts in the lower zone, locally common, 100–1200 m.

A. pachycaulon Bolle. Like *A. punctatum* but very robust, completely glabrous, the leaf-margins usually without dark crenulations. Flowers small.

Fuerteventura: Pico de la Zarza, on wet cliffs on the N. face, rare, 600 m; Tenerife: Vueltas de Taganana, deep shady banks in laurel woods, Agua Garcia, Madre del Agua, up to 1000 m, local.

AEONIUM

Perennial herbs or small shrubs. Stems woody at least at the base. Leaves rosulate, entire, fleshy, the margins usually ciliate. Flowers yellow, pink or white (1 sp. red). Calyx fleshy, 6- to 12-parted. Petals as many as calyx segments. Stamens twice as many as petals. Hypogynous scales usually present. Ovaries as many as petals, immersed in receptacle at the base. Seeds many, very small. ***Pastel del Risco, Oreja del Abad.***

A very difficult genus in the field as many species are very variable in size of rosettes and in the degree of red-pigmentation in the leaves. Several species-groups occur the members of which, without the aid of a hand-lens or microscope, are difficult to separate except by their geographical distribution.

1. Herbs, only the base woody, leaves green, usually without a red edge.
   2. Leaves hairy on both surfaces .. .. **A. canariense** group
      Gran Canaria .. .. **A. virgineum**
      Tenerife .. .. **A. canariense**
      La Palma .. .. **A. palmense**

Hierro .. .. **A. palmense** var. **longithyrsum**
Gomera .. .. **A. subplanum**

2. Leaves glabrous, sometimes glandular.
3. Leaves without linear glands.
4. Rosette flat, leaves densely imbricate, green, the margins long-ciliate .. .. **A. tabuliforme**
4. Rosette more or less cup-shaped, leaves glaucous, margins entire .. .. **A. cuneatum**
3. Leaves with linear glands .. .. **A. simsii**

1. Shrubs or subshrubs, stems woody.
5. Stems unbranched.
6. Flowers red, leaves very fleshy, suborbicular .. .. **A. nobile**
6. Flowers pink, white or yellow, leaves succulent, much longer than broad.
7. Flowers yellow, leaves green, glossy (Gran Canaria) .. **A. undulatum**
7. Flowers pink or white, leaves blue-glaucous.
8. Leaves narrow (Tenerife, Gomera) .. .. .. **A. urbicum**
8. Leaves broad (Hierro) .. .. **A. hierrense**

5. Stems branched.
9. Leaves smelling very strongly of balsam (Lanzarote & Fuerteventura) .. .. **A. balsamiferum**
9. Leaves odourless or if smelling slightly of balsam then plants very small, less than 20 cm, twiggy.
10. Flowers yellow.
11. Tall robust shrubs with long leaves and dense, conical inflorescences.
12. Inflorescence finely pubescent (Gran Canaria) .. **A. manriqueorum**
12. Inflorescences glabrous.
13. Several years' dead leaves persistent (La Palma) .. **A. vestitum**
13. Dead leaves not persistent for long periods.
14. Stems leafless in flowering period (Gomera) .. **A. rubrolineatum**
14. Stems leafy in flowering period .. .. .. **A. holochrysum**
11. Small, twiggy shrubs with short leaves and small diffuse inflorescences.
15. Leaves with linear glands.
16. Branches thick, with white hairs below the leaf-rosettes (Tenerife) .. .. **A. smithii**
16. Branches thin, without white hairs .. .. **A. spathulatum**
15. Leaves without linear glands.
17. Leaves very fleshy with red stripes (W. Tenerife) .. **A. sedifolium**
17. Leaves without red stripes.
18. Leaves glabrous, sticky (Gomera) .. .. **A. viscatum**
18. Leaves pubescent.
19. Leaves more or less rhomboidal, thick, pubescent (Tenerife, La Palma) .. .. .. **A. lindleyi**
19. Leaves ovate, flat, glandular-pubescent (Gomera) .. .. **A. saundersii**

10. Flowers pink or white.
20. Leaves pubescent on both surfaces (Hierro) .. .. **A. valverdense**
20. Leaves glabrous.
21. Rosettes 2–3 cm. across.
22. Flowers rose-pink (N. & W. Tenerife) .. .. **A. haworthii**
22. Flowers pale-pink or white.

23. Leaves sticky, green, margins not ciliate (La Palma) **A. goochiae**
23. Leaves blue-glaucous, margins ciliate.
 24. Densely branched twiggy shrublets.
  25. Leaves without red edges (Gomera) .. .. **A. castello-paivae**
  25. Leaves suffused with red .. .. **A. decorum**
 24. Laxly branched shrublet with few branches (E. Gomera) .. .. **A. gomerense**
21. Rosettes over 5 cm across.
 26. Leaves with slender straight marginal cilia (Tenerife) **A. ciliatum**
 26. Leaves with curved or broadly conical cilia.
  27. Cilia curved, calyx pubescent (Gran Canaria) .. **A. percarneum**
  27. Cilia broadly conical, calyx glabrous (Lanzarote) .. **A. lancerottensis**

A. canariense (L) Webb & Berth. Thick, short-stemmed herb, somewhat branched at base. Leaves in dense rosette, the inner suberect forming a cup, ovate-spathulate, subsessile, glandular pubescent, the margins finely ciliate; apex rounded to mucronate. Flowering stem 50–70 cm, leafy, with leaves of decreasing size towards tip. Inflorescence ovate-pyramidal. Flowers sulphur-yellow, 8- to 10-parted, shortly pedicellate. Calyx segments lanceolate; floral scales subquadrate.

TENERIFE: Common on rocks and cliffs of the N. coast from Anaga to Teno, from near sea-level to the forest zones 600–1300 m.

A. subplanum Praeger. Like *A. canariense* but the rosettes flat with broad-tipped overlapping spathulate leaves. Strongly smelling of balsam. Flowers bright yellow, 8- to 12-parted, subsessile. Floral scales fan-shaped. **(148)**

GOMERA: Very common on the N. side of the island from San Sebastian to Vallegranrey, on rocks, banks, walls and forest cliffs from 200–1100 m.

A. palmense Webb ex Christ. Similar to *A. canariense* but with loosely rosulate, obovate to ovate-rhomboidal leaves with very short dense, glandular hairs. Flowers usually 9-parted, bright lemon yellow. Floral scales expanded at apex.

LA PALMA, HIERRO: In the lower and forest zones in similar habitats to *A. canariense*, common in the Barranco de Las Angustias on LA PALMA and in the El Golfo region of HIERRO.

A. virgineum Webb & Christ. Like *A. canariense* but plants forming dense clumps, rosettes small, freely branching at base, leaves with distinct reddish tinge, densely hairy on both surfaces. Floral scales expanded and rounded at apex.

GRAN CANARIA: Frequent along the N. coast at mid-altitude 500–1500 m, in similar habitats to the previous species, locally common at Moya, Barranco de la Virgen, Arucas etc.

A. tabuliforme (Haw.) Webb & Berth. Rosettes usually single or occasionally with a few offsets, up to 30 cm in diameter, very dense, flat, plate-like. Leaves more or less spathulate, the margins with long cilia. Flowering stems erect, arising from the centre of the rosette. Flowers yellow, 8-parted. Floral scales oblong, bluntly bilobed at tip. **(151)**

TENERIFE: N. coast from Teno to Taganana, usually below 500 m, in cliff crevices, locally abundant near San Juan de la Rambla, Roque de las Animas, Bajamar, Roque del Fraile, usually on N. or N.W.-facing cliffs.

A. cuneatum Webb & Berth. Rosette plant with horizontal offsets. Leaves rigid, glabrous, blue-glaucous, more or less oblong, the apex mucronate; margins finely ciliate. Flowering stem leafy, up to 1 m tall. Inflorescence conical. Flowers golden yellow, 8- to 10-parted, flat. Floral scales subquadrate or rounded. **(147)**

TENERIFE: Confined to the old mountain blocks at either end of the island, Anaga and Teno, forest cliffs and earthy banks, locally common along the Anaga ridge, Cumbres de Taganana, El Bailadero 600–800 m.

A. manriqueorum Bolle. Branched shrub with thick fleshy stem up to 1 m. Leaf rosettes flat, 12–20 cm across. Leaves elongate-spathulate, shining green, streaked with purple; margins with forward-pointing cilia. Inflorescence conical, the branches pubescent. Flowers yellow, the petals blunt. Floral scales small, quadrate. (**152**)

GRAN CANARIA: Cliffs and ravines on the N. coast and in the central mountain region 300–1200 m, Moya, Galdar, Santa Brigida, locally very common.

A. holochrysum Webb & Berth. Like *A. manriqueorum* but leaves generally narrower and longer. Inflorescence glabrous and the floral scales larger and forming a conspicuous ring. (**150**)

TENERIFE, GOMERA, LA PALMA, HIERRO: Common plant of the xerophytic zone 30–1200 m, especially on the N. and W. coasts.

A. rubrolineatum Svent. Like *A. holochrysum* but the leaves narrowly lanceolate-spathulate and the summer resting rosettes very small. Inflorescences rather small, formed in late summer. Calyx margins purple. Floral scales trapeziform, small.

GOMERA: W. and C. region, in or just below the forest zone, 800–1200 m, Roque Agando, Garajonay, Lomo de Carretón, Chorros de Epina, very local but often frequent.

A. vestitum Svent. Like *A. manriqueorum* but the leaves small, obovate to narrowly obrhomboidal in tight rosettes with dead leaves very persistent on the stems and the inflorescences very small and ovate.

LA PALMA: N.W. part of the island 600 m between Puntallana and Barlovento, locally frequent at La Galga and Los Sauces, on walls and cliffs in the upper xerophytic zone.

A. balsamiferum Webb & Berth. Shrub, smelling strongly of balsam. Branches erect. Leaf-rosettes small, cup-shaped. Leaves glabrous, glandular, obovate, acute. Inflorescences small. Flowers yellow. Floral scales club-shaped.

LANZAROTE: Very rare, on cliffs in the Famara and Haria regions 500–700 m. Occasionally cultivated near the coast; FUERTEVENTURA: commonly cultivated near Antigua and La Oliva. The sticky balsam was once used to preserve fishing lines.

A. undulatum Webb & Berth. Unbranched shrub with massive stem. New stems arise subterraneously from the rootstock of the parent stem. Leaf-rosette very large. Leaves dark green, red-edged, often undulate at margins, oblong-spathulate. Inflorescence very large, leafy, conical. Flowers yellow, usually 10-parted. Scales quadrate, whitish. (**149**).

GRAN CANARIA: Locally frequent, 400–600 m. Moya, Tenteniguada, Santa Brigida etc, often a feature of the landscape when in flower.

A. urbicum (Chr. Sm.) Webb & Berth. Unbranched, monocarpic shrub, 60–200 cm. Leaves in broad, dense apical rosette, narrowly oblanceolate, glabrous, pale green to blue-glaucous, red-edged; margins ciliate. Inflorescence very large, pyramidal, up to 80 cm long. Flowers 9- to 10-parted pink to greenish white. Styles purplish (**154**)

TENERIFE: N. coast from Anaga to Teno, Valle de Santiago del Teide, Masca, very frequent in the La Laguna area, rocks, walls and even rooftops in the lower and forest zones from sea-level to 1100 m. GOMERA: Fortaleza de Chipude in the S.W. of the island 800 m, locally frequent.

A. hierrense (Murray) Pit. & Proust. Like *A. urbicum* but shorter in stature with broader leaves and the flowers 8- rather than 9- to 10-parted.

HIERRO: Locally common round Valverde and in the El Golfo region on dry rocky slopes in the lower zone up to 700 m.

A. ciliatum (Willd.) Webb & Berth. Branched shrub up to 1 m. Branches woody, rough with prominent leaf-scars. Leaves glabrous, green with a red margin, shortly spathulate, the apex apiculate; margin with broad-based cilia. Inflorescences leafy, broadly pyramidal, pubescent. Buds conical, twisted. Flowers 7- to 8-parted, greenish-white or pink.

TENERIFE: Frequent in the Anaga region from sea-level to the forest zone, Vueltas de Taganana, Pico Inglés, El Bailadero, Taganana etc. Sporadic along the N. coast to Teno; LA PALMA: common in the E. and S. especially below Fuencaliente. The La Palma form is distinct from the Tenerife one in several characters, it is much less branched and has reddish-brown tinged glaucous leaves and white flowers.

A. valverdense Praeger. Like *A. ciliatum* but usually more compact and the leaves densely and finely pubescent on both surfaces. Flowers pink, 7- to 8-parted. **(156)**

HIERRO: Valverde region, between the town and the coast 20–500 m, rocky open ground and dry cliffs.

A. gomerense Praeger. Like *A. ciliatum* but lax with a few slender, ascending branches. Leaves glaucous, very succulent, edges red-tinged. Flowers white, 8-parted, campanulate.

GOMERA: Barranco de la Villa between San Sebastian and Hermigua on steep, rocky slopes 600–1100 m, rare.

A. decorum Webb ex Bolle. Small, much-branched, spreading, glabrous subshrub. Stems rough, with leaf-scars and scales prominent. Leaves in loose rosettes, smooth, suffused with red especially in dry conditions, keeled on the back, acute; margins scarcely to densely ciliate. Inflorescence lax with reddish, pubescent branches. Flowers 6- to 8-parted. Petals pinkish white, sometimes greenish on the back.

GOMERA: Locally very frequent in the valley above San Sebastian and Barranco de la Laja on the E. side of the island and in the Vallegranrey region on the W. coast, 100–500 m. TENERIFE: A small form of this species with the leaf-margins not or scarcely ciliate is found on the cliffs below the village of Masca on the W. side of the island at about 350 m.

A. castello-paivae Bolle. Like *A. decorum* but with slender, smooth stems and very fleshy glaucous leaves with purplish streaks on the back and coarse, forward-directed cilia on the margins. Inflorescence glandular-pubescent. Flowers 8-parted, petals pink to greenish-white.

GOMERA: Very common on the N. side of the island from San Sebastian to Vallehermoso and beyond, usually on dry, rocky slopes, 100–1000 m.

A. haworthii (Salm-Dyk.) Webb & Berth. Very similar to the preceding species but with rough branches and broader, glaucous, red-edged leaves. Flowers 8-parted. Petals pale yellow to rose-pink.

TENERIFE: Restricted to the N.W. of the island where it is locally very common, Icod, Buenavista, Teno, Montaña de Taco to Masca usually on rocks and rocky slopes below 500 m.

A. percarneum (Murray) Pit. & Proust. Stout, erect shrub up to 1 m. Branches thick, divaricate, smooth. Leaves fleshy glaucous-green to purplish, usually with a red margin with strong, forward-curving cilia. Flowering stem densely leafy. Flowers 9-parted. Petals pink or white with pink streaks.

GRAN CANARIA: N. and C. region of the island, rocky slopes in the xerophytic and forest zones, 200–1500 m, Cuesta de Silva, Bandama, Santa Brigida, etc, locally very common.

A. lancerottense Praeger. Glabrous, erect, much-branched subshrub. Branches silvery. Leaves pale glaucous green with the edges often pale brown-orange to reddish; margins with short, blunt cilia. Flowers 8-parted. Petals pink. (**155**)

LANZAROTE: N. part of the island on the cliffs of the Famara Massif, Peñitas de Chache, Haria, El Valle, Riscos de Famara, 300–600 m, locally frequent.

A. goochiae Webb & Berth. Much-branched, sticky shrub with tortuous thin, woody branches. Leaves limp, flat, glandular-pubescent, rhomboidal, the margins entire. Flowers usually 8-parted, petals pink.

LA PALMA: N.E. region on shady cliffs and forest banks and on exposed dry slopes, 150–850 m, locally common, La Galga, Los Sauces, Barlovento.

A. viscatum Webb ex Bolle. Like *A. goochiae* but with glabrous, sticky succulent leaves and yellow, 7- to 8-parted flowers.

GOMERA: N.E. region of the island from the coast to 500 m, locally very common, San Sebastian, La Laja, Agulo, Roque Cano, Puerto de Vallehermoso.

A. lindleyi Webb & Berth. Like *A. goochiae* but with very thick, succulent, glandular-pubescent, blunt leaves and 8- to 9-parted golden-yellow flowers.

TENERIFE: Along the N. coast from Anaga to Orotava, frequent on rocks in the lower zone, Taganana, San Andrés, Bajamar, sea-level to 500 m, also reported from the W. side of LA PALMA.

A. smithii (Sims) Webb & Berth. Small shrub. Branches succulent, densely white-hairy towards apex. Leaves with reddish, linear glands and irregularly ciliate, crispate margins. Flowers usually 10-parted, large, golden-yellow. (**160**)

TENERIFE: Very local species of S. part of Las Cañadas and the mountain slopes to the S., Tamaimo, Vilaflor, Ladera de Guimar, Montaña de Diego Hernandéz etc, 500–2200 m, rather rare and probably declining species.

A. spathulatum (Hornem.) Praeger. Small, dense, erect, much-branched subshrub with thin woody branches. Leaves spathulate, very small and sticky with linear glands; margins with bead-like cilia. Young leaves forming a tight, globose bud during the dry season. Flowers 8- to 10-parted. Petals golden-yellow. (**157**)

TENERIFE, GRAN CANARIA, GOMERA, LA PALMA, HIERRO: Cliffs in forest regions and subalpine rocks, locally very common, 800–2000 m, Agua Mansa, Tenteniguada etc.

A. saundersii Bolle. Similar in habit to *A. spathulatum* but the leaves much larger, suborbicular, densely pubescent on both surfaces. Inflorescences simple, few-flowered. Flowers 12- to 16-parted, large. Petals pale yellow.

GOMERA: Very distinctive species known only from a few localities on the E. side of the island, locally frequent on dry cliffs near La Laja, 400–500 m.

A. sedifolium (Webb) Pit & Proust. Dense, twiggy, erect subshrub. Leaves fleshy, club-shaped, green with red lines, very sticky. Flowers few, large, 9- to 11-parted, bright yellow. (**153**)

TENERIFE: W. region of the island from Los Silos to Tamaimo in dry exposed rocky areas and on cliffs, Punta de Teno, Masca, Hoya de Malpais, El Retamal,

Montaña de Taco etc. 100–600 m, very common in some areas. (The hybrid of this species and *A. urbicum* which occurs in the Masca and Tamaimo regions has been described as a distinct species *A. burchardii*. It is recognisable by its smooth, reddish-brown stems and thick, green leaves with orange-red edges.)

A. simsii (Sweet) Stearn. Tufted, caespitose perennial. Rosettes more or less sessile, dense. Leaves lanceolate to narrowly oblong with prominent linear glands beneath; margins with long, hyaline cilia. Flower stems arising from lateral rosettes. Flowers golden yellow, 8-parted, about 1 cm across. **(158).**

GRAN CANARIA: Mountains of the C. and S. regions from 600–2000 m, rock-faces, cliffs, walls etc, Crux de Tejeda, Tenteniguada, Tirajana valley, Barranco de Fataga etc, locally very common.

A. nobile Praeger. Robust, unbranched, up to 60 cm. Rosette on a short stem. Leaves very large, thick, fleshy, broadly obovate to suborbicular, red-tinged, the margins with a few coarse cilia. Inflorescence very large. Flowers bright red. **(159)**

LA PALMA: Local, N. of Santa Cruz, Los Llanos, Barranco de las Angustias, Tijarafe etc, 200–800 m, sometimes frequent.

GREENOVIA

Like *Aeonium* but leaves usually glaucous, glabrous, the flowers 18- to 32-parted and the carpels embedded to about half their length in the hypogynous disc.

1. Leaves glabrous, glaucous.
  2. Plants without offsets, leaf-margins very finely ciliate .. **G. diplocycla**
  2. Plants with offsets, leaf-margins usually without cilia.
    3. Flowers 28- to 32-parted, offsets few .. .. .. **G. aurea**
    3. Flowers 18- to 22-parted, offsets many .. .. .. **G. dodrentalis**
1. Leaves glandular-pubescent, green .. .. .. **G. aizoon**

G. aurea (Chr. Sm.) Webb & Berth. Stems very short, rosettes usually several. Rosette-leaves densely imbricate, cuneate to spathulate, glaucous, fleshy. Flowering stems leafy. Flowers 30- to 35-parted, bright yellow, up to 2·5 cm across. **(161)**

TENERIFE, GRAN CANARIA, HIERRO, GOMERA, LA PALMA: Common on rocks, cliffs, walls and dry slopes from the lower zone to the subalpine, 400–2000 m.

G. diplocycla Webb ex Bolle. Unbranched rosette plant resembling *G. aurea* but the leaves with finely ciliate margins and the flowers 20-parted.

GOMERA, HIERRO, LA PALMA: Common in the lower zone and on forest cliffs, 50–1700 m.

G. dodrentalis (Willd.) Webb & Berth. Like *G. aurea* but with small, globular rosettes and 18- to 22-parted flowers.

TENERIFE: Very local, Anaga region, Barranco de Bufadero, Igueste de San Andrés, Teno region, Valle de Masca, 150–1200 m.

G. aizoon Bolle. Small, much-branched rosette plant. Old leaves persistent. Leaves oblong-spathulate, densely glandular-pubescent. Flowers 20-parted, golden-yellow, about 1–2 cm across.

TENERIFE: Mountain regions in the S. of the island, Ladera de Guimar, Igueste de Candelaria etc, locally frequent, 600–1600 m.

ROSACEAE

Trees, shrubs or herbs. Leaves alternate, stipulate. Flowers regular, hermaphrodite, monoecious or dioecious. Sepals 5, sometimes with epicalyx. Petals usually 5, free, sometimes absent. Stamens usually many. Receptacle flat or concave. Carpels 1-many, free or connate. Styles usually free. Fruit of achenes, drupes, follicles or a pome.

1. Leaves entire or trifoliate.
 2. Scrambling shrubs with trifoliate leaves and prickly stems .. .. **Rubus**
 2. Trees with entire leaves, stems not prickly .. .. **Prunus**
1. Leaves pinnate.
 3. Fruits fleshy, globose or more or less pear-shaped .. **Bencomia**
 3. Fruits dry, winged or 4-angled.
  4. Fruit a winged samara .. .. **Marcetella**
  4. Fruit a 4-angled nutlet .. .. **Dendriopoterium**

BENCOMIA

Shrubs. Leaves pinnate. Inflorescence spikes all male, all female or plants dioecious. Receptacle fleshy or spongy, globular to pear-shaped, enclosing the achenes.

1. Leaves with up to 11 leaflets, fruits pear-shaped.
 2. Leaflets sessile, leaves without stipules (Tenerife, subalpine zone) .. .. **B. exstipulata**
 2. Leaflets shortly petiolate, stipules present.
  3. Inflorescences usually branched, leaflets lanceolate .. **B. brachystachya**
  3. Inflorescences usually simple, leaflets ovate .. .. **B. caudata**
1. Leaves with 13–25 leaflets, fruits globose (Hierro, La Palma) .. .. **B. sphaerocarpa**

B. caudata (Aiton) Webb & Berth. Small shrub up to 2 m. Leaves with 7–11 ovate, dentate, shortly petiolate leaflets. Inflorescences usually simple. Fruits pear-shaped to subglobose about 4–5 mm across.

TENERIFE: Sierra Anaga, Vueltas de Taganana, Agua Mansa, Los Silos, Adeje, usually on forest cliffs 600–1500 m, rare; GRAN CANARIA: N. region, Los Tiles de Moya, Teror etc, rare.

B. sphaerocarpa Svent. Like *B. caudata* but the leaves with 13–15 lanceolate, serrate, densely sericeus leaflets and globose, rather larger fruits.

HIERRO: Forest cliffs of the El Golfo region above Frontera, Fuente de Tinco etc, 500–1000 m, very rare; LA PALMA: (subsp. *sventenii*) Los Tilos, Barranco del Agua, 600–800 m, very rare.

B. brachystachya Svent. Like *B. caudata* but with narrower leaflets and long, branched inflorescences. **(26)**

GRAN CANARIA: Valle de Tirajana, cliffs in the mountain region, 1600 m, Tenteniguada, cliffs of El Rincŏn, 800–1000 m, rare.

B. exstipulata Svent. Small shrub with small, sessile leaflets. Fruit flattened, subglobose.

TENERIFE: Cliffs in the S. region of Las Cañadas de Teide, Boca de Tauce, Ucanca valley, 2100–2200 m, on shady rocks, in crevices, very rare.

MARCETELLA

Dioecious shrubs. Leaves pinnate, glabrous, glaucous. Inflorescences lax, spicate. Receptacle dry, compressed, 4-angled. Fruit a samara with 2 wings.

M. moquiniana (Webb & Berth.) Svent. Shrub. Upper stem with red, glandular hairs. Leaves in terminal rosettes; leaflets petiolate, 7–15, crenate. Male spikes long, few-flowered, pendulous. Female spikes pendulous. Fruits dry, winged, resembling elm (*Ulmus campestris*). **(163)**

TENERIFE: Sporadic, Los Silos, Cuevas Negras, Los Realejos, Barranco del Infierno at Adeje, Masca, 300–600 m; GRAN CANARIA: Very rare in the W. region.

DENDRIOPOTERIUM

Like *Marcetella* but with very large stipules and leaflets. Inflorescences monoecious, female flowers in the upper part of the spike and males below. Receptacle dry, 4-angled. Fruit a small, hard, brown 4-angled nutlet.

D. mendendezii Svent. Small shrub with glaucous leaves and very large leaf-like stipules. Inflorescences erect or ascending. Fruits about 2 mm. **(162)**

GRAN CANARIA: W. & S. regions, Agaete, Juncallilo, Tamadaba, Guayedra, Tirajana, locally frequent but sporadic on basalt cliffs 400–1500 m.

RUBUS

Scrambling woody shrubs with prickly stems and trifoliate leaves. Flowers in terminal and axillary clusters. Epicalyx absent. Sepals and petals 5. Stamens numerous. Fruit an aggregate of numerous, 1-seeded fleshy carpels on a domed receptacle. ***Zarza.***

The Canarian blackberries, which are common on the W. & C. islands, belong to or are closely related to the very variable *R. ulmifolius* Schott. A large-leaved forest form has been described as *R. bollei* Focke and is said to be endemic. Two other 'species', *R. canariensis* Focke and *R. palmensis* Hansen are also reported from the islands.

PRUNUS

Trees. Leaves leathery, glossy, glabrous, the margins crenate or dentate. Flowers 5-parted, in elongated axillary racemes. Fruit a drupe.

P. lusitanica L. subsp. ***hixa*** (Willd.) Franco. Tree up to 10 m. Twigs and petioles reddish. Leaves oblong-lanceolate to elliptic; margins regularly crenate or dentate; apex long-acuminate forming a 'dripping point'. Inflorescences pendunculate, suberect. Petals white. Fruit ovoid to subglobose, purplish black. ***Hija.***

TENERIFE: Sierra de Anaga, laurel forests, Las Mercedes to Taganana 600–900 m, locally common; GOMERA: El Cedro, rare; GRAN CANARIA: Barranco de la Virgen, Los Tiles de Moya, rare; LA PALMA, HIERRO.

LEGUMINOSAE ——— Pea Family

Trees, shrubs or herbs with alternate, entire, pinnate or trifoliate leaves. Flowers hermaphrodite. Sepals united into a tube. Petals free or connate. Stamens 10. Ovary superior. Fruit a dehiscent, 2-valved legume or indehiscent, sometimes lomentaceous.

1. Scrambling plant with tendrils and pinnate leaves .. **Vicia**
1. Shrubs or herbs without tendrils, leaves entire or trifoliate.
  2. Leaflets with serrate margins .. .. **Ononis**
  2. Leaflets with entire margins.
    3. Flowers white.
      4. Leaflets lanceolate to ovate, large .. .. .. **Chamaecytisus**
      4. Leaflets linear, small, caducous.
        5. Fruit a flattish, dehiscent black legume .. .. **Spartocytisus**

5. Fruit small, ovoid, indehiscent .. .. .. **Retama**
3. Flowers yellow, pink or creamish with red lines.
6. Tall shrubs usually over 1 m in height.
7. Flowers creamish usually with red lines .. .. **Dorycnium**
7. Flowers yellow or pink.
8. Legume lomentaceous .. .. **Anagyris**
8. Legume not lomentaceous.
9. Legume with glandular, black papillae, seeds without a strophiole .. .. **Adenocarpus**
9. Legume sericeous, seeds strophiolate .. .. **Teline**
6. Herbs or dwarf, woody-based perennials, less than 20 cm .. .. **Lotus**

ONONIS

Shrubs or herbs. Leaves trifoliate or simple. Veins of leaflets ending in teeth on the margins. Calyx deeply divided, the teeth more or less equal. Fruit ovoid or oblong, usually more or less equalling or somewhat longer than the persistent calyx.

1. Shrubs, flowers pink or yellow.
2. Flowers pink .. .. **O. christii**
2. Flowers yellow .. .. **O. angustissima**
1. Herbs, flowers yellow .. .. **O. hebecarpa**

The W. Mediterranean *O. natrix*, a viscid, yellow-flowered shrub of sand-dunes is generally common on the E. islands and in S. regions of Tenerife.

O. christii Bolle. Shrub up to 90 cm. Leaves 3-foliate. Racemes pedunculate, 1- to 3-flowered. Corolla rose-pink, glabrous. Fruiting peduncle reflexed. Pod linear or oblong. Seeds tuberculate.

FUERTEVENTURA: Jandia region in xerophytic scrub, Pico de la Zarza, 600 m, local.

O. angustissima Lam. subsp. ***ulicina*** Webb & Berth. Like *O. christii* but with long, coarsely toothed leaflets and a yellow corolla which is sometimes red-lined. **(174)**

GRAN CANARIA: Locally very common, Caldera de Tirajana, Mogan, Agaete valley, N. coast etc, in *Euphorbia* and *Echium* communities; TENERIFE: Ladera de Guimar, 600–700 m, sporadic in the upper part of the Guimar cliffs.

O. hebecarpa Webb & Berth. Small annual. Peduncles 2-flowered. Corolla glabrous, yellow. Pod short. Seeds smooth.

LANZAROTE, FUERTEVENTURA: Locally abundant in the Jandia and Famara regions.

LOTUS

Herbaceous annuals to woody-based perennials or dwarf shrubs. Leaves with five leaflets. Flowers usually yellow (1 sp. red, 1 pink), axillary or in clusters at the end of long peduncles. Fruits narrow, linear, with 2 valves, many seeded. About 15 species, a rather difficult genus in the Canary Islands. **Corazoncillo.**

1. Keel petal with long beak, at least the tips of the petals bright red.
2. Petals deep red .. .. **L. berthelotii**
2. Petals yellow with bright red tips .. .. .. **L. maculatus**
1. Keel petal without a long beak.
3. Flowers pink .. .. **L. glinoides**

3. Flowers yellow.
  4. Leaflets linear.
    5. Plants densely covered with long silky hairs (S. Gran Canaria) .. .. **L. holosericeus**
    5. Plants glabrous or with short hairs.
      6. Small shrubs with suberect branches and petiolate leaves.
        7. Calyx with long hairs (Valle de Masca, Tenerife) .. **L. mascaensis**
        7. Calyx with short hairs and purplish stripes (Gran Canaria) .. .. **L. spartioides**
      6. Procumbent perennials with short woody stock and more or less sessile leaves .. .. **L. sessilifolius**
  4. Leaflets lanceolate to orbicular.
    8. Leaflets orbicular, petals usually purplish-tipped (La Gomera) .. .. **L. emeroides**
    8. Leaflets lanceolate to obovate, petals without purplish tips.
      9. Flowers borne singly or in pairs on axillary peduncles **L. glaucus**
      9. Flowers in heads of 3–8, on axillary or terminal peduncles.
        10. Peduncles usually less than 4 cm, leaflets obovate (Lanzarote & Fuerteventura) .. .. .. **L. lancerottensis**
        10. Peduncles usually more than 4 cm, leaflets narrowly lanceolate to oblanceolate (La Palma & Tenerife).
          11. Leaflets 4–10 mm; peduncles 1·5 cm or less; habit procumbent (Tenerife) .. .. .. **L. campylocladus**
          11. Leaflets 2–4 mm; peduncles usually more than 1·5 cm; habit erect (La Palma) .. .. .. **L. hillebrandii**

L. campylocladus Webb & Berth. Woody-based, procumbent perennial. Leaflets narrowly lanceolate to oblanceolate, obtuse, 4–10 mm long with dense, appressed hairs. Peduncles short. Flowers about 1–1·5 cm, in 3- to 6-flowered clusters. Calyx hirsute.

TENERIFE: Subalpine zone 1800–2200 m, common in open pine forest and lava desert of Las Cañadas. Occasionally found at lower levels on the S. side of the island particularly in the Chio region at about 400–600 m.

L. hillebrandii Christ. Like *L. campylocladus* but erect or suberect, leaflets shorter (2–4 mm), peduncles very long. **(177)**

LA PALMA: Montane regions of the Gran Caldera and Cumbrecita region, in pine forests 500–1700 m, reaching almost to sea-level at Fuencaliente in the S.

L. spartioides Webb & Berth. Dwarf shrub with slender, suberect branches. Leaflets more or less linear, 5–10 mm, covered with short, whitish hairs. Flowers bright yellow, 2–5 per cluster. Calyx with purple stripes. **(175)**

GRAN CANARIA: Pine forests and montane scrub in the mountain zone above 1000 m, locally common in the pinar at Tamadaba.

L. holosericeus Webb & Berth. Like *L. spartioides* but the whole plant covered with long, silky hairs and the clusters with 6–10 flowers.

GRAN CANARIA: Caldera de Tirajana 600–800 m in legume scrub, locally common on the San Bartolome side of Paso de la Plata.

L. glaucus Aiton. Very variable decumbent plant, usually perennial and woody at the base. Leaflets oblanceolate, 4–5 mm, sparsely hairy. Flowers borne singly or in pairs on long axillary peduncles, pale yellow. Calyx lobes narrow, acute. Pods slender, brown. **(33)**

TENERIFE, HIERRO, GRAN CANARIA, LA GOMERA: Frequent in the littoral region on coastal rocks and cliffs from sea-level to 200 m. The Hierro form of the species is much more densely hairy and has been referred to var. *villosus* Pitard. *L. glaucus* is also found on Madeira.

The very similar ***L. dumetorum*** R. P. Murray **(176)**, is found only on the south side of Tenerife. It is a densely branched shrublet with reddish-brown pods and pale yellow flowers in axillary clusters of 3–4.

L. emeroides R. P. Murray (*L. borzii* Pitard). Like *L. glaucus* but more robust, leaflets orbicular, flowers in clusters of 3–4, petals usually purple-tipped.

LA GOMERA: Frequent on dry lowland slopes up to 500 m, common on the N. and E. coasts.

L. callis-viridis Bramwell & D. H. Davis, a very rare species from the W. side of GRAN CANARIA (Andén Verde) differs from *L. emeroides* by its larger flowers, very short calyx tube and bright yellow petals lacking a purplish tip.

L. sessilifolius DC. Decumbent perennial with woody stock. Leaves sessile to very shortly petiolate. Leaflets linear, 3–9 mm, covered with short, silky hairs. Flowers in terminal or axillary clusters of 3–5; peduncles up to 7 cm. Calyx purplish at base. Pods linear, brown. **(32)**

TENERIFE: Frequent in coastal regions from sea-level to 150 m. Common on the N. coast from Teno to Puerto de la Cruz. A very narrow, filiform-leaved variety, var. *petaphyllos* (Link) D. H. Davis, is found along the S. coast of the island from Guimar to Medano and on the island of Hierro in the Golfo region.

L. leptophyllus (Lowe) Larsen appears to be a local variant of this species found in a few coastal areas of GRAN CANARIA (Arinaga, Cuesta de Silva). It has very short, round leaflets and a compact habit.

L. kunkelii (Esteve) Bramwell & D. H. Davis also belongs to this group of species. It has succulent leaves with long silky hairs and is known only from a single locality at Playa de Jinama on the E. coast of GRAN CANARIA.

L. lancerottensis Webb & Berth. (incl. *L. erythrorhizos* Bolle). Prostrate perennial with woody stock. Leaflets obovate, 3–5 mm, silky-hairy. Flowers in terminal or axillary clusters of 3–5, peduncles 2–4 cm. Pods blackish.

LANZAROTE, FUERTEVENTURA: Common in lowland and coastal regions up to 500 m.

L. mascaensis Burchard Dwarf, silvery shrub. Leaflets linear, 12–15 mm; petiole 1–3 mm. Flower clusters terminal, 3- to 7-flowered, peduncles 2–5 cm. Calyx with long hairs.

TENERIFE: endemic to the Valle de Masca on the W. side of the island where it is locally frequent 400–600 m.

L. glinoides Desf. Dwarf annual with glaucous, retuse leaflets, pinkish purple flowers and dark brown twisted pods.

LANZAROTE, FUERTEVENTURA: Frequent in coastal regions and dunes; GRAN CANARIA, TENERIFE: rare in S. coast areas. (Canaries & N. Africa.)

L. berthelotii Masf. (*L. peliorhynchus* Webb) Pendulous cliff plant with woody rootstock. Leaflets linear, 10–18 cm, silky hairy. Flowers scarlet with long-beaked keel and wing petals. ***Pico Paloma.*** **(34)**

TENERIFE: Forest cliffs at La Florida in the Orotava Valley and in the Barranco de Tamadaya on the S. side of the island 700–1200 m. Virtually extinct in the wild but often cultivated in gardens.

L. maculatus Breitf. Like *L. berthelotii* but the leaves broader and the petals yellow with orange-red tips. **(178)**

TENERIFE: Extremely rare on the N. coast. Cultivated in the Jardin Canario at Tafira, Gran Canaria.

ANAGYRIS

Tall shrubs with large yellow flowers and trifoliate leaves. Legumes large, lomentaceous.

A. latifolia Brouss. Shrub up to 3·5 m. Leaflets lanceolate to narrowly ovate, about 5–6 cm. Flowers large, yellow with reddish lines on the standard. Legume large, light brown, indehiscent with the seeds separated by cross-walls. **(27)**

TENERIFE: Rare shrub of the lower zone, Guia de Isora, San Juan de la Rambla, Masca, Punta de Hidalgo, 200–500 m; GRAN CANARIA: Known only from a single locality which is now included within the Jardin Canario in Barranco de Guiniguada at Tafira.

TELINE

Shrubs. Leaves trifoliate or rarely unifoliate, usually petiolate, usually stipulate. Flowers borne in axillary or terminal racemes. Calyx tubular-campanulate, 2-lipped; upper lip bifid, lower with 3 teeth. Corolla yellow. Legume narrowly oblong, compressed, sericeus, 2- to 8-seeded. Seeds strophiolate.

1. Standard petal ± uniformly sericeus; leaves subsessile or very shortly petiolate .. .. **T. linifolia**
1. Standard petal more or less glabrous or pubescent only towards apex; leaves distinctly petiolate.
  2. Standard petal glabrous.
    3. All leaves trifoliate; leaflets elliptical, more than 4 mm wide .. .. **T. stenopetala**
    3. Upper leaves unifoliate; leaflets more or less linear, not more than 2 mm wide .. .. **T. osyroides**
  2. Standard petal pubescent towards apex.
    4. Twigs leaves and calyx woolly-tomentose; leaflets ± elliptical, involute (Gran Canaria) .. .. **T. microphylla**
    4. Twigs, leaves and calyx sericeus; leaflets obovate, usually not involute .. .. **T. canariensis**

T. canariensis (L.) Webb & Berth. Erect or spreading, much branched shrub up to 3 m. Twigs sericeus. Leaflets small, obovate to rotund, subglabrous above, densely hairy beneath. Racemes terminal. Legume shortly villous. **(29) (168)**

TENERIFE: Common in the Anaga region in and below the forest zone and along the N. coast to Icod, 500–1500 m. GRAN CANARIA: Forest zone of the N. coast, Moya, Barranco de la Virgen etc.

T. microphylla (DC.) Gibbs & Dingwall. Like *T. canariensis* but a much more densely branched and compact shrub. Leaflets involute. Infloresences short. Legume densely tomentose. ***Retama amarilla.*** **(169)**

GRAN CANARIA: Common in the C. and S. mountain regions 250–1900 m, San Mateo, Tejeda, Tirajana, Santa Brigida etc, locally dominant.

T. stenopetala Webb & Berth. Tall shrub up to 6 m. Branches glabrous. Leaves petiolate. Leaflets elliptical to lanceolate, 3–18 mm across, glabrous to sericeus above, densely sericeus beneath. Racemes terminal, elongate, 10- to 26-flowered. Standard more or less glabrous. Legume narrowly oblong, densely villous. ***Gacia.*** **(30)**

LA PALMA: Locally frequent in and below the laurel forest zone, Los Tilos, Barlovento; TENERIFE: Valle de Orotava near Agua Mansa; HIERRO: Laurel woods of the El Golfo region; GOMERA: El Cedro, Vallehermoso, Arure, forest and *Erica* heath regions. 600–1500 m.

T. osyroides (Svent.) Gibbs & Dingwall. Small shrub up to 1·5 m. Leaflets linear to narrowly elliptical, sessile; upper leaves unifoliate. Inflorescences terminal, 4- to 15-flowered. Standard glabrous.

TENERIFE: Valle de Masca, locally frequent near Masca village below shady cliffs. S. valleys between Arona and Escabonal, sporadic, 450–600 m.

T. linifolia (L.) Webb & Berth. Very variable small shrub up to 2 m. Leaves sessile or very shortly petiolate. Leaflets narrowly oblanceolate to elliptical, sericeus, the margins revolute. Racemes terminal, short, dense, 4- to 20-flowered. Standard densely sericeus. **(167)**

TENERIFE: Taganana, Roque de las Animas, rare on coastal cliffs, 50–100 m; GOMERA: Cumbres de Hermigua, Roque de Vallehermoso, in *Erica* heath, 400–600 m; LA PALMA: Barranco del Rio near Santa Cruz, in laurel woods, La Cumbrecita, cliffs in pine forest 1800 m, frequent; GRAN CANARIA: Risco Blanco region of the Tirajana valley, rare (the form from this island with linear-oblong leaves may best be considered as a distinct species *T. rosmarinifolia* Webb & Berth.).

ADENOCARPUS

Shrubs with petiolate, trifoliate leaves, usually in fascicles. Flowers in terminal racemes. Calyx tubular, 2-lipped, the upper deeply bifid, the lower with 3 teeth. Flowers yellow. Legume narrowly oblong with glandular papillae. Seeds without a strophiole.

1. Calyx with glandular papillae; standard petal subglabrous **A. viscosus**
1. Calyx without glands; standard petal sericeus.
   2. Petioles long (1 cm); leaflets linear-lanceolate; legume very densely glandular .. .. **A. ombriosus**
   2. Petioles short (1–3 mm); leaflets lanceolate or obovate; legumes sparsely glandular .. .. **A. foliolosus**

A. viscosus (Willd.) Webb & Berth. Densely leafy, viscous shrub. Leaves densely fasciculate; leaflets involute. Calyx with glandular papillae. Standard petal subglabrous. Legume with glandular papillae and sparse hairs. ***Codeso del Pico.*** **(164)**

TENERIFE: High mountain zone of Tenerife, especially Las Cañadas. Volcanic slopes and screes 1800–2100 m, common; LA PALMA: Cumbres region of the outer rim of the Caldera de Tabouriente 1800–2000 m, frequent.

A. foliolosus (Ait.) DC. Erect, leafy shrub. Leaves fasciculate; leaflets not or only slightly involute, the upper surface glabrous. Calyx densely hairy. Standard petal sericeus. ***Codeso.*** **(165) (166)**

GRAN CANARIA: C. region, Cumbres de San Mateo, Santa Brigida, *Erica* heath and woodland 500–1000 m, locally frequent; TENERIFE: Locally rather common in the forest zones of the N. coast and on dry slopes in the S. and W. part of the island and in the pine forests of the central region, up to 1500 m, Las Mercedes, Agua Mansa, Icod, Santiago del Teide; GOMERA: Forest region of the centre, Agando, El Cedro, Arure, 600–1000 m, locally abundant.

A. ombriosus Ceb. & Ort. Like *A. foliolosus* but a more lax shrub with long petioles, narrow leaflets and a densely glandular legume.

HIERRO: San Salvador, in pine woodland and *Erica* heath, 900 m, now very rare and probably almost extinct.

DORYCNIUM

Differs from *Lotus* by its shrubby habit and dense inflorescence of white to cream flowers which often have dark red veins and tips.

1. Stipules petiolate, free at the base.
  2. Calyx with short hairs .. .. **D. broussonetii**
  2. Calyx glabrous .. .. **D. spectabile**
1. Stipules sessile, united at the base .. .. **D. eriophthalmum**

D. broussonetii (Choisy) Webb & Berth. Shrub up to about 1·5 m. Leaflets obovate to elliptical, 2–4 cm. Stipules suborbicular, shortly petiolate, free at base. Flowers whitish with pink or purple tips, in dense clusters of 5–8. Calyx campanulate, pilose, the teeth unequal. **(35)**

TENERIFE: N. coast cliffs, Anaga region, 200–400 m, rare.

D. spectabile (Choisy) Webb & Berth. Like *D. broussonetii* in its free, petiolate stipules but differing by its glabrous calyx with equal teeth.

TENERIFE: S. coasts, very local, from Guimar to Masca. N. W. region in the mountains above Los Silos, rare.

D. eriophthalmum Webb & Berth. Differs from both the above species because of its sessile stipules which are united at the base. The calyx is glabrous and the teeth more or less equal.

LA PALMA: N.E., shady cliff in the Puntallana region. W. coast, Barranco de Las Angustias 400–800 m, rather rare.

VICIA

Annual to perennial herbs with tendrils. Leaves pinnate with 1 to 8 pairs of leaflets. Flowers solitary or in racemes. Calyx usually not 2-lipped. Legume compressed.

1. Flowers white to pale pink.
  2. Leaflets 1–2 pairs, flowers usually solitary .. .. **V. filicaulis**
  2. Leaflets 2–6 pairs, flowers in racemes.
    3. Racemes less than 12-flowered, calyx teeth very small, upper absent .. .. **V. cirrhosa**
    3. Racemes 16- to 30-flowered, calyx teeth prominent, unequal .. .. **V. scandens**
1. Flowers pale yellow with a purplish keel, solitary .. **V. chaetocalyx**

V. cirrhosa Chr. Sm. Slender, scrambling herb. Stems intricate, filiform. Leaflets in 2–4 pairs, linear to linear-lanceolate. Tendrils coiled. Peduncles 3- to 12-flowered. Calyx teeth small, the upper absent. Corolla white or pinkish.

TENERIFE: Shady, rocky places and open slopes in the S. part of the island. Adeje, Arico, Guimar etc, 400–600 m, locally frequent; LA PALMA: Coastal zone near Breña Baja; GOMERA: N. coast, Agulo to Vallehermoso, 400 m; HIERRO: Risco de Jinamar in the forest zone, 800 m; GRAN CANARIA: S. region Arguineguin, Fataga, 400 m.

V. filicaulis Webb & Berth. Like *V. cirrhosa* but the leaflets in 1 or 2 pairs and the flowers usually solitary with equal calyx teeth.

GRAN CANARIA: Tejeda region of the Cumbres, rare.

V. scandens Murray. Like *V. cirrhosa* but more robust with many-flowered inflorescences. **(31)**

TENERIFE: Agua Mansa, laurel woods and *Erica* heath below Los Organos, 1200–1400 m, locally frequent.

V. chaetocalyx Webb & Berth. Scrambling herb. Leaflets in 6–7 pairs. Flowers solitary, subsessile, calyx teeth long, keel with purplish blotch.

GRAN CANARIA: Telde region near the coast but not found recently.

SPARTOCYTISUS

Erect, much branched shrubs with small, caducous, trifoliate leaves and white flowers. Calyx 2-lipped, teeth very short, obscure. Legume black, somewhat hairy, 4- to 6-seeded.

1. Stems slender, flexuous; leaves shortly petiolate; leaflets oblanceolate to narrowly obovate; pedicels longer than calyx .. .. **S. filipes**
1. Stems stout, erect; leaves subsessile; leaflets linear; pedicels shorter than or equalling calyx .. .. .. **S. supranubius**

S. filipes Webb & Berth. Small shrubs with dense, flexuous branches and long, interrupted clusters of white, strongly scented flowers. **(171)**

TENERIFE: N. coast region to Masca, very local, 200–500 m. La Rambla, Icod, Masca; GOMERA: Locally frequent on the N. side of the island, Vallehermoso, Hermigua, Vallegranrey, 500–800 m; LA PALMA: Common in some parts, Fuencaliente, Los Llanos, El Paso, lower and forest zones; HIERRO.

S. supranubius (L.) Webb & Berth. Stout, erect shrub with thick, glaucous stems. Leaflets linear. Flowers in dense clusters along the upper parts of stems, shortly pedicellate, very fragrant. ***Retama del Pico.*** **(172)**

TENERIFE: Abundant and often dominant in the high mountain zone of Las Cañadas 1900–2200 m; LA PALMA: Cumbres de Garafia, tops of the mountains circling the Caldera de Tabouriente, 1700–2000 m.

RETAMA

Like *Spartocytisus* but with a small calyx with a deeply 2-toothed upper lip and small, ovoid, indehiscent 1- to 2-seeded fruits.

R. monosperma (L.) Boiss. subsp. ***rhodorrhizoides*** Webb & Berth. is abundant in the lower zones of all the islands especially on S. facing slopes near coasts. ***Retama.*** **(173)**

CHAMAECYTISUS

Tall shrubs with trifoliate leaves. Flowers in axillary clusters, white. Calyx tubular, 2-lipped, teeth small. Legume black. Seeds strophiolate.

1. Leaves glabrous, broadly elliptical-ovate (La Palma) .. **C. palmensis**
1. Leaves pubescent to silky hairy, lanceolate .. .. **C. proliferus**

C. proliferus (L.) Link. Tall, very variable shrub. Leaves petiolate; leaflets lanceolate, acute or mucronate. Flowers in axillary fascicles of 1–4. Calyx deeply bilobed, pubescent to densely sericeus. Legume compressed, black when ripe. ***Escabon.*** **(28) (170)**

TENERIFE: Very abundant in the forest zones up to 1800 m especially along the N. coast and in the Anaga region. Local in the S., Agua Mansa, Esperanza, Portillo, Santiago del Teide, Orotava to Icod, Masca, Arico, Guimar, Las Mercedes, Taganana etc; GRAN CANARIA: Common in the C. region up to 1500 m, Cruz de Tejeda, Tenteniguada, Paso de la Plata, Fataga etc; GOMERA: C. region, Agando, El Cedro, Chipude, 600–1200 m; HIERRO: Locally frequent in the El Golfo area.

C. palmensis (Christ) Hutch. Like *C. proliferus* but with glabrous, broader leaves and very densely hairy pods.

LA PALMA: Common in the N. region, La Galga to Barlovento, Roque del Faro etc, 400–1200 m.

GERANIACEAE ——— Geranium Family

Herbs. Leaves alternate, deeply lobed or compound. Flowers regular. Sepals and petals 5. Stamens 10 or 15, more or less fused at the base. Ovary of 3–5 fused carpels ending in a slender beak formed from the fused styles. Beak in fruit splitting into coiled sections projecting the 1-seeded carpels outwards.

GERANIUM

Leaves compound. Flowers regular. Sepals and petals 5. Stamens 10 or 15, fused at base. Ovary of 5 fused carpels ending in a long beak formed by the fused styles and which splits on maturing into coiled, spring-like sections carrying the 1-seeded carpels.

G. canariense Reut. Robust perennial herb often with a woody stock or short woody stem. Leaves in a large rosette, broadly ovate, deeply lobed and dissected; middle lobe sessile. Inflorescence branched. Flowers 2–3 cm across. Petals clawed, oblong, pink with whitish back. Filaments white or pink; anthers red. ***Pata de Gallo.* (179)**

TENERIFE: Sierra Anaga, frequent in the laurel woods of Taganana, Las Mercedes and El Bailadero, Agua Garcia, Icod, 500–700 m; LA PALMA: Breña Alta, Cumbre Nueva to Barlovento, common in cloud forest zones; GOMERA: Central forest region, El Cedro, Barranquillos de Vallehermoso, Chorros de Epina etc; HIERRO: Laurel forests and pine woods of the Cumbre and El Golfo region, 500–1000 m.

ZYGOPHYLLACEAE

Herbs or shrubs. Leaves usually opposite, compound, stipulate, often fleshy. Flowers solitary or in cymes, usually hermaphrodite, regular, 5-parted. Ovary superior. Fruit a capsule or drupe.

ZYGOPHYLLUM

Small shrublets. Leaves and branches fleshy. Leaves opposite, 2-foliate. Flowers solitary. Petals and sepals 5. Stamens 10. Fruit 4-angled, breaking into 5-valves.

Z. fontanesii Webb & Berth. Curious succulent shrublet of coastal regions. Leaves and stems glaucous to yellowish. Fruits whitish, corky when ripe.
Halophyte of dunes and coastal rocks. TENERIFE: El Medano, Los Cristianos etc, S. coast region; GOMERA: Puerto de Vallehermoso, Alojera, S. coasts; GRAN CANARIA: La Isleta, Baha de Confital, Maspalomas, Melenara, Arinaga; LANZAROTE: La Caleta, Arrecife, Puerto Naos, Graciosa etc; FUERTEVENTURA: Jandia, Matas Blancas, Cotillo, Cofeté, Lobos.

EUPHORBIACEAE ——— Spurge Family

Trees, shrubs or herbs with milky latex. Leaves usually simple. Inflorescence usually compound. Flowers more or less regular, small. Ovary usually 3-celled, superior. Fruit a 3-parted capsule. Seeds usually with a caruncle.

EUPHORBIA

Trees, shrubs or herbs with milky juice. Leaves alternate or absent. Flowers axillary or in terminal umbels. The flower-heads consisting of a small, cup-shaped involucre with 4 or 5 yellow or brownish glands. Stamens 1-several (each in fact representing a single male flower). Female flowers consisting of a single, central 3-celled ovary on a pedicel which elongates in fruit. Styles 3. Fruit a 3-celled capsule.

1. Trees with terminal panicles of flowers (Tenerife, La Palma) .. .. **E. mellifera**
1. Shrubs.
 2. Stems cactus-like with spines or thorns.
  3. Stems 4- to 5-sided, spines usually short, curved .. **E. canariensis**
  3. Stems 8- to 14-sided, spines long, straight .. .. **E. handiensis**
 2. Stems unarmed.
  4. Stems leafless, inflorescences sessile .. .. .. **E. aphylla**
  4. Stems with at least a terminal rosette of leaves, inflorescences pedunculate.
   5. Inflorescences consisting of a single, terminal flower .. **E. balsamifera**
   5. Inflorescences many-flowered.
    6. Inflorescences purple-red.
     7. Leaves spathulate to oblong, inflorescence bracts more than 1 cm across (Tenerife) .. .. .. **E. atropurpurea**
     7. Leaves linear-lanceolate, inflorescence bracts less than 0·5 cm across (Gomera) .. .. .. **E. bravoana**
    6. Inflorescences green-yellow or only the ripe capsules reddish.
     8. Bracts below flower-heads persistent in fruiting stage.
      9. Bracts 1–2 cm long, fused at least at base.
       10. Floral glands crescent-shaped, bracts united only at the base (Tenerife) .. .. .. **E. bourgaeana**
       10. Floral glands dentate, bracts united for $\frac{2}{3}$ of their length (Gomera) .. .. .. **E. lambii**
      9. Bracts less than 1 cm long, free at base .. .. **E. obtusifolia**
     8. Bracts below the flower-heads deciduous before the fruit matures.. .. .. **E. regis-jubae**

The perennial, herbaceous European/Mediterranean ***E. paralias L.*** **(183)** is frequent on dunes in coastal regions of S. TENERIFE, GRAN CANARIA, LANZAROTE and FUERTEVENTURA.

E. canariensis L. Tall succulent shrub with square or 5-sided cactus-like stems. Spines paired, curved, borne along the stem ridges. Flowers green-reddish, solitary on short peduncles. Capsules red or brown. ***Cardon.*** **(180) (181)**

ALL ISLANDS: Common in the coastal regions of the W. islands and GRAN CANARIA on dry rocky slopes and screes, lava fields, cliffs, etc, up to 900 m. Rather rare on the E. islands.

E. handiensis Burchd. Small shrub up to 80 cm, often densely branched. Stems cactus-like with 8–14 ribs and long, paired, straight spines, up to 2–3 cm long and a tuft of spines at the apex. Flowers reddish. Capsules brown or red. **(182)**

FUERTEVENTURA: Jandia region between Moro Jable and El Faro de Jandia in the coastal region up to 150 m, this species is now a very rare and endangered plant.

E. mellifera Aiton. Tree up to 15 m. Bark grey, smooth. Leaves crowded towards the apices of branches, subsessile, narrowly lanceolate, obtuse to acute, dark green. Inflorescences terminal, paniculate. Capsules large. Seeds with a sessile, plate-like caruncle. ***Tabaiba Silvestre.*** **(189)**

TENERIFE: Sierra de Anaga in laurel forests, Cumbres above Taganana, Chinamada; LA PALMA: Cumbre Nueva above Las Breñas, Monte de Barlovento, extremely rare, almost extinct species.

E. bourgaeana Gay. Shrub up to 1·5 m. Stems light brown. Leaves lanceolate to oblanceolate, apex retuse. Inflorescences more or less simple umbelliform, sometimes compound, greenish-yellow. Floral bracts large, fused at base. Floral glands crescent-shaped. Capsules light-brown or rarely dark-reddish. **(188)**

TENERIFE: Montes de Teno, Buenavista, El Fraile, Teno Bajo, Ladera de Guimar, 100–600 m, locally frequent only at El Fraile, otherwise extremely rare.

E. lambii Svent. Like *E. bourgaeana* but taller with more slender branches; leaves narrowly lanceolate; floral bracts very large, fused for at least ½ their length; floral glands dentate; capsules light brown or yellowish.

GOMERA: Forest margins in the N.W. and C. regions of the island, Benchijigua, Chorros de Epina, Lomo de Carreton, 600–800 m, rare.

E. obtusifolia Poiret. Shrub up to 2 m. Stems light brown. Leaves narrowly oblong, apex acute or obtuse. Inflorescences umbelliform, simple or rarely compound, with 5–8 rays. Floral bracts large, persistent in fruit, free at the base, greenish-yellow. Floral glands usually crescent-shaped or rarely dentate. Capsules light-brown or red. Seeds with a stalked caruncle. ***Tabaiba.*** **(186)**

EASTERN ISLANDS: Common in the lower zones of GRAN CANARIA, LANZAROTE and FUERTEVENTURA, ascending to 1500 m on GRAN CANARIA; GOMERA: S.W. region between Arure and Chipude, Argaga etc; HIERRO: Valverde, El Golfo, S. region; TENERIFE: Teno, very rare.

E. regis-jubae Webb & Berth. Like *E. obtusifolia* but the umbels usually compound, the floral bracts small, falling well before the fruits ripen, the floral scales ovoid, dentate and the seeds with a more or less sessile caruncle. ***Tabaiba.*** **(187)**

TENERIFE: Very common on the N. coast from Anaga to Teno, San Andrés, Guimar, Adeje, Masca etc, often dominant over wide areas, 0–1500 m; GOMERA: N. coast, San Sebastian, Agulo, Vallehermoso, Epina, Vallegranrey, very common; LA PALMA: Very common in the lower zone in all parts of the island, Santa Cruz, La Galga, Los Sauces, Fuencaliente, Los Llanos etc, up to 1500 m in the Caldera. The existence of this species on GRAN CANARIA and the E. ISLANDS has yet to be confirmed.

A low-growing form from the E. side of GOMERA with glaucous leaves reddish stems and dense, compound umbels has been described as a separate species *E. berthelotii* Bolle.

E. atropurpurea Brouss. Shrub with succulent brown stems and branches. Leaves crowded towards tips of stems, glaucous, oblong-spathulate, blunt. Inflorescence dark red-purple, 5- to 15-rayed. Capsules dark-red to brown. Seeds dark brown, wrinkled; caruncle stalked. **(36)**

TENERIFE: Locally very common in the lower zone of the W. and S., Teno, Masca, Valle de Santiago del Teide, Tamaimo, Guia de Isora, Adeje, Ladera de Guimar, 300–1200 m.

E. bravoana Svent. Like *E. atropurpurea* but less densely branched with very dark stems and branches. Leaves below the inflorescence purplish, umbels few-rayed (2–5). Capsules light reddish-brown. **(190)**

GOMERA: N.E. region in ravines of the lower zone, cliffs at Agulo, Barranco de Majona, up to 800 m, rare.

E. balsamifera Aiton. Shrub up to 2 m. Stems gnarled, greyish. Leaves rosulate at tips of branches, oblong-spathulate, pale green to glaucous, obtuse to acute. Flower-heads solitary, subsessile or on a short peduncle, campanulate. Floral glands oval to rounded, more or less entire. Capsules solitary, globose. Seeds brown, wrinkled, the caruncle absent. ***Tabaiba dulce.*** **(37) (184)**

ALL ISLANDS: Common in coastal regions especially near the sea or on dry S. slopes; TENERIFE: Teno, Los Cristianos, Medano etc; GOMERA: Vallegranrey, Puerto de Vallehermoso, S. region; LA PALMA: Fuencaliente etc; GRAN CANARIA: Very common S. of Telde.

E. aphylla Brouss. ex Willd. Small compact shrub up to 50 cm. Stems slender, pencil-like, leafless. Flowers more or less sessile in small clusters at the tips of stems. Fruits very small, light-brown or reddish. Seeds small, brown; caruncle sessile, plate-like. **(185)**

GRAN CANARIA: N. coast, La Isleta to Cuesta de Silva, Aldea de San Nicolas, locally frequent near the sea; TENERIFE: N.W., Buenavista region, Punta de Teno, coastal region below Los Silos, S. coasts, below Guimar etc; GOMERA: N. coast, Puerto de Vallehermoso, Alojéra, S. coasts, Santiago, Alajéro, halophyte always found on coastal rocks and slopes facing the sea between 0 and 150 m.

CNEORACEAE

Shrubs with medifixed hairs. Leaves usually alternate, simple. Flowers hermaphrodite, borne on the petioles, 3- to 4-parted. Ovary superior.

CNEORUM

Shrubs, leaves alternate. Inflorescences borne on petioles of leaves. Flowers 3- to 4-parted. Fruit of 2–4 (usually 3) cocci, each globose, drupaceous.

C. pulverulentum Vent. Up to 1·5 m. Stems and leaves densely clothed with short medifixed hairs. Leaves entire, linear to narrowly oblanceolate, obtuse. Flowers borne singly on petioles of upper leaves, yellow. Fruits of usually 2 or 3 grey cocci resembling a *Euphorbia* capsule. ***Leña blanca.*** **(38) (191)**

GRAN CANARIA, TENERIFE, GOMERA, LA PALMA, HIERRO: Locally very common and sometimes dominant in the lower zone especially in *Euphorbia* communities, S. GRAN CANARIA, TENERIFE, Teno region.

RUTACEAE ——————— Rue Family

Shrubs or herbs. Leaves alternate or opposite, usually compound, gland-dotted. Flowers hermaphrodite, regular. Sepals and petals 4–5. Stamens 8–10. Ovary superior with 4–5 cells. Style 1. Fruit a berry or capsule.

RUTA

Shrubs with pinnate leaves with translucent oil glands. Flowers regular, 4-parted. Stamens 8. Petals yellow, margins ciliate. Capsule 4- to 5-lobed.

1. Leaves glaucous-blue, petals erect (Gran Canaria) .. **R. oreojasme**
1. Leaves green, petals patent.
 2. Densely branched shrub, leaflets remotely toothed, fruits small (4–5 mm) (Gomera) .. .. **R. microcarpa**
 2. Lax shrub, leaflets entire, fruits about 6 mm across (Tenerife, La Palma) .. .. **R. pinnata**

R. oreojasme Webb & Berth. Small, strongly scented, procumbent to ascending shrub. Leaves glaucous-blue, pinnate; lobes obtuse. Flowers large, yellow. Petals erect, keeled. Fruits light brown, rugose. ***Ruda.*** **(193)**

GRAN CANARIA: S. region, Barrancos de Tirajana and Fataga, Mogan, cliffs in the lower zone, 200–600 m. Locally frequent.

R. pinnata L. fil. Tall shrub up to 1·5 (–2) m. Leaves light green, pinnate; lobes linear to rhomboidal, very variable. Flowers yellow. Petals small, flat. Fruits orange-brown. ***Tedera Salvaje.*** **(192)**

TENERIFE: Sporadic in the lower zone, 150–600 m, Buenavista, La Rambla, Montaña de las Arenas, Punta del Hidalgo, El Escabonal, Adeje; LA PALMA: Dry rocky areas towards the south, Mazo, 350 m, occasional.

R. microcarpa Svent. Like *R. pinnata* but smaller (80 cm) dense, much branched; leaflets remotely toothed; fruits small, yellowish.

GOMERA: Rocks and cliffs near Hermigua and in the S.W. region, very rare.

AQUIFOLIACEAE —————— Holly Family

Trees or shrubs. Leaves alternate, simple, often spiny. Flowers hermaphrodite or dioecious, regular, 4- to 5-parted. Sepals small. Petals free or united at base. Ovary superior. Style very short or absent. Fruit a small drupe with 3 or more stones.

ILEX

Evergreen shrubs or small trees. Dioecious with vestigial stamens in female flowers. Petals 4, slightly united or free at the base. Sepals 4, united. Fruit a small red or blackish drupe (holly berries).

1. Leaves very broad, orbicular, the margins with forward-pointing spines; apex sharply pointed .. .. .. **I. platyphylla**
1. Leaves ovate, the margins with few spines or spineless; apex blunt .. .. **I. canariensis**

I. platyphylla Webb & Berth. Tree up to 15 m. Bark greyish. Leaves glossy, up to 15 cm long, broadly ovate to orbicular; margins undulate with forward-pointing spines or occasionally more or less entire; apex with a sharp spine. Petals pinkish-white. Fruit about 6–9 mm, dark red to blackish. ***Naranjero salvage.***

TENERIFE: Fairly frequent in the laurel forests of the N. slopes and in the Anaga region, Las Mercedes, Agua Garcia; GOMERA: El Cedro.

I. canariensis Poiret. Shrub or small tree up to about 10 m. Bark greyish-brown. Leaves glossy, about 6–8 cm long, ovate; margins usually entire but occasionally with a few small spines; apex obtuse or rounded. Petals white. Fruit about 1 cm, red. ***Acebino.*** **(194)**

TENERIFE: Very frequent in forests and *Erica* heath up to 1800 m, Agua Mansa, Las Mercedes, Agua Garcia, Icod etc; GRAN CANARIA: In relict pockets of laurel forest especially on the N. coast, near Arucas, Moya, Tamadaba; GOMERA, HIERRO, LA PALMA: Very common in laurel woods and heath zones.

CELASTRACEAE

Trees or shrubs. Leaves simple. Inflorescence cymose. Flowers hermaphrodite, regular, 5-parted. Fruit a 3-locular capsule. Seeds arillate.

MAYTENUS

Shrubs with entire, coriaceous, remotely serrulate leaves. Flowers 5-parted, regular. Capsules 3-locular. Seeds brown with whitish aril at base.

M. canariensis (Loesl.) Kunk. & Sund. (*Catha cassinoides* Webb & Berth.) Up to 4 m. Densely branched. Leaves leathery, glossy. Inflorescence cymose. Flowers small, pale greenish-yellow. Fruit pale greenish to light brown, globose, 3-celled, dehiscent. Seeds chestnut-brown, oval, with a white, fluffy aril forming a basal cup. **(195)**

TENERIFE, GRAN CANARIA, LA PALMA, GOMERA, FUERTEVENTURA: Locally common on rocks and cliffs in the lower zone and forest regions 200–1500 m. Frequent on the N. coast of TENERIFE, Teno and at Guimar.

RHAMNACEAE

Trees or shrubs, sometimes spiny. Leaves alternate, simple. Flowers small, in axillary clusters, 5-parted, hermaphrodite or unisexual. Perigynous disc present. Ovary superior. Fruit a drupe or capsule.

RHAMNUS

Trees or shrubs. Leaves simple. Flowers 4-parted, unisexual. Stamens 4, alternating with calyx lobes. Ovary superior. Fruit a drupe.

1. Tree, leaves ovate with small glands .. .. .. **R. glandulosa**
1. Shrubs, leaves lanceolate to oblanceolate, without glands.
 2. Leaves entire, lanceolate .. .. **R. integrifolia**
 2. Leaves with toothed margin, narrowly oblanceolate to ovate .. .. **R. crenulata**

R. glandulosa Aiton. Small tree up to 10 m. Leaves ovate with small, round, prominent glands in the axils of the veins; margins bluntly serrate. Inflorescences shorter than the leaves. Fruits globose, black. ***Sanguino.***

TENERIFE: Sierra Anaga, locally common in laurel woods, Taganana, Las Mercedes, Orotava Valley, Sta. Ursula, Guimar, 600–800 m; LA PALMA: Forests of Cumbre Nueva and N.E. region, Breña Alta, Los Tilos 700–900 m; GOMERA: El Cedro woodland, Chorros de Epina to Arure.

R. crenulata Aiton. Shrub up to 2 m. Leaves narrowly oblanceolate to obovate, without glands; margins bluntly dentate. Inflorescences few-flowered, axillary. ***Espinero.* (40)**

TENERIFE, GOMERA, LA PALMA, HIERRO, GRAN CANARIA: Very frequent in *Euphorbia* scrub in the lower zone, especially on the N. coast of TENERIFE, rather rare on HIERRO.

R. integrifolia DC. Small shrub. Leaves lanceolate, acute, without glands; margins entire. Inflorescences axillary or terminal. Fruit more of less 4-lobed, reddish-black when ripe. ***Moralito.* (39)**

TENERIFE: Subalpine zone of Las Cañadas, on cliffs of the old crater wall, La Fortaleza, Montaña de Deigo Hernandez, Boca de Tauce, 2000 m. Occasionally found on cliffs in the deep valleys of the S., Guimar, Guia de Isora, Adeje, 400–800 m.

MALVACEAE —————— Hibiscus Family

Shrubs or herbs, often with stellate pubescence. Leaves palmately lobed, stipules present. Flowers in small panicles or solitary. Sepals 3 or 5, free or fused, often with an epicalyx. Petals 5, free. Stamens numerous, the filaments fused below. Ovary superior. Fruit schizocarpic by separation of 1-seeded nutlets.

LAVATERA

Woody shrubs. Flowers in small clusters or more or less terminal and solitary. Epicalyx segments 3, more or less caducous. Fruit a schizocarp split into numerous 1-seeded, indehiscent mericarps in a single whorl.

1. Flowers salmon-pink, petals narrowly oblong .. .. **L. phoenicea**
1. Flowers mauve, with darker bases to the ovate petals .. **L. acerifolia**

The weed *L. arborea* also occurs on roadsides and waste ground.

L. acerifolia Cav. Tall shrub up to 2·5 m. Leaves palmately lobed; lobes irregularly dentate; petioles long. Flowers in small terminal or axillary clusters or occasionally solitary, long-pedicellate. Petals mauve with dark bases or very occasionally white. **(197)**

TENERIFE: Locally frequent, Teno, Masca, Guimar, Adeje, Anaga region, cliffs of the lower zone, 200–500 m; GRAN CANARIA: N.E. region, Guiniguada, Montaña Lentiscal etc, 250–400 m; GOMERA: Barranco de la Villa, Argaga, Agulo.

L. phoenicea Vent. Like *L. acerifolia* but the leaves with narrower lobes, the petals narrower, oblanceolate, salmon-pink. **(196)**

TENERIFE: Cuevas Negras de los Silos, 250 m, N. coast of the Anaga region, Taganana, Bajamar, Punta del Hidalgo, very rare.

GUTTIFERAE ———— St. John's Wort Family

Shrubs or herbs with resinous juice. Leaves opposite, simple, usually entire, often gland-dotted. Flowers usually terminal, solitary or in cymes, hermaphrodite, regular. Sepals 5, free. Petals 5. Stamens numerous, often connate in bundles. Ovary superior. Styles free or connate. Fruit a capsule or berry.

HYPERICUM

Shrubs. Leaves opposite, simple, often gland-dotted. Flowers yellow, solitary or in branched cymes, terrestrial, regular. Sepals and petals 5. Stamens numerous often more or less joined at base in bundles. Ovary superior, styles 3 or 5. Fruit usually a capsule, sometimes fleshy. Seeds very small, brown, numerous.

1. Plants without glands on the margins of leaves or sepals.
   2. Leaves narrowly elliptical to linear-lanceolate; inflorescence many-flowered .. .. **H. canariense**
   2. Leaves broadly ovate; inflorescence 2- to 4-flowered .. **H. grandifolium**
1. Plants with glandular margins to leaves and/or sepals.

3. Leaves shortly petiolate . . . . **H. glandulosum**
3. Leaves sessile, subamplexicaul to perfoliate.
  4. Leaves perfoliate, hairy beneath . . . . . . **H. coadunatum**
  4. Leaves subamplexicaul at base, glabrous . . . . **H. reflexum**

H. canariense L. Glabrous shrub up to 2·5 m. Leaves linear-lanceolate to narrowly elliptical, 2–7 cm long, obtuse or acute. Flowers yellow, 2 cm across, in large, dense, terminal panicles. Sepals ovate, fused at base, without glandular margins. Petals elliptical, about 1 cm long. Fruit a somewhat fleshy capsule, becoming brown and hard when ripe. ***Granadillo.*** **(198)**

GRAN CANARIA, TENERIFE, LA PALMA, GOMERA, HIERRO: Xerophytic scrub and forest zones 150–800 m, locally very common.

H. grandifolium Choisy. Glabrous shrub up to 1 m. Stems reddish-brown. Leaves broadly ovate, subsessile, 4–7 × 2·4–5 cm, obtuse. Inflorescence lax 2- to 4-flowered. Flowers large, up to 4·5 cm across. Petals lanceolate, about 2 cm long. Fruit a hard, dark brown capsule when ripe. ***Malfurada.*** **(43)**

TENERIFE, GOMERA, GRAN CANARIA, LA PALMA, HIERRO, LANZAROTE: Common shrub in laurel and pine forest zones particularly on Gomera, 400–1500 m, Peñitas de Chache on LANZAROTE.

H. glandulosum Aiton. Small shrub up to 1 m. Stems glabrous or tomentose. Leaves elliptical to ovate, attenuate to short petiole at base, 4–5 × 1·5–2·5 cm, acute; margins subcrenate, glandular. Inflorescence dense, many-flowered. Flowers 1·5–2 cm across. Sepals with black glands. **(42) (199)**

GOMERA, TENERIFE, LA PALMA, GRAN CANARIA: Frequent on forest cliffs, occasionally extending into the upper xerophytic zone, 500–1500 m. Plants with brown, woolly hairs on the stems have been separated as *H. joerstadii* Lid.

H. reflexum L. fil. Like *H. glandulosum* but the leaves decussate, sessile and subamplexicaul at the base. ***Cruzadilla.*** **(44)**

TENERIFE, GOMERA, GRAN CANARIA: Locally common on basalt cliffs, in the lower zone, 200–800 m.

H. coadunatum Chr. Sm. Small spreading shrub of damp localities. Leaves perfoliate, ovate, hairy especially below, inflorescence dense. Flowers small, about 1 cm across. **(41)**

GRAN CANARIA: Very rare, on wet cliffs and walls in the central region above San Mateo, 1200 m.

TERNSTROEMIACEAE

Trees or shrubs. Leaves alternate, simple. Flowers solitary or fasciculate and axillary, hermaphrodite, 4- to 5-parted. Sepals fused at base. Petals more or less free. Ovary superior. Fruit capsulate.

VISNEA

Trees with alternate, leathery leaves. Petals and sepals 5. Stamens numerous. Fruit a dehiscent capsule.

V. mocanera L. fil. Up to 15 m. Leaves lanceolate, the margins serrate to subentire. Flowers in axillary clusters, more or less pendulous, campanulate. Calyx lobes blunt, pubescent. Petals cream-white. Fruit an oval, greyish-brown, dehiscent capsule. ***Mocan.*** **(45)**

TENERIFE: Laurel woods, Sierra Anaga, Guimar, Cuevas Negras, rare, 300–

1000 m; LA PALMA: Breña Alta, forests of Cumbre Nueva, Mazo; HIERRO: Locally frequent in the forests of El Golfo, large groves above Frontera; GOMERA: El Cedro, rare; GRAN CANARIA: Los Tiles de Moya, a declining species.

VIOLACEAE ——————— Pansy, Violet Family

Herbs or shrublets. Leaves simple, stipulate. Flowers usually solitary. Pedicels with 2 bracteoles. Flowers hermaphrodite, zygomorphic, 5-parted. Corolla spurred. Ovary superior, 1-celled. Style simple. Fruit a 3-valved capsule.

VIOLA

Herbs with alternate, simple leaves. Flowers usually solitary, irregular, usually spurred. Sepals, petals and stamens 5. Fruit a dehiscent, 3-valved, many-seeded capsule.

V. cheiranthifolia H.B. & K. Small perennial herb with densely pubescent, ovate to spathulate, dentate to entire leaves. Flowers pansy-like with a short spur, tricoloured, predominantly mauve with white and yellow markings. **(46)**

TENERIFE: Subalpine zone of Las Cañadas, Pico de Teide up to the Refugio, Montaña Blanca, Pico Viejo, 2100–2800 m.

V. palmensis Webb & Berth. Resembling *V. cheiranthifolia* but much more robust with larger pale mauve and yellow flowers with a longer, slender spur. LA PALMA: High mountain regions on the rim of the Caldera de Tabouriente, Cueva de Tamagantera, Pico del Cedro etc, 1900–2400 m, rare.

Several species of violets are frequent in the laurel forest regions of the islands including *V. odorata* with deep violet flowers and a pale violet spur, *V. reichenbachiana* with blue-violet flowers and a violet spur and *V. canina* with blue flowers. These are fairly widespread European species. *V. plantaginea* Webb ex Christ was described last Century from Gomera but is still very poorly known.

CISTACEAE

Shrubs or herbs. Leaves simple, usually opposite, mostly with stellate hairs. Flowers hermaphrodite, regular. Sepals 5 or 3. Petals 5, usually caducous. Stamens numerous. Ovary superior. Styles simple. Fruit a capsule with 3–10 valves.

1. Capsule with 3 valves, petals yellow .. .. .. **Helianthemum**
1. Capsule with at least 5 valves, petals pink or white .. **Cistus**

HELIANTHEMUM

Dwarf shrubs with small, entire leaves. Flowers usually small, less than 2 cm. Petals yellow (rarely pink or white). Capsule 3-valved.

1. Dwarf shrubs up to 25 cm, stems procumbent to ascending, leaves less than 1 cm.
  2. Leaves densely silvery-hairy, ovate .. .. .. **H. canariense**
  2. Leaves green, sparsely hairy, oblanceolate to obovate .. **H. thymiphyllum**
1. Shrubs 30–100 cm, erect, leaves 1·2–5 cm.
  3. Leaves up to 5 cm, ovate-lanceolate (Gran Canaria) **H. bystropogophyllum**
  3. Leaves less than 3 cm, oblong (Tenerife, La Palma).
    4. Sepals hispid, style curved .. .. **H. teneriffae**

4. Sepals densely and shortly white-tomentose, style straight .. .. **H. broussonetii**

H. canariense Pers. Densely branched, dwarf shrub up to 25 cm. Leaves 0·5–1 cm, oval, obtuse, densely silvery-grey pubescent. Sepals with 3-veins. Flowers about 1·2 cm across. Petals pale yellow.

ALL ISLANDS: Locally frequent in dry areas, Cuesta de Silva on GRAN CANARIA, Adeje region on TENERIFE, up to 600 m in the lower zone.

H. thymiphyllum Svent. Like *H. canariense* in habit but with smaller, oblanceolate-obovate, very shortly and sparsely hairy leaves.

LANZAROTE: Famara cliffs, 400–600 m, locally abundant in dry areas; FUERTEVENTURA: Betancuria region in dry valleys, 300–500 m, very local.

H. bystropogophyllum Svent. Erect shrub 50–100 cm. Leaves up to 5 cm, ovate-lanceolate, pale green, densely pubescent. Flowers sulphur-yellow, about 2 cm across.

GRAN CANARIA: San Nicolas valley on the W. side of the island, pine woodland at about 1400 m, very rare.

H. teneriffae Coss. Like *H. bystropogophyllum* but with small, oblong leaves. Inflorescence with 4–12 flowers.

TENERIFE: Ladera de Guimar, 800 m, very rare.

H. broussonetii Dun. ex DC. Resembling *H. canariense* but much taller and with oblong-lanceolate, obtuse leaves and the sepals 4-nerved, densely white-tomentose.
TENERIFE: Anaga region, Valle de Afur, Taganana region; LA PALMA: Los Sauces. (N. Africa).

CISTUS

Shrubs with entire leaves and large, showy flowers 2 cm or more with ephemeral white or pink petals. Fruit a capsule with 5–10 valves.

1. Flowers pink.
  2. Capsules sparsely hairy .. .. **C. symphytifolius**
  2. Capsules densely hairy (Tenerife, subalpine) .. .. **C. osbeckifolius**
1. Flowers white.
  3. Leaves linear, flowers small .. .. **C. monspeliensis**
  3. Leaves lanceolate, flowers large (Gran Canaria) .. **C. ladiniferus**

C. symphytifolius Lam. (incl. *C. candidissimus* Dun. ex DC.) Very variable shrub up to 1 m. Leaves broadly lanceolate to ovate, sparsely to densely hairy, acute or obtuse, the surface rugose; veins very prominent beneath. Flowers up to 5 cm. Petals pink. Stigma capitate. Capsules brown, sparsely hairy. **(200A)**

TENERIFE, LA PALMA, HIERRO, GOMERA, GRAN CANARIA: Very common in pine forest and montane zones from 800–1800 m.

C. osbeckifolius Webb ex Christ. Like *C. symphytifolius* but the leaves lanceolate to elliptic, much smaller, densely hairy with ciliate, silvery margins. Petals rose-pink. Capsules densely hairy. **(47)**

TENERIFE: Subalpine zone of Las Cañadas, La Fortaleza, Tope de la Grieta, ca. 2000 m, rare.

C. monspeliensis L. Small shrub with linear, sticky, revolute leaves and small white flowers. **(200)**

W. ISLANDS & GRAN CANARIA: Locally common in forest regions especially on GOMERA (Mediterranean region).

C. ladiniferus L. Shrub with lanceolate leaves and large white flowers with a brown blotch at the base of the petals.

GRAN CANARIA: Pine forests in the Tamadaba region. Very local and almost certainly introduced (Mediterranean region).

TAMARICACEAE ——— Tamarisk Family

Trees or shrubs. Leaves small, scale-like, alternate. Flowers hermaphrodite, regular, 4- to 5-parted, in slender spikes. Disc present. Petals free. Stamens 5, more or less free. Ovary superior. Fruit a capsule. Seeds with long hairs.

TAMARIX

Shrubs with scale-like, amplexicaul or sheathing leaves. Flowers small, white or pink, in spike-like, dense racemes. Flowers 5-parted with a nectar-secreting disc between stamens and ovary.

1. Bark black or purplish, flowers subsessile, white or pale pink .. .. **T. africana**
1. Bark reddish brown, flowers pedicellate, pink .. .. **T. canariensis**

T. africana Poir. Small tree or large shrub with black or purplish bark. Leaves up to 4 mm. Spikes 30–60 mm. Flowers sessile, pale pink or white.

GRAN CANARIA: S. region near Maspalomas, Arguineguin etc, locally frequent in beds of dry valleys; LANZAROTE, FUERTEVENTURA.

T. canariensis Willd. Shrub with reddish-brown bark. Leaves 1–3 mm. Spikes 15–45 mm. Flowers pedicellate, pink. ***Tarajal.***

ALL ISLANDS: Locally abundant, in coastal regions, dunes, dry rocky areas near sea-level, Maspalomas, Medano etc.

FRANKENIACEAE

Herbs or shrublets. Leaves opposite, heather-like. Flowers hermaphrodite, regular, solitary or cymose. Sepals 4–6, often joined at base. Petals 4–6, clawed. Stamens 6. Ovary superior. Capsule enclosed within the calyx, opening by valves.

FRANKENIA

Perennials. Flowers in cymes. Petals and sepals 5. Calyx more or less tubular. Petals imbricate. Outer whorl of stamens shorter than inner. Ovary sessile. Ripe capsule enclosed in persistent calyx.

1. Flowers in dense, terminal clusters, leaves linear, revolute .. .. **F. laevis** subsp. **capitata**
1. Flowers scattered on upper parts of stems and branches, at least some leaves lanceolate to oblanceolate, flat .. **F. ericifolia**

F. laevis L. subsp. ***capitata*** Webb & Berth. Woody-based, procumbent, mat-forming. Leaves linear, white-encrusted, margins revolute. Flowers in terminal clusters. Petals pink or whitish.

ALL ISLANDS: Very common in coastal regions close to sea-level, up to 100 m; Teno, Arinaga, Jandia etc. The annual *F. pulverulenta* with long slender prostrate branches is also present in similar habitats.

F. ericifolia Chr. Sm. Like *F. laevis* but more lax, with flat lanceolate to oblanceolate leaves and scattered flowers. Petals usually white.

TENERIFE: N. coast, Puerto de la Cruz, Playa de Castro, Teno; LA PALMA: Fuencaliente coastal region, N. coast between Garafia and Barlovento; GOMERA, HIERRO.

CUCURBITACEAE — Melon Family

Climbing or scrambling herbs with tendrils. Stems hispid, sappy. Leaves palmate. Tendrils usually coiled, axillary. Flowers unisexual, regular. Calyx with 5 narrow sepals. Petals 5, united at the base. Stamens 5, consisting of 2 fused pairs and 1 free stamen. Ovary inferior. Fruit a large, fleshy berry.

BRYONIA

Usually dioecious perennial, hispid herbs. Stems climbing by means of tendrils. Flowers in small axillary inflorescences, greenish-white to pale yellow. Corolla deeply 5-lobed. Fruits round, up to 2 cm across.

B. verrucosa Aiton. Leaves large, 5-angled or palmately lobed. Flowers about 1 cm across. Immature fruits greenish with pale stripes; mature fruits yellow-orange. **(48)**

TENERIFE: Locally common in the lower zone of the Anaga region, La Laguna, Bajamar, Punta del Hidalgo etc, 50–500 m, Teno, Los Silos; GRAN CANARIA: Barranco de Moya, Barranco de Guiniguada, Santa Lucia de Tirajana, locally frequent below 700 m; HIERRO: Pico Muertes, El Golfo, Sabinosa; LA PALMA: Barranco de las Angustias, Tijarafe etc.

ARALIACEAE — Ivy Family

Woody climbers. Leaves alternate, simple or lobed. Flowers small, in umbels, regular, hermaphrodite or unisexual, usually 5-parted. Sepals very small. Petals free. Stamens 3-many, disc present. Ovary inferior. Styles fused into a column. Fruit a berry.

HEDERA

Woody, evergreen climbers with simple leaves. Flowers globose, in umbels. Berries globose.

H. canariensis Willd. Leaves broader than long, those of non-flowering shoots more or less 3-lobed; those of flowering shoots suborbicular-cordate. Berries about 0·8 cm across, black with 3–5 whitish seeds. ***Yedra.*** **(49)**

TENERIFE, LA PALMA, GRAN CANARIA, HIERRO, GOMERA: Occasionally frequent in the laurel woods (commonly cultivated on old walls and in gardens).

UMBELLIFERAE — Carrot Family

Annual to perennial herbs or shrubs. Inflorescence a simple or compound umbel. Flowers 5-parted. Fruit dry, consisting of two lateral, indehiscent carpels, usually separating when ripe.

1. Leaves entire or shallowly lobed.
 2. Shrubs, leaves alternate, linear-lanceolate to narrowly ovate, glabrous .. .. **Bupleurum**
 2. Herbs, leaves opposite, more or less shallowly 3-lobed, hairy .. .. **Drusa**
1. Leaves ternately or variously pinnately divided.
 3. Coastal plant with thick, fleshy leaves .. .. .. **Astydamia**
 3. Leaves not thick or fleshy.
  4. Flowers yellow.

5. Plants very tall, up to 3 m, leaf-lobes usually linear-filiform .. .. **Ferula**
5. Plants up to 1 m, leaf-lobes flat.
  6. Leaves glabrous .. .. **Seseli**
  6. Leaves pubescent .. .. **Todaroa**
4. Flowers white.
 7. Fruits glabrous, blackish .. .. **Cryptotaenia**
 7. Fruits pubescent, light brown.
  8. Fruits ovoid to subglobose, villous to velutinous, scarcely ridged .. .. **Pimpinella**
  8. Fruits somewhat compressed, narrowing to a beak, pubescent with prominent ridges .. .. .. **Tinguarra**

DRUSA

Prostrate, slender herbs. Leaves petiolate, opposite. Inflorescences axillary, few-flowered, compact, simple umbels. Petals white. Fruits flattened, winged, the wings with glochidiate spines.

D. glandulosa (Poir.) Bornm. (*D. oppositifolia* DC.) Sparsely branched. Leaves membranaceous, more or less 3-lobed with simple or forked hairs. Stems and lower leaf surface occasionally with glochidiate hairs. Peduncle short, bracts absent. Fruits few, the lateral ribs extended into glochidiate wings. **(50)**

TENERIFE: Locally frequent in and below the laurel forest zone, Sierra Anaga, Bajamar, Agua Garcia, Orotava, Icod, Teno, Guimar, Adeje etc; GRAN CANARIA: Barranco de Moya, Tenteniguada, Agaete, Anden Verde, locally abundant; LANZAROTE: Famara Massif; FUERTEVENTURA: La Oliva, Jandia; LA PALMA: Los Llanos, San Andrés y Los Sauces; HIERRO: Valverde region; GOMERA: Hermigua valley.

BUPLEURUM

Leaves simple, often with more or less parallel veins. Sepals usually absent. Petals yellow. Fruits ovoid or oblong, blackish, the ridges more or less conspicuous.

1. Leaves broadly oblong-ovate, mucronate or obtuse .. **B. handiense**
1. Leaves linear to narrowly lanceolate, acute to acuminate **B. salicifolium**

B. salicifolium Soland. (incl. *B. aciphyllum* Webb & Berth.) Shrub up to 1·5 m. Stem woody. Leaves glaucous, linear to narrowly lanceolate, acute or apiculate; nerves 8–10, more or less parallel. Inflorescence lax. Rays 5–20. Umbellules 10- to 20-flowered. Fruits black or dark brown. **(54)**

TENERIFE: W. region, Masca, Teno, Guia de Isora, S. slopes, Guimar, Adeje etc, frequent on cliffs in the lower zone to 1000 m; GRAN CANARIA: C. mountain region, high cliffs between Tejeda and Roque Nublo; LA PALMA: Barranco de las Angustias 450 m, Los Tilos, forest cliffs; GOMERA: Roque Cano de Vallehermoso, basalt cliffs at Chipude, Valle de Santiago; HIERRO: El Golfo, Jinamar, Hoya de Tinco etc.

B. handiense Bolle. Small shrub. Leaves glaucous, broadly oblong-ovate, mucronate or obtuse, 5- to 7-nerved. Inflorescence 5- to 8-rayed. Umbellules 4- to 6-flowered.

FUERTEVENTURA:* Handia region on N. facing, exposed cliffs near the summit of Pico de la Zarza, 600 m; LANZAROTE: Upper part of the Riscos de Famara 550 m, rare.

***Ruthea herbanica** Bolle. Perennial herb with more or less pinnate leaves. Terminal umbel with about 20 rays; laterals 10- to 15-rayed. Involucral bracts present, linear. Flowes white. Fruits ovoid or oblong, weakly ribbed, somewhat compressed.

FUERTEVENTURA: Poorly known species which is said to be locally frequent in the central region of the island near La Oliva and Antequera.

ASTYDAMIA

Succulent-leaved herbaceous perennials. Flowers yellow. Petals small. Fruits ovoid, more or less fleshy, becoming dry, weakly ribbed.

A. latifolia (L. fil.) O. Kuntze. Leaves pinnate or deeply incised-dentate, very thick and fleshy with broad lobes, dying back in summer. Umbels compound. Rays up to 15. Flowers yellow. Mature fruits with a corky texture, light brown with 3 ribs and a somewhat expanded margin. Coastal regions of all islands, halophyte. **(202)**

TENERIFE: Frequent on N. coast cliffs, Bajamar, Puerto de la Cruz, Teno, S. coast, Playa de Viuda etc; GRAN CANARIA: N. coast Baha de Confital, Cuesta de Silva, La Isleta; LA PALMA: Tazacorte, coast below Barlovento etc, rather rare; HIERRO: Coasts of El Golfo, Roques de Salmar, Cuesta de Sabinosa; GOMERA: Puerto de Vallehermoso; LANZAROTE, FUERTEVENTURA: Very local in a few coastal areas.

Crithmum maritimum L. With pale yellow flowers and leaves with linear, succulent lobes is also frequent on coastal rocks on most of the islands (Mediterranean region–N.W. Europe).

FERULA

Tall, perennial herbs with 3- to 4-pinnate leaves with linear-filiform lobes. Petioles sheathing at base. Umbels large. Flowers yellow. Fruits oblong-elliptical, compressed dorsally with thin wings and ridges.

F. linkii Webb. Up to 3 m. Leaves with very conspicuous sheathing bases; lobes linear, flat to filiform. Terminal umbel large with up to 20 rays, usually surrounded by several smaller lateral umbels. Fruit about 1–1·5 cm. ***Cañaheja, Julan.*** **(201)**

TENERIFE: N. coast region from Anaga to Los Silos, Las Cañadas, Montaña de las Arenas Negras, La Fortaleza, Granadilla, Vilaflor etc, locally abundant, 200–2000 m; GRAN CANARIA: Los Tiles de Moya, Cruz de Tejeda, common in the Cumbres; GOMERA: Hermigua valley, Vallehermoso, Chipude; HIERRO: Miradero, Jinamar.

F. lancerottensis Parl. Like *F. linkii* but with broader, flat leaf-lobes.

LANZAROTE: Locally abundant in the Famara, Haria region 400–650 m, Peñitas de Chache.

SESELI

Perennial herbs with fibrous stock. Leaves 2-pinnate to ternately divided. Sepals absent or small. Petals yellow. Fruit oblong to ellipsoid, not strongly compressed. Ridges prominent.

S. webbii Cosson. Up to 50 cm, usually procumbent. Leaves glaucous, glabrous; lobes linear to ovate, very variable, apiculate. Umbels compound. Rays 10–25. Umbellules more or less subglobose.

TENERIFE: Punta de Teno, Los Silos, Playa de Masca etc, coastal region up to

200 m, locally frequent; GRAN CANARIA: Andén Verde, between Agaete and San Nicolas, 150 m, rare; also reported from HIERRO.

TINGUARRA

Perennial herbs. Leaves up to 4-pinnate or ternate. Sepals small. Petals white, somewhat pubescent. Fruits narrowly ovate, somewhat compressed, densely pubescent, narrowing to a beak; ridges prominent.

1. Leaves ternate, the lobes large, ovate, biserrate .. .. **T. cervariaefolia**
1. Leaves 2- to 4-pinnate, the lobes very small, usually lanceolate .. .. **T. montana**

T. cervariaefolia Parl. Robust perennial herb of cliff-ledges. Leaves long-petiolate, more or less glabrous, ternately divided; lobes ovate, biserrate. Umbels compound.

TENERIFE: Bandas del Sur, Guia de Isora, Adeje, Ladera de Guimar, up to 450 m, ledges on basalt cliffs, rare; GOMERA: Vallegranrey, Vallehermoso, Fortaleza de Chipude, up to 700 m; LA PALMA: Forest cliffs of the Cumbrecita region, Caldera de Tabouriente, 600–1600 m.

T. montana Webb. Pubescent perennial herb. Leaves 2- to 4-pinnate, densely pubescent; lobes small, lanceolate. Umbels compound, often in groups. **(52)**

TENERIFE: Forest and lower zone cliffs. Agua Mansa, Barranco Ruiz, Icod el Alto, Sierra Anaga, Malpais de Tamaimo, Masca etc, locally frequent. 400–1600 m; GRAN CANARIA: Rincón de Tenteniguada, Juncalillo etc, cliffs below the pine-forest zone; HIERRO: Laurel woods between Frontera and Fuente de Tinco, 800 m, El Golfo; LA PALMA: Laurel woods of N.E. region, Barranco del Rio, La Galga, Barranco de los Tilos, 600–800 m, rare.

TODAROA

Like *Tinguarra* but the flowers yellow and the fruits glabrous with the ridges produced into prominent wings.

T. aurea Parl. Leaves up to 4-pinnate, sparsely pubescent, long-petiolate. Umbels dense, many-rayed, compound. Fruits readily splitting into 2 mericarps, each with 5 wings.

TENERIFE: Punta de Teno, *Euphorbia* communities near the coast, 60–100 m, rare; coasts of the Anaga region, Bufadero, San Andrés, Taganana etc; GOMERA: Agulo, coastal rocks; LA PALMA: coastal zone below Fuencaliente.

PIMPINELLA

Perennial. Stock woody with abundant, scale-like remains of petioles. Basal leaves 1- to 2-pinnate. Bracts and bracteoles usually absent. Sepals minute. Petals white. Fruit ovoid to subglobose, densely villous to velutinous.

1. Basal leaves 1-pinnate.
   2. Stems and leaves densely pubescent .. .. .. **P. cumbrae**
   2. Stems and leaves glabrous .. .. **P. dendrotragium**
1. Basal leaves 2-pinnate.
   3. Lateral leaf-lobes deeply dissected almost to midrib, subglabrous to very sparsely pubescent; fruits oblong-ovoid .. .. **P. anagodendron**
   3. Lateral leaf-lobes coarsely dentate to laciniate, pubescent; fruits more or less subglobose .. .. .. **P. junionae**

P. cumbrae Buch ex DC. Branches pubescent, strongly ridged. Basal leaves

1-pinnate; segments ovate to rotund, coarsely dentate, densely white-pubescent. Umbels with 7–10 rays. Fruit subglobular, velvety-hairy.

TENERIFE: Subalpine zone of Las Cañadas de Teide, Los Roques, La Fortaleza etc, locally abundant, 1900–2100 m.

P. dendrotragium Webb & Berth. Branches glabrous. Basal leaves 1-pinnate; segments ovate, glabrous, dentate-laciniate. Cauline leaves pinnate or 3-lobed, reduced. Umbels with 7–9 rays. Fruit ovoid to subglobular, pubescent. ***Culantrillo.***

LA PALMA: Locally frequent in and below the forest zones, Barranco de las Angustias, Caldera de Tabouriente, La Cumbrecita, Barranco del Rio etc, 400–1200 m; TENERIFE: Mountains above Guimar, Barrancos de Badajoz, Añavigo, Agua etc, up to 800 m.

P. anagodendron Bolle (*P. rupicola* Svent.) Branches very finely pubescent. Basal leaves 2-pinnate, primary segments deeply dissected into very sparsely pubescent lobes. Umbels with 7–12 rays. Fruit oblong to ovoid, shortly pubescent.

TENERIFE: Sierra de Anaga, Punta del Hidalgo, Barranco del Rio, La Mina, Cruz del Carmen etc. 200–600 m, Valle de Masca, 400–1200 m, very local.

P. junionae Ceb. & Ort. Branches sparsely pubescent. Basal leaves 2-pinnate; segments with up to 3 pairs of serrate to laciniate, finely pubescent lobes. Umbels with 7–10 rays. Fruit ovate to subglobular, velvety-hairy. **(51)**

GOMERA: Locally common, Barranco de la Villa, Vallegranrey, Lomo de Carretón, La Fortaleza de Chipude, Roque de Agando etc, cliffs in the upper reaches of the xerophytic zone and in forests, 600–1200 m.

CRYPTOTAENIA

Herbs. Leaves ternate or 1- to 2-pinnate. Primary rays very slender, often wavy. Petals white. Fruit subglobose, weakly ridged.

C. elegans Webb. Stems ridged. Leaves with pinnatisect, lanceolate lobes, sparsely pubescent. Inflorescences diffuse, up to 10-rayed, compound. The umbellules with irregular rays. Fruits black, the surfaces between the weak ridges somewhat rugose. **(53)**

TENERIFE: Laurel forests of Sierra Anaga, Vueltas de Taganana, 500–600 m; LA PALMA: N.E. region in the laurel woodland from Cumbre Nueva to Barlovento, La Galga, Los Tilos etc; GOMERA: El Cedro forest, Laguna Grande, Roque de Agando; HIERRO: Forests of El Golfo near Fuente de Tinco.

SAPOTACEAE

Trees or shrubs with milky juice. Leaves alternate, simple. Flowers clustered in leaf-axils, usually 5-parted. Staminodes sometimes present between the corolla-lobes. Ovary superior. Style simple. Fruit a berry.

SIDEROXYLON

Trees or shrubs. Leaves petiolate, thick and leathery; venation reticulate with prominent lateral nerves; apex obtuse to retuse. Flowers congested in the axils of leaves, pedicellate. Sepal 5, overlapping at the margins. Corolla 5-lobed with a short tube. Stamens slightly longer than the corolla. Staminodes present. Fruit a fleshy, single-seeded, ovoid berry.

S. marmulano Banks. Up to 15 m. Bark dark greyish. Leaves elliptic to obovate, more or less glabrous. Corolla white. Fruit black when ripe. **(60)**

TENERIFE: N. coast cliffs, Sierra Anaga to Los Silos, Punta del Hidalgo, Icod, Monte del Agua, S. region, Barranco del Infierno, 200–600 m; GOMERA: Chorros de Epina, very rare.

ERICACEAE

Shrubs or trees. Leaves simple, without stipules. Flowers regular. Petals fused. Stamens twice as many as petals. Ovary superior. Style simple. Fruit a capsule or berry.

1. Trees with oblong-lanceolate leaves, fruit a 2–3 cm yellow-orange berry .. .. **Arbutus**
1. Shrubs or small trees, leaves linear, fruit a small, brownish capsule .. .. **Erica**

ARBUTUS

Trees. Leaves evergreen, coriaceous. Inflorescences paniculate. Sepals fused at the base. Corolla flask-shaped with short reflexed lobes. Fruit a globose berry.

A. canariensis Viell. Up to 15 m. Bark reddish-brown, peeling in flakes. Leaves oblong-lanceolate, margins toothed. Panicles drooping or suberect. Corolla white or greenish, often tinged with pink. Berries 2–3 cm, covered with papillae, yellow-orange when ripe. ***Madroño.*** **(55)**

TENERIFE: Laurel and pine forests, occasional, Barranco Añavigo above Guimar, El Tanque, Pinar de La Guancha, Orotava valley, Sierra Anaga; LA PALMA: Caldera de Tabouriente; GOMERA: El Cedro, Cumbres de Vallehermoso; HIERRO: Frequent in the El Golfo region above Frontera, Jinamar etc., 600–1200 m.

ERICA

Tall evergreen shrubs. Leaves whorled, linear. Flowers in clustered lateral racemes forming a 'thyrse'. Sepals 4, free, short. Petals united into a campanulate tube with short lobes, persistent in fruit. Fruit a capsule.

1. Flowers white, leaves more or less erect .. .. .. **E. arborea**
1. Flowers deep pink-reddish, leaves patent .. .. .. **E. scoparia**

E. arborea L. Shrub or small tree up to 15 m. Leaves up to 6 mm, linear, dark green, completely revolute. Inflorescence a many-flowered thyrsoid, Corolla white, broadly campanulate. ***Brezo.*** **(203)**

ALL ISLANDS: Forest zones of the western islands and GRAN CANARIA, often dominant and forming a tall shrubby or tree heath. Ecologically agressive and replacing felled laurel species in many places. Rare on LANZAROTE and FUERTEVENTURA.

E. scoparia L. subsp. ***platycodon*** Webb & Berth. Like *E. arborea* but never forming a small tree. Leaves slightly revolute, patent. Flowers in interrupted terminal racemes. Corolla companulate, deep pink or reddish. ***Techo.*** **(56)**

TENERIFE: Sierra Anaga, Las Mercedes, crests in laurel forests, Orotava valley, Icod 600–800 m; GOMERA: Roque de Agando, Barranco de La Laja, Vallehermoso, Montes de Arure, locally common 400–1200 m; HIERRO: El Golfo region, forests of the central region.

MYRSINACEAE

Trees or shrubs. Leaves alternate, simple, with pellucid glands. Inflorescences cymose or fasciculate. Flowers uni- or bisexual. Petals free or fused, 5. Ovary superior. Fruit a drupe or berry.

1. Petals united at base, inflorescences few-flowered .. **Pleiomeris**
1. Petals free, inflorescences many-flowered .. .. .. **Heberdenia**

HEBERDENIA

Trees. Inflorescences axillary. Flowers hermaphrodite, 5-parted. Sepals joined at base. Petals free. Stamens attached to base of petals. Ovary superior. Fruit a globose, 1-seeded, more or less hard, berry.

H. bahamensis (Gaertn.) Sprague. Up to 10 m. Leaves entire, leathery, obovate to oblong. Shortly petiolate, obtuse. Inflorescences subcorymbose. Petals white or pinkish, strongly scented, glandular. Fruit about 5 mm across. ***Aderno.*** **(58)**

TENERIFE: Laurel forests and forest cliffs, Sierra Anaga, Las Mercedes, Vueltas de Taganana, Monte de los Silos etc, 500–1000 m, becoming very rare due to forest exploitation; GOMERA: Central forest region, El Cedro, Agando etc.

PLEIOMERIS

Like *Heberdenia* but the inflorescences small, subsessile, axillary. Petals united at base.

P. canariensis (Willd.) DC. Trees up to 15 m. Leaves elliptical to oblong, prominently veined, obtuse. Inflorescence few-flowered, sessile. Petals united at base to form a short, campanulate tube. Fruit globose or flattened at apex, about 5–7 mm across. ***Marmulan.*** **(59)**

TENERIFE: Laurel forests 600–1000 m, Monte de las Mercedes, Cuevas Negras de los Silos, Barranco de Castro, Vueltas de Taganana, Pico Inglés, rare; GOMERA: El Cedro forest; LA PALMA: Cumbre Nueva, Barlovento, very rare.

PLUMBAGINACEAE

Shrubs or herbs. Leaves entire or lobed. Inflorescence a panicle. Calyx tubular, often enlarged and petal-like at the top. Corolla tubular, small, Stamens 5. Ovary with a single ovule. Styles 5. Fruit a capsule enclosing a single seed.

LIMONIUM

Perennial herbs or shrubs. Leaves in basal or terminal rosettes. Flowers in 1- to 5-flowered, cymose spikelets aggregated to spikes. Calyx funnel shaped, expanded above into a scarious, coloured limb. Corolla white, small, with a short tube. Styles free to base. ***Siempreviva.***

1. Main flowering stems winged.
 2. Leaves always deeply sinuately or pinnately lobed.
  3. Leaves sinuately lobed, terminal lobe very large, 10–15 cm .. .. **L. brassicifolium**
  3. Leaves pinnately lobed, terminal lobe not more than 4–5 cm .. .. **L. imbricatum**
 2. Leaves entire or rarely very shallowly lobed towards base.
  4. Leaves long-petiolate, ovate to rhomboidal.
   5. Flowering stems broadly winged (Gomera) .. .. **L. redivivum**
   5. Flowering stems very narrowly winged (Tenerife).

6. Tall shrub with long woody stem up to 1·5 m .. **L. arborescens**
6. Small shrub with short stem, less than 50 cm .. **L. fruticans**
4. Leaves sessile, broadly oblanceolate .. .. .. **L. macrophyllum**
1. Main flowering stems not winged.
7. Leaves pinnately lobed .. .. **L. spectabile**
7. Leaves entire or very shallowly lobed towards base.
8. Leaves pubescent (Lanzarote).
9. Plants tall, up to 40 cm, leaves shallowly sinuately lobed near base.. .. .. **L. bourgaeii**
9. Plants small, 10–15 cm, leaves entire .. .. .. **L. puberulum**
8. Leaves glabrous, sometimes glaucous.
10. Plants small, coastal, leaves spathulate or oblanceolate, sessile, greyish, up to 5 cm long .. .. **L. pectinatum**
10. Plants large, inland, leaves lanceolate or rhomboidal, green, 15–30 cm long.
11. Arborescent, leaves lanceolate to elliptical (Gomera) .. .. **L. dendroides**
11. Frutescent, leaves rhomboidal (Gran Canaria).
12. Petiole twice as long as leaf-blade .. .. **L. rumicifolium**
12. Petiole more or less equalling leaf-blades .. **L. preauxii**

L. arborescens (Brouss.) O. Kuntze. Shrub up to 1·8 cm, with cylindrical, smooth, woody stem. Leaves ovate, glabrous, glaucous, long-petiolate, margins erose; apex obtuse; petiole expanded at base. Inflorescence branched, dense, the stem and lower branches narrowly winged. Wings of peduncles with two curved, acute extensions above. Calyx blue-mauve, persistent. Flowers small, white. **(208)**

TENERIFE: Extremely rare, endangered species from a few localities on the N. coast of the island. Cultivated in some parks and gardens (Jardin Botanico, Orotava).

L. fruticans (Webb) O. Kuntze, Like. *L. arborescens* and perhaps only a local form of it but differing in its shorter stature, broader leaves and narrower peduncle-wings with blunt projections. **(205)**

TENERIFE: N. coast, El Fraile, Los Silos, etc, very rare and in danger of extinction and therefore should not be collected.

L. macrophyllum (Brouss.) O. Kuntze. Leaves entire, sessile or shortly petiolate, large (30 cm) oblanceolate to narrowly ovate. Flowering stems narrowly winged, alternately branched; peduncles with triangular bracts and broad wings. Calyx blue-mauve.

TENERIFE: N.E. region, along the N. coast of Anaga 50–100 m, Taganana.

L. brassicifolium (Webb & Berth.) O. Kuntze. Woody-based rosette plant with stock up to 20 cm. Leaves petiolate, sinuately lobed; terminal lobe very large, up to 15 cm long; apex acuminate. Flowering stems broadly winged (wings up to 1·5 cm wide). Calyx pale mauve.

GOMERA: N. coast region between Agulo and Valleherrmoso, Los Organos; HIERRO: Sabinosa region of El Golfo, up to 500 m, rare. Often cultivated in gardens on Tenerife. The Hierro form has been referred to a distinct species *L. macropterum* (Webb & Berth.) O. Kuntze.

L. redivivum (Svent.) Kunkel & Sund. Resembles *L. brassicifolium* but the leaves long-petiolate, obovate to spathulate and entire not sinuately lobed.

GOMERA: E. region, 600 m. Very rare, Chipude, on steep cliffs, 1000 m.

L. rumicifolium (Svent.) Kunkel & Sund. Like *L. arborescens* but lacking the long woody stem. Leaves with very long petioles, twice as long as blade, rhomboidal, entire, obtuse or mucronulate. Inflorescences large, the main flowering stem unwinged. Calyx deep mauve. **(206)**

GRAN CANARIA: Rocky slopes on the S. side of the island, Tirajana region, Valle de Fataga 400–600 m. Locally frequent.

The poorly known *L. perezii* Stapf from the western end of Tenerife appears to closely resemble this species.

L. preauxii (Webb & Berth.) O. Kuntze. Like *L. rumicifolium* but much smaller in all its parts and perhaps just a local form.

GRAN CANARIA: W. coast near Agaete.

L. bourgaeii (Webb) O. Kuntze. Subsessile rosette plant with woody stock. Leaves pubescent, long-petiolate, broadly ovate-rhomboidal, often shallowly sinuate-lobed towards the base. Inflorescence branches pubescent, unwinged, bracteate. Calyx deep violet. **(207)**

LANZAROTE: Famara mountains, especially on the N.W.-facing cliffs above Playa de Famara, 500 m. Locally abundant.

L. puberulum (Webb) O. Kuntze. Like *L. bourgaeii* but much smaller with shortly petiolate, ovate leaves with ciliate margins. Inflorescences short, about 6 cm, flowering stems and peduncles not winged.

LANZAROTE: Frequent on the cliffs and rocks above El Rio on the N. coast especially on the cliff-tops of Mirador del Rio, 500–600 m.

L. spectabile (Svent.) Kunkel & Sund. Small subshrub. Leaves pinnate, the lobes linear-lanceolate with ciliate margins, the terminal lobe triangular. Flowering stem more or less unwinged. Peduncles narrowly winged. Calyx mauve. **(204)**

TENERIFE: Very rare, known only from the sea-cliffs of the Barranco de Masca.

L. imbricatum (Webb) Hubbard. Rosette plant. Leaves pinnately lobed, the lobes broadly lanceolate to ovate, pubescent. Flowering stem and inflorescence branches broadly winged. Calyx deep mauve. **(57)**

TENERIFE: Coastal rocks west of Buenavista, near sea-level, rare; LA PALMA: On the coast near Puntagorda, also rare.

L. pectinatum (Aiton) O. Kuntze. Very variable dwarf-shrub of coastal regions. Leaves usually spathulate, obtuse to retuse, very variable in size, but usually 2–4 cm, densely rosulate. Inflorescences unwinged, often with groups of leaf-like bracts along the main flowering stems. Flowers borne unilaterally on upper side of inflorescence branches. Calyx pale mauve. **(209)**

ALL ISLANDS: Very common on coastal rocks and on the foreshore of all the islands.

L. dendroides (Svent.) Kunkel & Sund. Arborescent, up to 3 m, on long woody stem. Leaves rosulate, lanceolate, acute, glabrous and sometimes more or less coriaceous. Inflorescences paniculate, the branches unwinged. Calyx pinkish.

GOMERA: Barranco de Argaga on sea-cliffs, Barranco de Cabrito, near Pico de Aragan, 50–400 m, extremely rare endangered species.

OLEACEAE ——— Olive and Jasmine Family

Trees or shrubs, usually glabrous. Leaves opposite or alternate. Flowers hermaphrodite 4- to 5-merous. Calyx campanulate, sometimes small. Corolla with at least a short tube. Stamens 2. Fruit a drupe or berry.

1. Leaves pinnate, corolla yellow .. .. **Jasminum**
1. Leaves entire, corolla white or greenish.
 2. Corolla with a tube, fruit fleshy .. .. **Olea**
 2. Corolla tube almost absent, fruit dryish with a crustaceous endocarp .. .. **Picconia**

OLEA

Shrubs or small trees. Leaves opposite, simple, lanceolate. Flowers in axillary panicles usually hermaphrodite. Calyx small with 4 teeth. Corolla tubular, 4-lobed. Fruit a fleshy drupe.

O. europaea L. var. ***cerasiformis*** Webb & Berth. Shrub up to 6 m. Bark grey. Leaves green, glossy above, white-scaly beneath. Corolla white. Fruit ellipsoid, fleshy, green becoming brown or black when ripe. ***Acebuche.***

ALL ISLANDS: Rocky places in the lower zone, on cliffs, GOMERA: Epina; FUERTEVENTURA: Betancuria; GRAN CANARIA: Lentiscal etc, up to 600 m, very local.

PICCONIA

Shrubs or trees with whitish bark. Leaves opposite, simple, entire or rarely serrulate, glabrous, leathery. Flowers hermaphrodite. Corolla and calyx 4-lobed. the corolla without a tube. Fruit a dryish drupe with a crustaceous endocarp.

P. excelsa (Aiton) DC. Up to 10 m. Leaves lanceolate to obovate. Corolla white, deeply 4-lobed. Fruit oblong-ovoid, black. ***Palo blanco.* (61)**

TENERIFE: Locally frequent in laurel forests and on forest cliffs, Sierra Anaga, Guimar, Los Silos, Orotava valley etc, up to 1000 m; LA PALMA: N.E. region, Barlovento, Los Tilos, La Galga; GOMERA: El Cedro, Monte de Arure; HIERRO: Very frequent in the forests of El Golfo; GRAN CANARIA: N. coast, Los Tiles de Moya, Barranco de la Virgen, rare.

JASMINUM

Shrubs with alternate, pinnate leaves. Inflorescences 1- to 4-flowered, axillary cymes. Calyx campanulate with 5 lobes. Corolla with a long tube and 5 patent lobes. Fruit an oblong, black berry.

J. odoratissimum L. Shrub up to 4 m. Leaflets usually 3, oblong to obovate. Flowers yellow, corolla lobes shorter than the tube.

TENERIFE: Locally frequent in rocky places in the lower zone and occasionally in forest regions 400–1000 m, Guia de Isora, Punta de Teno, Icod, Guimar, Sierra de Anaga; LA PALMA: Mazo, Los Sauces, Barranco de Carmen, Barranco de las Angustias; GOMERA: Barranco de la Villa, Vallehermoso, Lomo de Carreton, Argaga; HIERRO: El Golfo, Fuente de Tinco, Jinmar, Frontera.

GENTIANACEAE —— Gentian Family

Herbs. Leaves opposite, entire, usually sessile. Flowers hermaphrodite, regular, 4- or 5-parted. Calyx fused, the lobes imbricate. Ovary superior. Fruit a capsule with numerous small seeds.

IXANTHUS

Perennial herbs. Leaves opposite, entire. Calyx enclosed in 2 leaf-like bracteoles, short, campanulate, 5-lobed, the lobes acute. Corolla tube 5-lobed. Stamens 5. Capsule 2-valved. Seeds numerous, small.

I. viscosus Griseb. Tall viscous herb up to 80 cm. Leaves lanceolate to ovate, acute, with 3–5 more or less parallel veins. Inflorescence branched, bracteose. Flowers large, yellow. Capsule oblong, greenish-black, somewhat fleshy. Seeds many, black. ***Reina del Monte.*** **(62)**

TENERIFE: Locally common in laurel woodland particularly in the Anaga region, Las Mercedes, Pico Inglés, El Bailadero etc, Orotava valley; GRAN CANARIA: Los Tiles de Moya, rare; LA PALMA: Common in the woodland of the N.E., La Galga, Los Sauces, Barranco del Agua, Barlovento etc; GOMERA: El Cedro forest; HIERRO: Fuente de Tinco, Frontera, 600–1000 m.

ASCLEPIADACEAE

Shrubs or perennial herbs, sometimes succulent or with twining stems. Leaves usually opposite, sometimes scale-like or more or less absent. Flowers regular, 5-merous. Corolla usually with a tube and 5 lobes; corona present. Fruit a pair of long follicles. Seeds flat with a long 'coma' or 'pappus'.

1. Succulent-stemmed herbs.
  2. Stems 4-angled, corolla-tube very short .. .. .. **Caralluma**
  2. Stems more or less circular in cross section, corolla-tube much longer than the lobes .. .. **Ceropegia**
1. Shrubby climbers or scrambling plants with a short corolla tube .. .. **Periploca**

CEROPEGIA

Succulents with erect or sprawling, round to oval stems. Corolla with a slender tube, much longer than the lobes, yellow or reddish-brown. Fruits usually more or less erect.

1. Flowers reddish-brown, stems grey-brown .. .. **C. fusca**
1. Flowers yellow, stems green, grey or olive.
  2. Flowers in dense fascicles of 10–50.
    3. Stems sprawling, oval in section, corolla whitish with yellow lobes (Tenerife) .. .. **C. chrysantha**
    3. Stems erect, circular in section, corolla yellow (Gomera).
      4. Inflorescences 20- 50- (70)-flowered, mature corolla lobes not joined at tip .. .. **C. krainzii**
      4. Inflorescences 10- to 20-flowered, mature corolla lobes joined at tip .. .. **C. ceratophora**
  2. Flowers in small groups of 2–7.
    5. Inflorescences on long, slender, reddish terminal branches of current year's growth (La Palma) .. **C. hians**
    5. Inflorescences on short, greenish stems of previous year's growth (Tenerife) .. .. **C. dichotoma**

C. fusca Bolle. Stems erect or sprawling, up to 75 cm, greyish-white (Gran Canaria) or brownish (Tenerife), cylindrical. Inflorescences axillary, borne towards the tips of older stems, 2- to 5-flowered. Pedicels reddish-brown, 2–5 mm. Corolla brownish-red, the lobes joined at tips only in young flowers. Fruits 10–13 cm, apex acute. ***Cardoncillo.*** **(210)**

TENERIFE: Locally common in dry rocky areas of the S, Medano, El Roque Adeje, near sea-level at Montaña Roja to 600 m; GRAN CANARIA: Rather rare, dry cliffs in the S. and E. (Valle de Fataga, Agaete 100–300 m).

C. dichotoma Haw. Stems erect or sprawling, up to 60–75 cm, green or greyish olive. Leaves ephemeral. Inflorescences axillary on the upper sections of older stems, 2- to 5-flowered. Flowers 2–4 cm long, pale yellow with darker yellow lobes. Fruits 10–12 cm long, greenish-brown. ***Cardoncillo.*** **(64) (211)**

TENERIFE: Locally common along the N. coast, Teno (very common), San Juan de la Rambla to Taganana, up to 500 m, local in the S. from Guimar to Adeje. A very thick-stemmed, succulent form is a feature of the coastal vegetation at Punta de Teno.

C. hians. Svent. Stems densely branched from the base, erect, cylindrical, up to 70 cm, whitish to olive green. Inflorescences axillary on long, slender, reddish terminal branches of current year, 2- to 7-flowered. Pedicels about 6–10 mm, reddish. Corolla 3–4 cm, yellow, the narrow lobes usually not joined at the tip when fully open. Fruits 8–10 cm. **(212)**

LA PALMA: Locally common in the S. of the island in the vicinity of Fuencaliente, 100–700 m. A tall, pale yellow-flowered form from near Tigarafe and Barranco de las Angustias on the W. side of the island has been referred to var. *striata* Svent.

C. chrysantha Svent. Stems few, sprawling, brownish to grey-olive, oval in section. Inflorescences subterminal, subgobose, up to 20-flowered. Peduncles long, up to 8 mm, reddish. Corolla 3–3·5 cm, the tube usually somewhat curved, whitish, the lobes yellow, joined at the tip.

TENERIFE: Very rare and severely depleted, known only from a single locality near Adeje at about 150 m.

C. krainzii Svent. Stems erect, 40–60 cm, cylindrical, 1–2·5 cm across at base, light greyish-olive. Pedicels greyish. Flowers in fascicles of 20–70 towards apex of old stems, lemon-yellow, 3–3·5 cm long; corolla lobes not joined at tips in mature flowers. Fruits about 10 cm, greyish-brown.

GOMERA: N. and W. coast regions up to 600 m (Barranco Majona), rare.

C. ceratophora Svent. Stems branching from base, erect, up to 150 cm, brown-olive. Flowers in several 10- to 20-flowered whorls on upper nodes of the stems. Pedicels 7–8 mm, reddish or green. Corolla about 3 cm long, pale greenish yellow, the lobes joined at the tips. Fruits 10–14 cm long, with blunt tips.

GOMERA: Known only from cliffs on the S.W. side of the island 600–1000 m, between Epina and Vallegranrey, very rare.

PERIPLOCA

Shrubs with erect or twining stems. Leaves opposite. Flowers in axillary cymes. Corolla lobes green outside, purple-brown and white inside. Fruits 5–10 cm long, subpatent.

P. laevigata Aiton. Shrub with at least the tips of the stems twining. Leaves lanceolate, obtuse or acute, glabrous. Corolla about 1 cm across. ***Cornical*** **(63)**

ALL ISLANDS: Common in the *Euphorbia* communities of the lower zone up to 700 m. (Also S. Morocco. The European form of this species is probably best referred to *P. angustifolia* Labill.)

CARALLUMA

Succulents with subterranean stolons. Stems 4-angled, the angles with leaf-scars. Corolla 5-lobed. Fruits a pair of long, horn-like follicles. Seeds brown, flat with a pappus of long white hairs.

C. burchardii N.E. Br. Stems about 10–15 cm, ascending or erect, the faces flat. Inflorescences of about 6 small flowers. Corolla purplish-brown with densely white-ciliate margins. Fruits about 10 cm long, slightly curved, erect. **(213) (214)**

FUERTEVENTURA: La Oliva and Isla de Lobos, locally frequent on dry volcanic slopes; LANZAROTE: rare in the northern part of the island (var. *sventenii* Lamb).

RUBIACEAE — Madder Family

Shrubs or herbs. Leaves opposite or whorled, stipulate. Flowers usually in cymes, corymbs or panicles, hermaphrodite or male. Corolla 3- to 7-lobed, regular. Ovary inferior, 2-locular. Fruit fleshy or splitting into 2 dry mericarps.

1. Scrambling perennials, leaves with prickly margins .. **Rubia**
1. Shrubs, leaf-margins not prickly.
   2. Leaves filiform, pendulous, fruit a small black berry .. **Plocama**
   2. Leaves lanceolate, patent or suberect, fruit indehiscent, splitting into 2 small, dry black, 1-seeded sections .. **Phyllis**

RUBIA

Rough, prickly, scrambling perennials. Leaves stiff, whorled. Flowers small. Corolla yellow-cream; lobes 5. Fruit a black or whitish fleshy berry.

1. Herb, leaves linear-lanceolate, long .. .. .. **R. angustifolia**
1. Shrub, leaves elliptic-ovate, short .. .. **R. fruticosa**

R. angustifolia L. Prickly, scrambling herb. Stems 4-angled. Leaves linear-lanceolate, long. Flowers in axillary clusters. Fruits black.

TENERIFE: Laurel forests, Agua Garcia, Taganana, Icod el Alto, 600–800 m; LA PALMA: Cumbre Nueva, La Galga, Los Tilos; GOMERA: El Cedro, Laguna Grande; HIERRO: Forests above Frontera; GRAN CANARIA: Los Tiles de Moya, locally common.

R. fruticosa Aiton. Very variable scrambling or climbing woody-based shrub. Leaves whorled, elliptic to ovate, very prickly on the margins and beneath. Flowers pale yellow, in axillary or terminal clusters. Fruit a globose berry about 4–6 mm, black or translucent-whitish. ***Tasaigo.*** **(216)**

TENERIFE: Sierra Anaga, lower zone, N. coast, Teno, Masca, S. region, Adeje, Guimar, very common in *Euphorbia* communities; GRAN CANARIA: N. coast region, Cuesta de Silva, S. zone, Fataga, Tirajana etc, locally common; HIERRO, GOMERA, LA PALMA: very common throughout the lower zones; LANZAROTE: N. region, Haria, Cueva de los Verdes; FUERTEVENTURA: La Oliva, Betancuria.

PHYLLIS

Leaves opposite or verticillate. Flowers hermaphrodite or polygamous. Calyx 5-dentate in male flowers; 2-dentate in hermaphrodite. Corolla 5-parted. Style 2-fid. Fruit dry, black, splitting into 2 indehiscent, 1-seeded sections.

1. Leaves lanceolate to ovate, not sticky, inflorescence lax .. **P. nobila**
1. Leaves narrow-lanceolate, very sticky, inflorescence dense, short .. .. **P. viscosa**

P. nobila L. Small glabrous or pubescent subshrubs. Leaves entire, lanceolate to ovate, acute. Flowers whitish, small, in lax terminal and axillary panicles. Fruiting pedicels pendulous. ***Capitana.*** **(66)**

TENERIFE: Laurel forest cliffs and banks, Sierra Anaga, Las Mercedes to Vueltas de Taganana, Agua Mansa etc, locally very common, 600–1200 m; LA PALMA: Cumbre Nueva, El Paso, Barlovento etc; GOMERA: Monte del Cedro, Arure,

Chorros de Epina; HIERRO: Forest regions of El Golfo; GRAN CANARIA: Pinar de Tamadaba, pine forest cliffs, 1000 m, rare.

P. viscosa Webb & Berth. Like *P. nobila* but smaller with narrow, sticky leaves and short, dense inflorescences.

TENERIFE: W. region, Los Silos, Cuevas Negras, Montes de Teno, Masca etc, cliffs in the lower zone up to 1000 m at Cumbre de Masca; LA PALMA.

PLOCAMA

Shrubs with pendulous branches. Leaves filiform. Corolla 5- to 7-lobed. Stamens 5–7; filaments short. Fruit a small, globose berry, black when ripe.

P. pendula Aiton. Very strong-smelling shrub up to 2 m. Flowers minute, axillary or terminal, crowded towards the tips of the slender branches. ***Balo.*** **(65) (215)**

TENERIFE: Common in the lower zone of many parts of the island, particularly in the S. and W., Punta de Teno, Guimar, San Andrés, Igueste, dry slopes in *Euphorbia* communities; GRAN CANARIA: Locally very common, N. coast, Agaete, S. region, Maspalomas, Mogan etc; GOMERA: San Sebastian, Vallehermoso, Vallegranrey to Playa de Santiago; LA PALMA: Rather rare, Puerto de Naos, Tazacorte, Punta Gorda, coast below Fuencaliente; HIERRO: El Júlan, S. region, rare; LANZAROTE, FUERTEVENTURA.

CONVOLVULACEAE ——— Bindweed Family

Herbs or shrubs. Stems often climbing. Leaves alternate. Flowers hermaphrodite, regular. Sepals 5, usually free. Corolla funnel or bell-shaped, shallowly 5-lobed. Stamens 5. Ovary superior. Style terminal. Fruit a capsule.

CONVOLVULUS

Shrubs or lianas with alternate, entire leaves. Flowers in axillary or terminal corymbs or panicles. Sepals 5. Corolla more or less broadly campanulate. Stamens 5, included. Style 1, bifid. Fruit a capsule with few seeds in 2 locules.

1. Erect shrubs or cushion plants, flowers white.
 2. Erect shrubs, flowers in 3- to many-flowered corymbs.
  3. Leaves oblong, inflorescences many-flowered, paniculate .. .. **C. floridus**
  3. Leaves linear-filiform, inflorescences few-flowered .. **C. scoparius**
 2. Cushion plants, flowers solitary .. .. **C. caput-medusae**
1. Scrambling, woody lianas, flowers pink or mauve.
 4. Leaves ovoid-oblong, densely long-hairy (Laurel forests) **C. canariensis**
 4. Leaves elliptic-lanceolate to narrowly oblong, finely pubescent to glabrous (lower zone).
  5. Corolla pink, leaves more or less glabrous.
   6. Inflorescences 1- to 2-flowered, leaves 2–4 cm, densely glandular (Gran Canaria) .. .. **C. glandulosus**
   6. Inflorescences 3- to 6-flowered, leaves 4·5–6 cm, sparsely glandular (Lanzarote) .. .. .. **C. lopez-socasi**
  5. Corolla pale mauve-violet, leaves finely pubescent.
   7. Leaves elliptical with short auricles at base, corolla violet (Gomera) .. .. **C. subauriculatus**

7. Leaves linear-lanceolate to oblong, base without auricles, corolla pale blue-mauve.
  8. Leaves narrow, linear-lanceolate, 5–7·5 cm long .. **C. diversifolius**
  8. Leaves oblong-lanceolate or lanceolate, 2–4 cm long.
    9. Leaves and stems densely grey-pubescent, leaves oblong-lanceolate inflorescences 2- to 3-flowered .. **C. perraudieri**
    9. Leaves and stems greenish or brown, finely pubescent, leaves lanceolate, flowers usually solitary .. **C. fruticulosus**

C. floridus L. fil. Tall shrub 2–4 m. Leaves oblong-linear to oblong or rarely spathulate, 2–14 cm long, shortly and densely pubescent. Inflorescence paniculate, terminal, many-flowered. Corolla white or very pale pinkish, about 1 cm across. ***Guaydil.*** **(217)**

ALL ISLANDS: Locally very common in the lower zone especially in *Euphorbia* communities. Commonly cultivated in the Canaries as a garden shrub.

C. scoparius L. fil. Shrub about 1 (-2) m. Leaves linear, filiform, caducous, 0·5–4·5 cm; indumentum hairs very short, some glandular. Inflorescences axillary or terminal, 5- to 6-flowered, Corolla white or pinkish. ***Lena Noel.*** **(221)**

TENERIFE: Locally frequent, S. region, Candalaria, Guimar 50–200 m. Teno, coastal region up to 150 m; LA PALMA: Barranco de la Herradura, rare; GRAN CANARIA: S. region, on dry slopes, Arguineguin to Mogan, up to 300 m, locally common.

C. diversifolius Mendoza-Heuer. Subshrub or liana with woody, scrambling stems. Leaves linear-lanceolate 5–7·5 cm long, acute to more or less obtuse, glabrescent to pilose, glandular. Inflorescences axillary, 1- to 3-flowered. Corolla about 2 cm long, bluish.

TENERIFE: Rocks of the coastal region near Taganana, rare.

C. perraudieri Coss. Like *C. diversifolius* but leaves oblong-lanceolate, grey, densely and finely pubescent, glandular. Inflorescences axillary 2- to 3-flowered. Corolla blue. **(67)**

TENERIFE: S. and S.W. region, Valle de Masca, 300–400 m, Arico, up to 1700 m rare; GRAN CANARIA: S. region, Mogan valley, 200–300 m.

C. subauriculatus (Burchd.) Lindinger. Like *C. diversifolius* but the leaves ovate-elliptical with short auricles at the base. Corolla violet.

GOMERA: N. region, up to 400 m. Roque Cano, Vallehermoso, Riscos de Agulo, 300 m, rare.

C. fruticulosus Desr. Like *C. diversifolius* but the leaves lanceolate, 1–2 cm, long, indumentum short, fine, tomentose. Inflorescences axillary 1- (3)-flowered. Flowers pale mauve. **(218)**

TENERIFE: Local in the lower zone in *Euphobia* communities, Anaga region, 50–500 m; LA PALMA: Barranco de las Angustias, 400 m, rare.

C. caput-medusae Lowe. Dwarf cushion shrublet, 10–15 cm. Branches terminating in spines. Leaves small, spathulate, grey, densely pubescent. Flowers solitary, white or sometimes pale pink on the outside. ***Chaparro.*** **(220)**

FUERTEVENTURA, LANZAROTE: very rare in coastal areas; GRAN CANARIA: Windswept coastal rocks in the S.E., 10–30 m, Arinaga, Melenara.

C. canariensis L. Scrambling shrub, at least the lower portions of the stems woody. Leaves ovoid-oblong, 4–9 cm long, densely hairy, nerves prominent on lower surface. Inflorescences axillary, cymose, 4- to 7-flowered. Sepals densely hairy. Corolla pale blue. ***Correguelon.***

GRAN CANARIA: Los Tiles de Moya, in laurel woods, rather scarce; TENERIFE: Common in the forest regions of the Anaga region, Vueltas de Taganana, El Bailadero 600–800 m; LA PALMA: N.E. region in laurel forest zone, La Galga, Los Tilos, Barlovento etc, locally frequent; GOMERA: Forests of El Cedro; HIERRO: Locally common in the forests of the Golfo region.

C. lopez-socasi Svent. Like *C. canariensis* in habit but leaves elliptic-lanceolate, 4–6 cm long, more or less glabrous, glandular. Inflorescences with up to 6 flowers. Calyx more or less glabrous. Corolla pink.

LANZAROTE: Cliffs in the Famara Massif, 600 m, rare.

C. glandulosus (Webb) Hallier. Subshrub with densely leafy, woody, scrambling stems. Leaves lanceolate, 2–4 cm long, more or less glabrous, very densely glandular. Inflorescences axillary, 1- to 2-flowered. Corolla pale pink, small, outer surface pubescent. **(219)**

GRAN CANARIA: Cliffs and rocks of the S. region, Caldera de Tirajana, Valle de Fataga, 400–1000 m, locally abundant.

BORAGINACEAE ——— Viper's Bugloss Family

Shrubs or herbs, usually hispid. Leaves alternate, simple. Flowers in scorpioid cymes, usually more or less regular. Calyx 5-lobed. Corolla 5-lobed, usually with a distinct tube. Stamens 5, inserted on the corolla. Ovary superior 2- or 4-locular. Style simple or bilobed, usually arising from between the 4 ovary-lobes (gynobasic). Fruit of 2 or 4 nutlets.

1. Fruit splitting into 2 mericarps, flowers inconspicuous, greenish-white .. .. **Messerschmidia**
1. Fruit of 4 nutlets, flowers conspicuous, blue, pink or white.
   2. Corolla with a short tube and patent lobes, nutlets smooth .. .. **Myosotis**
   2. Corolla with a long tube and erect lobes, nutlets rough .. **Echium**

ECHIUM

Annual to perennial herbs or shrubs with an indumentum of tuberculate setae and stiff hairs. Leaves entire. Inflorescence thrysoid with lateral scorpioid cymes. Calyx 5-lobed, enlarging in fruit. Corolla more or less funnel-shaped, usually hairy on the outside; corolla tube with a ring of tissue at the base, sometimes divided into 5 or 10 small scales. Stamens 5, often unequal, included or variously exserted from the corolla. Style longer than the corolla, bilobed at the tip. Ovary superior. Fruit of 4 rough nutlets. ***Taginaste.***

1. Annual to perennial herbs.
   2. Corolla pale pink, leaves linear .. .. **E. triste**
   2. Corolla blue, leaves lanceolate to ovate.
      3. Indumentum of long, yellowish setae, root stock woody, (Tenerife, subalpine) .. .. **E. auberianum**
      3. Indumentum of short, white setae, root stock not woody.
         4. Corolla over 20 mm, stem leaves sessile .. .. **E. plantagineum**
         4. Corolla less than 20 mm, stem leaves with a short stalk.
            5. Leaves ovate, densely hairy (N. Lanzarote) .. .. **E. pitardii**

5. Leaves lanceolate, hispid .. .. **E. bonnetii**

1. Branched or unbranched shrubs.

6. Unbranched shrubs with a dense leaf-rosette and a single, very large inflorescence.

7. Leaves linear-lanceolate, corolla red .. .. .. **E. wildpretii**

7. Leaves broadly lanceolate, corolla blue or white.

8. Corolla white, leaf-rosette with a short stem .. .. **E. simplex**

8. Corolla blue, leaf-rosette with a long stem .. .. **E. pininana**

6. Branched shrubs with several inflorescences.

9. Calyx lobes more or less equalling the tube, indumentum of flat discs (La Palma) .. .. .. **E. gentianoides**

9. Calyx divided almost to the base, indumentum of setae or hairs.

10. Corolla laterally compressed, white, lobes unequal.

11. Leaves with the margins and midrib densely spiny.

12. Calyx segments as long as the corolla .. .. **E. aculeatum**

12. Calyx segments shorter than the corolla (La Palma) **E. brevirame**

11. Leaf-margins with few or no spines.

13. Inflorescence conical, without basal branches, leaves at least 1·5 cm wide .. .. .. **E. giganteum**

13. Inflorescence a flattish dome, usually with lateral, basal branches, leaves less than 1·5 cm wide .. **E. leucophaeum**

10. Corolla not laterally compressed, usually pink or blue or if white then with bluish stripes, lobes more or less equal.

14. Inflorescence lax with long internodes and few lateral cymes, leaves ovate-lanceolate to ovate .. **E. strictum**

14. Inflorescence dense, with long internodes and many lateral cymes, leaves linear to lanceolate.

15. Inflorescence conical, very broad at the base, corolla white with bluish stripes (Gran Canaria, Lanzarote, Fuerteventura) .. .. .. **E. decaisnei**

15. Inflorescence cylindrical, scarsely broadening at base, corolla pink or blue.

16. Upper leaf-surface with dense, large setae, lower surface with simple, stiff hairs (Gran Canaria).

17. Leaf-margins revolute, corolla pale pinkish-white .. .. **E. onosmifolium**

17. Leaf-margins not revolute, corolla deep blue .. **E. callithyrsum**

16. Leaf-hairs similar on both surfaces.

18. Leaves broadly ovate-elliptical, corolla deep blue (Fuerteventura) .. .. **E. handiense**

18. Leaves linear to lanceolate, corolla pale blue or pink.

19. Leaves linear, corolla pale pink, appearing 4-lobed (the 2 ventral lobes more or less fused), (Tenerife) .. .. **E. sventenii**

19. Leaves lanceolate, corolla pink or blue, 5-lobed.

20. Nutlets very ornate, leaves large, up to 10 cm wide (Gomera) .. .. **E. acanthocarpum**

20. Nutlets papillate or shortly echinulate, leaves less than 3 cm wide.

21. Calyx segments lanceolate, longest leaves less than 8 cm long (Hierro) .. .. **E. hierrense**

21. Calyx segments linear to linear-lanceolate, longest leaves more than 8 cm long.
  22. Lateral cymes bifid, corolla pink to pale blue (Tenerife) .. .. **E. virescens**
  22. Lateral cymes simple, corolla blue (La Palma) .. **E. webbii**

E. bonnetii Coincy. Small annual. Basal leaves lanceolate to ovate. Stem leaves reduced. Calyx lobes linear. Flowers blue, up to 1·5 cm. Lower pair of stamens longer than the corolla. **(68)**

TENERIFE: W. and S. regions, very dry rocky areas, cinder cones etc, usually near the coast, Punta de Teno, Adeje, Volcan de Guimar 0–300 m; GRAN CANARIA: W. coast near Andén Verde; FUERTEVENTURA: Locally common, La Oliva, Betancuria, Jandia.

E. pitardii A. Chev. Like *E. bonnetii* but rather larger, the basal leaves ovate and the stamens all included within the corolla.

LANZAROTE: N. region, locally frequent in the Famara, Haria and Mirador del Rio regions, roadsides and cliff ledges up to 700 m.

E. plantagineum L. Like *E. bonnetii* but more robust and often with a branching stem. Flowers about 2–3 cm, much larger than *E. bonnetii.*

WESTERN ISLANDS: Locally very common, weedy, up to 1800 m on Tenerife (Europe, Mediterranean region).

E. giganteum L. fil. Tall shrub up to 2·5 m. Leaves lanceolate to oblanceolate, up to 20 cm long, hispid to hirsute. Inflorescence conical. Corolla white, laterally compressed, the lobes unequal. Stamens longer than the corolla. **(222)**

TENERIFE: N. coast region from near sea-level to 700 m, Orotava Valley, Mirador de Humboldt, Icod el Alto, San José, Icod de los Vinos, Los Silos, usually in *Erica* heath or forest relict communities.

E. aculeatum Poiret. Like *E. giganteum* but smaller, the leaves linear and very spiny and the calyx lobes as long as the corolla. **(224)**

TENERIFE: W. and S.W., Teno, Masca, Valle de Santiago del Teide, Tamaimo, Barranco del Infierno de Adeje etc, locally very common in the lower zone to 1000 m above Masca; GOMERA: Frequent in the lower zone of many parts of the island, Barranco de la Villa, Tagamiche, Vallegranrey, Chipude, Argaga, Barranco de la Laja, Agando, Vallehermoso; HIERRO: El Golfo region, Valverde.

E. brevirame Spr. & Hutch. Like *E. aculeatum* but generally with shorter, broader calyx segments and very short style lobes. **(223)**

LA PALMA: Widespread and variable species of the lower zone especially in the S., Barranco Carmen, Santa Cruz, Fuencaliente, Volcan de San Antonio, Tazacorte, Las Angustias etc, up to 600 m.

E. leucophaeum Webb ex Spr. & Hutch. Tall shrub up to 2 m. Leaves linear-lanceolate, usually blunt, leaves with short, stiff hairs but marginal spines usually absent or very few. Inflorescence dome-shaped, flattish, usually with several lateral branches arising near the base. Flowers white. Corolla laterally compressed, the lobes unequal. **(233)**

TENERIFE: Sierra de Anaga, upper part of the dry S. coast zone, Valle de San Andrés, El Bailadero, Igueste, N. coast, Bajamar, Punta del Hidalgo, 300–600 m, locally frequent.

E. triste Svent. Annual to perennial herb. Leaves in a basal rosette, usually linear, spiny. Inflorescence simple with a few lateral cymes. Corolla with a narrow tube, pinkish. Stamens slightly longer than the corolla.

TENERIFE: S. region, Adeje to Puerto de San Juan, Igueste de Candelaria, dry slopes amongst rocks; GRAN CANARIA: S. and W. regions in similar habitats, Valle de Mogan, Agaete; GOMERA: Valle de Argaga, Alojera, sea-level to 350 m.

E. simplex DC. Unbranched short-lived perennial. Rosette leaves elliptic-lanceolate, densely silvery-hispid. Inflorescence up to 2 m, cylindrical. Flowers white. stamens longer than the corolla. ***Arrebol.*** **(228)**

TENERIFE: Sierra Anaga, N. coast rocks, Bajamar, Punta del Hidalgo, Taganana, 50–350 m, rare and very local.

E. pininana Webb & Berth. Like *E. simplex* but with much larger rough leaves, a longer stem and a very tall inflorescence up to 4 m. Flowers blue. ***Pininana.*** **(229)**

LA PALMA: Laurel forests of the N.E. region, Cubo de la Galga 600 m, Barranco de los Tilos, Monte de Barlovento, extremely rare.

E. wildpretii Pearson ex Hook. fil. Resembling *E. simplex* but with linear leaves with long dense setae and a tapering inflorescence of red flowers. ***Taginaste roja.*** **(227)**

TENERIFE: Subalpine zone of Las Cañadas, 2000 m, La Fortaleza, Montaña de la Grieta, Vilaflor, Valle de Ucanca, Parador Naciónal etc, locally frequent; LA PALMA: High mountains of the Cumbre Nueva and Cumbrecita regions above El Paso, 1600–1800 m, extremely rare.

E. virescens DC. Densely branched shrub up to 2 m. Leaves lanceolate, acute, silvery-hispid. Inflorescence a long slender spike. Flowers pink or bluish. Corolla-lobes equal. Stamens longer than the corolla.

TENERIFE: Very common in forest regions and in the lower zone of S. slopes, usually on cliffs, Agua Mansa, Ladera de Guimar, Sierra Anaga, Barranco del Infierno, Orotava Valley, Guia de Isora, Lomo de Pedro Gil, Monte de Los Silos, Masca etc, 500–1900 m.

E. sventenii Bramwell. Like *E. virescens* but taller and very densely branched, the leaves very narrow with revolute margins and the corolla pale pink and more or less 4-lobed. **(70)**

TENERIFE: Mountain regions E. of Adeje, 350–500 m, very rare.

E. webbii Coincy. Resembling *E. virescens* in habit but with rather broader leaves and simple lateral cymes. Flowers blue. **(235)**

LA PALMA: Frequent in the forest zones and below, Barranco del Rio, Los Tilos, Barlovento, Fuencaliente, La Cumbrecita, Barranco de las Angustias, Cumbres de Garafia etc, 500–1800 m, laurel and pine woodland.

E. acanthocarpum Svent. Tall shrub with reddish-brown bark. Leaves ovate-lanceolate, large, hispid. Inflorescence a long, dense thyrse. Flowers blue. Nutlets very ornate and spiny.

GOMERA: Central forest region, Roque Agando, El Cedro, laurel forest cliffs 800–1000 m, rare.

E. hierrense Webb ex Bolle. Compact shrub with lanceolate to ovate, short, silvery leaves. Inflorescence a rather short, dense cylindrical thyrse. Corolla pink or blue. **(225)**

HIERRO: El Golfo region from Roques de Salmar to Sabinosa, in the forests and below, locally frequent from 400–800 m.

E. onosmifolium Webb & Berth. Shrub up to 1 m. Leaves linear to lanceolate, somewhat revolute at the margins, the upper surface with dense, large-based setae; lower surface usually with simple hairs, the setae confined to the midrib.

Inflorescence narrow, cylindrical. Corolla whitish, pale pink or very rarely blue, the tube narrow; lobes short. **(230) (231)**

GRAN CANARIA: S. and S.W. regions, Tejeda, Caldera de Tirajana, Barranco de Fataga, locally very common, lower zone and montane scrub, usually on dry S. slopes, 400–1500 m.

E. callithyrsum Webb ex Bolle. Shrub. Leaves lanceolate to ovate, the upper surface with large-based setae; lower surface with simple hairs and prominent, pubescent veins. Inflorescence a rather short, more or less ovate thyrse. Corolla deep blue.

GRAN CANARIA: Rather local, Cumbre de Tenteniguada, El Rincón, cliffs below Juncalillo, 800–1500 m.

E. handiense Svent. Small shrub resembling *E. callithyrsum* but the leaves more or less elliptical, densely hispid on both surfaces; inflorescence longer and lax; corolla blue, densely pilose. **(69)**

FUERTEVENTURA: Cliffs and dry slopes near the summit of Pico de la Zarza in the Jandia region, very rare.

E. strictum L. fil. Very variable shrub up to 1 m. Leaves broadly lanceolate to ovate, shortly hispid. Inflorescence very lax with widely spaced lateral cymes. Corolla pink or pale blue, sometimes with deep blue lines. **(226)**

TENERIFE: Locally frequent in the lower zone and forest regions, Sierra Anaga, Bajamar, Cruz de Taborno, Taganana, Icod el Alto, Guimar, Barranco de Badajoz, Orotava valley, Barranco del Infierno, etc, Buenavista, Punta de Teno (subsp. *exasperatum* (Webb) Bramwell); GRAN CANARIA: N. region, Santa Brigida, Los Tiles de Moya, Barranco de la Angostura, Guiniguada, Monte Lentiscal, Agaete etc; HIERRO: El Golfo, Frontera, Fuente de Tinco, Riscos de Jinamar; LA PALMA: Barranco del Carmen, Monte de Barlovento; GOMERA: Hermigua valley, Roque Cano de Vallehermoso, Barranquillos de Vallehermoso (subsp. *gomerae* (Pitard) Bramwell).

E. auberianum Webb & Berth. Perennial with short woody stock. Stems erect. Leaves in a basal rosette, linear to narrowly oblanceolate, both surfaces covered with long, yellowish setae. Inflorescence lax. Corolla blue. Stamens usually included or the two lower ones slightly longer than the corolla. ***Taginaste picante*****. (232)**

TENERIFE: Subalpine zone of Las Cañadas, Montaña Blanca, Los Azulejos, Montaña de las Arenas Negras, etc, 2100 m, rare.

E. decaisnei Webb & Berth. Shrub up to 2 m. Leaves lanceolate, the upper surface with evenly distributed, very short, large-based spines or flat discs; lower surface with spines confined to the margins and midrib. Inflorescence large, dense, broadly conical. Corolla white with pale blue stripes, broadly funnel-shaped, the lobes equal. **(234)**

GRAN CANARIA: Locally very common, lower and montane zones, Caldera de Tirajana, Tejeda, N. region. Cuesta de Silva etc, 100–1000 m; LANZAROTE: Famara Massif, Haria, Peñitas de Chache, ca. 700 m; FUERTEVENTURA: Jandia region, cliffs of the N. coast, Pico de la Zarza, local. The plants from the Eastern Islands are considered to be a distinct subsp. *purpuriense* Bramwell.

E. gentianoides Webb ex Coincy. Shrub up to 70 cm. Leaves lanceolate, the upper surface with short, large-based spinules, the lower glabrous. Calyx tube longer than the lobes. Corolla blue-violet. Stamens more or less equalling or slightly longer than the corolla, often unequal.

LA PALMA: High mountains of the rim of the Caldera de Tabouriente, Cumbre de Garafia, Gran Pared, Roque de los Muchachos, Topo Alto de Los Corralejos, 1100–1900 m, very rare.

MESSERSCHMIDIA

Shrubs. Flowers in terminal, branched cymes. Calyx lobes equalling the short tube. Corolla tube long, narrow, pubescent, the lobes acute, patent. Fruit somewhat fleshy, splitting into two separate mericarps.

1. Leaves stalked, lanceolate .. .. **M. fruticosa**
1. Leaves sessile, linear .. .. **M. angustifolia**

M. fruticosa L. fil. Shrub up to 3 m. Stem greyish-green, sometimes pubescent or with small setae. Leaves petiolate, lanceolate to ovate-lanceolate, covered with pustular-based setae. Inflorescences with small bracts. Calyx lobes about 2 mm. Corolla whitish, rather inconspicuous. Fruits somewhat fleshy, wrinkled, black when ripe. ***Duraznillo.*** **(71)**

ALL ISLANDS: Common shrub in drier parts of the lower zone, in *Euphorbia* communities.

M. angustifolia Lam. Like *M. fruticosa* but with long, linear, sessile leaves which are often sickle-shaped. **(72)**

GRAN CANARIA, TENERIFE (Teno, Adeje); GOMERA: Locally frequent on dry slopes. Very distinct from *M. fruticosa* on TENERIFE and GOMERA but apparently merging with it on GRAN CANARIA.

Heliotropium

H. erosum Lehm. a grey-leaved annual herb with dense cymes of small white flowers is frequent on dry S. slopes of all the islands especially TENERIFE (Guimar to Medano and Los Cristianos), LANZAROTE and FUERTEVENTURA, (N. Africa).

MYOSOTIS

Annual to perennial herbs. Flowers in paired scorpioid cymes. Calyx with hooked hairs. Corolla blue or white; tube short; lobes flat. Style included. Stigma capitate.

1. Perennial, basal leaves large, ovate. Corolla blue or pink **M. latifolia**
1. Annual, basal leaves lanceolate. Corolla whitish, becoming pink or blue later .. .. **M. discolor**

M. latifolia Poiret (*M. macrocalycina* Cosson). Perennial with rhizome. Stems up to 60 cm, often woody at the base. Basal leaves large, ovate. Corolla up to 1 cm, usually blue but sometimes pink. **(236)**

TENERIFE, GRAN CANARIA, LA PALMA, HIERRO, GOMERA: Common in laurel forest zones. The Hierro populations often have deep pink flowers.

M. discolor Pers. subsp. ***canariensis*** (Pitard) Grau. Slender annual up to 30 cm, often branched at base. Leaves lanceolate, obtuse; the upper often broader and acute. Corolla cream at first, becoming pink or blue later.

GRAN CANARIA, TENERIFE, HIERRO: Damp places in the xerophytic zone, locally common.

LABIATAE —— Sage Family

Herbs or shrubs, often aromatic. Leaves usually opposite. Flowers zygomorphic, usually aggregated into modified cymes known as 'verticillasters' which are, in turn, grouped into the inflorescence. Inflorescences usually with bracts or 'floral leaves'. Calyx 4- to 5-lobed, often 2-lipped, the upper lip with 3 teeth, the lower with 2. Corolla with a tube and 5 lobes, usually 2-lipped, the upper lip with 2 lobes, the lower with 3, sometimes the upper lip is absent Stamens 4 or 2. Ovary superior. Fruit of 4, 1-seeded nutlets.

1. Leaves trifoliate .. .. **Cedronella**
1. Leaves simple or pinnately lobed.
  2. Leaves pinnately lobed .. .. **Lavandula**
  2. Leaves simple.
    3. Corolla red, 1-lipped, the upper lip absent .. .. **Teucrium**
    3. Corolla blue, white or pink, 2-lipped.
      4. Fertile stamens 2 .. .. **Salvia**
      4. Fertile stamens 4.
        5. Shrubs with minute (1–2 mm) flowers in more or less globular verticillasters .. .. **Bystropogon**
        5. Shrubs or herbs with flowers over 4 mm long, verticillasters not globular.
          6. Shrubs, calyx 10-veined, corolla white or yellowish often with coloured lips .. .. **Sideritis**
          6. Herbs or small shrublets, calyx 13- to 15-veined, corolla usually pinkish or blue without brown or reddish lips.
            7. Corolla tube curved, leaves coarsely toothed .. **Nepeta**
            7. Corolla tube very short, straight, leaves with more or less entire margins.
              8. Calyx 10-veined (Lanzarote) .. .. .. **Thymus**
              8. Calyx 13- to 15-veined .. .. **Micromeria**

MICROMERIA

Dwarf shrubs. Calyx tubular to campanulate, 13- to 15-veined, more or less 2-lipped with unequal or sometimes more or less equal teeth. Corolla 2-lipped with a straight tube. Stamens shorter than corolla. Style subulate. ***Tomillo.***

A very complex and difficult genus especially on Gran Canaria where hybrids are common and species appear to intergrade and on the W. islands where *M. varia* is represented by many local races and ecotypes. The account outlined below should be considered as a very provisional one.

1. Erect robust shrub up to 75 cm, leaves large, up to 2 cm (Gran Canaria, pine forests) .. .. **M. pineolens**
1. Dwarf shrubs up to 30 cm, leaves less than 1·5 cm.
  2. Plants white-lanate.
    3. Stems procumbent, leaves opposite, not overlapping (Gran Canaria) .. .. **M. lanata**
    3. Stems erect, leaves fasciculate, overlapping.
      4. Leaf-hairs long coarse, pale pinkish-red tinged (Gran Canaria) .. .. **M. polioides**
      4. Leaf-hairs white-woolly (Gran Canaria) .. .. **M. linkii**
  2. Plants pubescent or glabrous.
    5. Calyx-tube at least as long as leaves, wide at base, with a narrow neck, corolla about 8 mm across (S. Gran Canaria) .. .. **M. helianthemifolia**
    5. Calyx tube shorter than leaves, campanulate or cylindrical, corolla less than 5 mm across.

6. Leaves glabrous, more or less suborbicular (S. Tenerife) .. .. **M. teneriffae**
6. Leaves pubescent, linear to broadly lanceolate.
7. Densely branched, more or less erect shrublets, leaves with revolute margins.
8. Leaves sparsely pubescent to subglabrous (Gomera) **M. densiflora**
8. Leaves densely pubescent to white-lanate beneath.
9. Individual flowers long-pedicellate (Gomera) .. **M. lepida**
9. Individual flowers subsessile (Gran Canaria) .. **M. benthamii**
7. Lax, branched shrublets, leaves usually without revolute margins or margins only slightly revolute.
10. Verticillasters sessile (Lanzarote, Gran Canaria) .. **M. bourgaeana**
10. Verticillasters pedunculate.
11. Calyx campanulate with long teeth (Tenerife, La Palma) .. .. **M. herphyllimorpha**
11. Calyx tubular with short teeth.
12. Peduncles as long as verticillasters, inflorescences with long internodes between the verticillasters (Tenerife, La Palma, subalpine) .. .. **M. julianoides**
12. Peduncles much shorter than verticillasters, inflorescences with short internodes (W. islands **M. varia**

M. pineolens Svent. Tall shrub up to 75 cm. Leaves up to 2 cm, ovate, densely long-hairy on both surfaces, the margins revolute. Inflorescence a terminal spike. Verticillasters sessile. Calyx with long hairs, cylindrical with obtuse teeth. Corolla large, pink. **(253)**

Gran Canaria: N.W. region, shrub in the pine forests of Tamadaba where it is extremely common, 1000–1400 m.

M. lanata (Chr. Sm.) Benth. Densely branched, procumbent shrublet. Leaves opposite, narrowly elliptical, white-woolly, the margins revolute. Calyx woolly, the teeth acute. Corolla white or pink.

Gran Canaria: Pinar de Tamadaba, pine forest floor. Juncalillo, sporadic in the C. region of the island, Pozo de las Nieves, 700–1900 m.

M. linkii Webb & Berth. Erect shrublet. Leaves fasciculate, overlapping, densely white-grey pubescent, erect. Inflorescences few-flowered. Flowers small, unobtrusive. Calyx densely woolly, cylindrical, the teeth very short, more or less equal. Corolla pink.

Gran Canaria: Coastal regions and dry S. slopes, Cuesta de Silva, Fataga, Mogan, 150–300 m.

M. polioides Webb. Resembling *M. linkii* but much less densely hairy and the hairs pinkish to reddish-brown.

Gran Canaria: S., W., and C. regions, locally rather common, Juncalillo, San Nicolas, Agaete, Arguineguin, Barranco de Fataga, etc, lower zone on dry slopes up to 700 m.

M. helianthemifolia Webb & Berth. Procumbent subshrub with long, lax, leafy branches. Leaves large, lanceolate-oblong, pubescent. Inflorescences short, dense. Calyx tube very long and slender with a narrow neck; teeth long, filiform. Corolla very large, pink. **(75)**

Gran Canaria: S. region, rather rare on cliffs in the lower zone, Barranco de Fataga, Arguineguin, Mogan, etc.

M. teneriffae Benth. Small procumbent shrublet with opposite to more or less fasciculate, suborbicular, acute, glabrous leaves. Calyx tubular, pubescent, the teeth obtuse. Corolla white or pink. **(74)**

TENERIFE: S.E. region on cliffs up to about 300–400 m, Ladera de Guimar, Escabonal, Igueste de Candelaria, Barranco Herques to Arico, locally abundant.

M. densiflora Benth. Very densely branched, more or less erect shrublet. Leaves opposite, tightly revolute, subglabrous to sparsely hairy. Calyx long, (6 mm); teeth acute. Corolla 1½ times as long as calyx, pink.

GOMERA: E. & C. region, Fuente de Agando, Tagamiche, Barranco de la Laja, generally in forest zones, El Cedro, etc, 600–1200 m, locally abundant.

M. lepida Webb & Berth. Like *M. densiflora* in habit but densely pubescent. Leaves linear to lanceolate, very revolute, densely hairy beneath. Verticillasters lax; individual flowers long-pedicellate. Corolla pale-pink or whitish. Calyx tubular with spreading, acute teeth.

GOMERA: S. and W. of the island, Alajeró to Chipude, Vallegranrey, Barranco de Argaga, etc, 400–1000 m, in more xeric habitats than the previous species.

M. benthamii Webb & Berth. Lax, twiggy shrublet with erect or ascending branches. Leaves opposite, the margins revolute; upper surface sparsely pubescent, the lower white-lanate. Inflorescence of several interrupted whorls. Verticillasters pedunculate; individual flowers subsessile. Calyx tube long. Corolla pink.

GRAN CANARIA: Locally very common in many regions of the island, Moya, Teror, San Bartolomé de Tirajana, Cruz de Tejeda, Valle de Tejeda, etc, 500–1900 m, open rocky places.

M. bourgaeana Webb. Small, compact, very densely leafy shrublet. Leaves ovate, the margins pubescent. Inflorescence dense, the verticillasters sessile. Calyx more or less tubular; teeth purplish. Corolla small, deep pink.

LANZAROTE: Famara region, on steep cliffs nr. Peñitas de Chache, Mirador del Rio, Haria, 500–700 m, sporadic; GRAN CANARIA; FUERTEVENTURA: Jandia, Pico de la Zarza, central region nr. Betancuria.

M. herphyllimorpha Webb. Lax, erect or procumbent. Leaves fasciculate, linear-lanceolate, subglabrous above, pubescent beneath. Inflorescence lax. Verticillasters pedunculate. Calyx campanulate with long teeth. Corolla deep pink to reddish. **(255)**

TENERIFE: Local, usually in pine or laurel woodland, Agua Mansa, La Esperanza, Granadilla, Icod, Lomo de Pedro Gil, Anaga etc, 400–1500 m; LA PALMA: El Paso, La Cumbrecita, Barlovento, La Galga, Las Breñas, 200–1400 m; HIERRO: Pine forest region of the central region, locally abundant.

M. julianoides Webb & Berth. Small, compact shrublets. Leaves fasciculate, linear, light green, the margins somewhat revolute. Inflorescence long, with long internodes between the verticillasters. Peduncles equalling the verticillasters. Calyx tubular hairy, the teeth more or less unequal, pink-tipped. Corolla pale pink or whitish.

TENERIFE: Las Cañadas de Teide, Vilaflor, El Sombrerito, Los Roques, Valle de las Arenas Negras, La Fortaleza etc, 1900–2200 m, locally common; LA PALMA: High regions of the Cumbres and outer walls of the Caldera (the La Palma plant has deeper pink flowers and short peduncles and is often separated as *M. palmense* (Bolle) Lid).

M. varia Benth. Subglabrous to densely pubescent, dense to lax shrublet. Leaves opposite or fasciculate, linear to lanceolate, sometimes with a slightly revolute margin. Inflorescence lax or dense. Verticillasters with a short peduncle. Calyx tube cylindrical, the teeth acute. Corolla small, pink or white. ***Tomillo.*** **(254)**

ALL ISLANDS: with the probable exception of FUERTEVENTURA: Locally very common on the W. islands, rather less frequent on GRAN CANARIA, up to 2000 m but most common below 1500 m, extremely variable.

THYMUS

Calyx campanulate, 2-lipped, 10-veined. Corolla weakly 2-lipped. Stamens 4, 2 usually exserted.

T. origanoides Webb. Small procumbent or creeping shrublet. Leaves small, lanceolate to ovate, glandular-punctate. Heads few-flowered. somewhat interrupted below. Flowers very small, pinkish. **(238)**

LANZAROTE: Rocks and cliffs in the Famara Massif, Riscos de Famara, Haria.

SIDERITIS

Shrubs. Leaves entire. Verticillasters many-flowered. Calyx campanulate, 10-veined, 5-toothed, the teeth equal. Corolla yellow or white often with reddish-brown or brown lips. Stamens included in corolla-tube.

1. Inflorescences more or less erect or ascending, bracts linear to lanceolate.
  2. Stems and branches densely white-felted.
    3. Inflorescences very dense, internodes between the verticillasters very short.
      4. Inflorescence a very dense spike with the ends of the bracts projecting, leaves 7 to over 10 cm long, ovate .. .. **S. macrostachys**
      4. Inflorescence with at least a few distant verticillasters in the lower part, leaves usually less than 7 cm.
        5. Leaves grey above, apex blunt .. .. .. **S. argosphacelus**
        5. Leaves green above, apex acute .. .. .. **S. lotsyi**
    3. Inflorescences lax with long internodes between the verticillasters.
      6. Corolla lips reddish-brown (Tenerife) .. .. .. **S. candicans**
      6. Corolla lips pale yellow (Gran Canaria) .. .. **S. dasygnaphala**
  2. Stems and branches pubescent or yellowish-felted.
    7. Verticillasters crowded at least in the upper part of the inflorescence.
      8. Tall shrub up to 2 m, leaves lanceolate with a cordate base, acute .. .. **S. dendro-chahorra**
      8. Small shrubs less than 75 cm, leaves ovate or heart-shaped, usually obtuse.
        9. Nerves very prominent on lower surface of leaf.
          10. Leaves ovate (N.W. Tenerife) .. .. .. **S. nervosa**
          10. Leaves narrow, heart-shaped (Lanzarote, Fuerteventura) .. .. **S. massoniana**
        9. Nerves not prominent beneath, leaves large, ovate, membranous (N.W. Tenerife, forest zone) .. .. **S. kuegleriana**
    7. Verticillasters distant, internodes at least as long as verticillasters.
      10. Leaves glandular-pubescent, the surface more or less rugose (S.W. Tenerife, Adeje region) .. .. **S. infernalis**

10. Leaves pubescent but not glandular, surface more or less smooth.
  11. Tall shrubs with large, heart-shaped or broadly oblong leaves.
    12. Indumentum very yellowish, leaves heart-shaped (W. islands) .. .. **S. canariensis**
    12. Indumentum cream-coloured, leaves broadly oblong to more or less heart-shaped (Gran Canaria) .. .. **S. discolor**
  11. Dwarf shrubs with small, more or less lanceolate leaves (S.W. Tenerife) .. .. **S. cystosiphon**
1. Inflorescences pendulous, bracts ovate, surrounding the verticillasters (Gomera).
  13. Leaves glandular-hairy, greenish (W. Gomera) .. **S. nutans**
  13. Leaves white-felted to lanate (N.E. & E. Gomera).
    14. Calyx pubescent, leaf-blade equalling or shorter than petiole .. .. **S. gomeraea**
    14. Calyx subglabrous, leaf-blade longer than petiole .. **S. cabrerae**

S. argosphacelus (Webb & Berth.) Clos. Small subshrub. Stem branched, white-woolly. Leaves heart-shaped, grey-canescent above, densely white-felted beneath; margins crenate. Inflorescence a short, dense spike with the lower whorls usually somewhat distant. Corolla yellowish with brown lips. Spike sometimes nodding in fruit. ***Chahorra.*** **(248)**

TENERIFE: N. coast from Orotava to Teno, locally frequent, Barranco de Ruiz, El Fraile, Teno Bajo, Los Silos, 50–250 m; GOMERA: Locally frequent, Barranco de la Villa, La Laja, Agulo, Vallehermoso, on N. coast cliffs to 300 m.

S. lotsyi (Pitard) Ceb. & Ort. Shrub up to 80 cm. Leaves ovate to elliptical; base cordate; lower surface densely white or rarely yellowish-felted, upper surface greenish; margins crenate; apex acute. Inflorescence often branched at base. Lower whorls distant. Corolla white or pale yellow with pronounced reddish or brown lips, **(80)**

GOMERA: Roque Agando, Roque Cano de Vallehermoso, Fortaleza de Chipude, Lomo de Carretón, etc, 600–1000 m, usually on cliffs, locally abundant; TENERIFE: Valle de Masca, 400 m, rare (var. *mascaensis* Svent.).

S. candicans Aiton. Tall, erect to spreading, much-branched shrub up to 80 cm. Lower leaves very variable, lanceolate to narrowly ovate, often cordate at the base, very densely white-woolly on both surfaces, the margins crenate, upper leaves and bracts linear-lanceolate. Inflorescence a long, interrupted, erect spike, sometimes branching from the base. Corolla pale yellow, somewhat pubescent, the lips light brown to orange-red. ***Chagorro.*** **(250)**

TENERIFE: Montane and pine forest zones, Vilaflor, Las Cañadas, La Fortaleza, Ucanca valley, Montaña de Diego Hernandez, locally very common, Agua Mansa, frequent in pine woods, 1300–2100 m.

S. dasygnaphala (Webb) Clos. Like S. *candicans* but the calyx teeth with an apical spine and the yellow corolla without coloured lips. **(249)**

GRAN CANARIA: Central montane region in leguminous scrub, Cruz de Tejeda, Roque Nublo, Paso de la Plata, Pozo de las Nieves, etc, 1500–1900 m, locally common.

S. macrostachys Poiret. Shrub up to 1 m. Mature leaves very large, ovate, cordate at the base, very densely white-felted beneath, grey-green above; margins crenate. Inflorescence a very dense, erect, cylindrical to quadrangular, spike, often branched from the base. The calyx and bracts white-woolly. Corolla whitish with brown lips. (**244**)

TENERIFE: N. coast, laurel forest cliffs and humid rocky areas in the lower zone. Sierra Anaga, El Bailadero, Las Mercedes, Vueltas de Taganana, etc, Bajamar, Punta Hidalgo, Realejo Alto, Icod el Alto, 200–700 m, local, rather rare.

S. infernalis Bolle. Small shrub with thin, rather membranous, rugose, glandular-hairy, green leaves. Inflorescence lax with few-flowered, distant verticillasters. Corolla white with reddish-brown lips. (**245**)

TENERIFE: S.W. region in the mountains of Adeje, Barranco del Infierno, Valle Seco, etc, wet cliffs, very rare.

S. cystosiphon Svent. Like *S. infernalis* but much-branched with narrow, lanceolate, densely hairy leaves. Inflorescences with distant verticillasters and linear, paired bracts. Corolla pale yellow or white, expanded and bulbous near tip, the lips minute, brownish. (**78**) (**246**)

TENERIFE: S.W. region on dry rocks S. of Santiago del Teide, 450 m, locally frequent in its only known locality.

S. canariensis L. Tall shrub with long, lax, erect inflorescences. Leaves ovate, cordate at the base; margins with small teeth; indumentum of yellowish, branched hairs. Inflorescence of large, many-flowered, distant verticillasters, each subtended by a pair of linear bracts. Corolla yellowish with darker lips.

TENERIFE: Common in the laurel forests of the Anaga region, Las Mercedes, Vueltas de Taganana, N. coast forest zone, Agua Garcia, Agua Mansa, Icod el Alto, 500–1200 m; HIERRO: Forest cliffs between Jinamar and Frontera, Fuente de Tinco, 600 m; LA PALMA: Locally frequent in the forests of the N.E. region, Breña Alta, Cumbre Nueva, Los Sauces, Barlovento, etc, 650–1000 m.

S. discolor Webb. Like *S. canariensis* but the leaves green, pubescent above, greyish-cream tomentose below. Inflorescence often branched, corolla white, the lower lip rather long.

GRAN CANARIA: Laurel forests of the N. coast, now almost extinct except for a small population at Los Tiles de Moya.

S. dendro-chahorra Bolle. Tall shrub. Leaves lanceolate to narrowly ovate, acute, indumentum of lower surface very dense, white-creamish, leaf-margins more or less entire. Inflorescences short, dense, often branched from the base; the lower verticillasters somewhat remote. Corolla yellow, the lips acute. (**247**)

TENERIFE: Sierra Anaga, abundant in the higher regions of the coastal zone. Valle de San Andrés, El Bailadero, Tunel de Taganana, Punta del Hidalgo, Batán, etc, 350–800 m; LA PALMA: N.E. region, pine forests above Los Sauces, S.W., between Fuencaliente and Los Llanos at the lower margins of the pine woods of Cumbre Vieja; GOMERA: S. side of the Agando region, Benchijigua; HIERRO: Central region in *Erica* heath, El Julán, Mocanal at the N. end of El Golfo etc.

S. nervosa Christ. Small shrub up to 50 cm. Leaves narrowly ovate, obtuse, the base cordate; nerves very prominent beneath; the indumentum yellowish. Inflorescences short, usually branched at the base. Corolla yellow. (**79**)

TENERIFE: Punta de Teno, in *Euphorbia* scrub vegetation, 100 m, rare.

S. kuegleriana Bornm. Very like *S. nervosa* but more with thin, rather membranous leaves with very short hairs. Inflorescence branched, often pendulous in fruit. Corolla lemon-yellow.

TENERIFE: N. coast, basalt rocks in the forest zone between Icod de los Vinos and Los Silos, 250–500 m, rare.

S. massoniana Aiton. Small, compact shrublet with small, heart-shaped leaves; the lower leaf-surface felty, the upper green, pubescent. Inflorescences short, few-flowered. Corolla pale yellow.

FUERTEVENTURA: Jandia Mountains, Pico de la Zarza, 600 m, local; LANZAROTE: Riscos de Famara, 400–650 m, rare.

S. nutans Svent. Cliff plant with short woody stem. Leaves greenish, lanceolate, to oblong, crenate, densely glandular-hairy, obtuse, base cordate. Inflorescence an unbranched, pendulous, dense, cylindrical spike. Corolla white with brown lips. **(252)**

GOMERA: W. region, Lomo de Carretón, Vallegranrey, Valle de Argaga, etc, on dry basalt cliffs in the lower zone, occasional 200–700 m.

S. gomeraea De Noe. Like *S. nutans* but densely white-woolly all over the vegetative parts and with a much longer inflorescence with ovate bracts. ***Tajora.*** **(251)**

GOMERA: E. region, Barranco de la Villa, Cumbre de Hermigua, Barranco de la Laja, Agando, etc, locally frequent on cliffs in the lower and forest zones, 200–900 m.

S. cabrerae Ceb. & Ort. Closely resembling *S. gomeraea* but the leaves long-petiolate and the calyx subglabrous.

GOMERA: Poorly known, rare species, Barranco de Cabrito, Benchijigua, up to 1000 m.

Two further species of *Sideritis*, *S. marmorea* Bolle (GOMERA) and *S. bolleana* Bornm. (LA PALMA) are reported. These are extremely rare plants of the forest zones known only from a few collections and may be local forms of *S. canariensis* and *S. dendro-chahorra* respectively.

BYSTROPOGON

Strongly scented shrubs. Leaves entire, sometimes with crenate margins. Flowers small, white or pink, in dense inflorescences. Calyx with 5 more or less equal teeth. Corolla 2-lobed; upper lobe 2-fid or with 2 small teeth, the lower 3-lobed. Stamens included.

1. Leaf-margins more or less entire; leaves silvery above.
   2. Leaves lanceolate, calyx exceeding the corolla .. .. **B. origanifolius**
   2. Leaves ovate-rhomboidal, calyx shorter than the corolla **B. plumosus**
1. Leaf-margins crenate; leaves green above .. .. **B. canariensis**

B. canariensis (L.) L'Hér. Very variable tall shrub up to 2·5 m. Leaves ovate-lanceolate, green, glabrescent above, sparsely to densely hairy beneath; margins coarsely crenate. Inflorescence consisting of small, round glomerules. Calyx teeth narrow, acute, densely setose, shorter than the white or pink corolla; ***Poleo de Monte.*** **(81)**

TENERIFE: Fairly frequent shrub in laurel forest areas 600–1500 m, Las Mercedes, El Bailadero, Vueltas de Taganana, Agua Mansa; GOMERA: El Cedro, Chorros de Epina, Agando, in laurel and *Erica/Myrica* woodland; LA PALMA: Monte de los Tilos, Cumbre Nueva above Breña Alta, Barlovento, La Galga, etc; GRAN CANARIA: Los Tiles de Moya, laurel woods; HIERRO: Forest regions of El Golfo, Frontera, Fuente de Tinco.

B. plumosus (L. fil.) L'Hér. Small robust shrub up to 80 cm. Leaves ovate-subrhomboidal, silvery, the margins entire or very occasionally remotely crenate;

upper surface silvery, sparsely hairy, the lower densely white-lanate. Flowers in large, lax glomerules forming a diffuse inflorescence. Calyx villous, shorter than the corolla; the teeth subequal. Corolla white. **(242)**

TENERIFE, GRAN CANARIA, LA PALMA, HIERRO, GOMERA: common in the lower zone 100–600 m and occasionally in exposed areas in the forest zone, usually in *Euphorbia* communities.

B. origanifolius L'Hér. Like *B. plumosus* but with lanceolate leaves and very long, slender calyx teeth exceeding the white corolla.

TENERIFE: Orotava valley, Garachico to Los Silos, dry rocky areas from the lower to forest zones up to 1200 m; LA PALMA: Barrancos above Los Sauces and near Tijarafe in *Erica* heath and pine woodland, exposed rocks; GRAN CANARIA: Barranco de la Virgen, rare.

SALVIA

Calyx campanulate, 2-lipped, upper lip entire or 3-lobed; lower lip 2-lobed. Corolla 2-lipped, upper lip hooded, the lower 3-lobed. Stamens 2.

1. Leaves narrow, base sagittate .. .. **S. canariensis**
1. Leaves broadly ovate, base cordate .. .. .. **S. broussonetii**

S. canariensis L. Shrub up to 2 m. Leaves lanceolate, 5–15 cm long, white-lanate at least below; base sagittate. Inflorescence branched. Bracts conspicuous, ovate, obtuse, papery, purple, longer than the calyx. Calyx pubescent, with prominent veins. Corolla pink. ***Salvia.*** **(239)**

GRAN CANARIA, TENERIFE, LA PALMA, GOMERA, HIERRO: common in the xerophytic, lower zone and locally frequent even up to the montane region on GRAN CANARIA but not usually found in the laurel forest, 50–1600 m.

S. broussonetii Benth. (*S. bolleana* Noe). Small shrub up to 75 cm. Leaves broadly ovate, 10–20 cm; margins coarsely crenate; base cordate. Inflorescence usually branched. Bracts hairy with an apical spine. Corolla white to pinkish. **(77)**

TENERIFE: Confined to cliffs at either end of the island, Anaga region between San Andrés and Igueste, 200 m, locally abundant, Teno region, very rare, Valle de Masca, 400 m.

TEUCRIUM

Shrubs up to 2 m. Corolla 2-lipped; upper lip very short.

T. heterophyllum L' Hér. Leaves lanceolate to ovate, densely hairy especially below; margins crenate, serrate or subentire. Flowers axillary in clusters of 1 to 4. Corolla pink to red, 2-lipped; upper lip very short, bifid; lower lip subentire to 3-lobed. Stamens and style exserted, twice as long as corolla. ***Jocama.*** **(73) (237)**

GRAN CANARIA, TENERIFE, LA PALMA: Dry rocky slopes 500–600 m. Locally frequent at Teno and Tamaimo. Doubtfully recorded from Las Cañadas.

CEDRONELLA

Herbaceous perennials, sometimes woody at the base. Leaves 3-foliate. Inflorescence a dense terminal head or spike. Corolla 2-lipped. Stamens equalling or slightly exceeding the corolla.

C. canariensis (L.) Webb & Berth. Scented. Stems up to 150 cm. Leaves 3-foliate; leaflets lanceolate, glabrous above, pubescent beneath; the margins serrate. Corolla pink ***Algaritofe.*** **(241)**

TENERIFE: Common in forest regions, Las Mercedes, Vueltas de Taganana, El Bailadero, Agua Garcia, Agua Mansa 500–1500 m; GOMERA: El Cedro,

Hermigua, Arure, Barranquillos de Vallehermoso, very common; LA PALMA: Laurel woods, Los Sauces, Barlovento, Garafia, etc; GRAN CANARIA: Los Tiles de Moya, N. coast region, Cumbre de San Mateo.

LAVANDULA

Shrubs. Flowers blue or purple in a short, bracteose, rather dense simple or branched spike. Corolla 2-lipped, the upper lip 2-lobed, the lower 3-lobed. Calyx ovoid, 13- or 15-veined, teeth short. Stamens 4, equalling or shorter than the corolla.

1. Leaves green, usually at least some 2-pinnatifid, stems glabrous .. .. **L. canariensis**
1. Leaves greyish, 1-pinnatifid, stems pubescent to lanate.
  2. Plants with woolly hairs on stems and leaves, calyx equal to or shorter than the subtending bract .. .. **L. minutolii**
  2. Plants densely covered with very short white hairs, calyx longer than bract .. .. **L. pinnata**

L. canariensis (L.) Mill. Shrub up to 1·5 m. Stems glabrous. Leaves ovate, 1- or more usually 2-pinnatifid, pubescent; lobes rounded, flat. Inflorescences branched, the spikes up to 10 cm. Bracts bluish towards the tip. ***Mato Risco.***

TENERIFE: Common, Sierra de Anaga, Igueste de San Andrés to Santa Cruz, Icod, Los Silos, S. region, Guimar, Arico, Granadilla, Adeje, etc, up to 600 m, in *Euphorbia* communities; GOMERA: Barranco de la Villa, Hermigua, Agulo, Vallehermoso, etc; HIERRO: Puerto de Estaca, Valverde, lower zone of El Golfo; LA PALMA: very common in the lower zone; GRAN CANARIA: frequent along the N. coast, Cuesta de Silva, Agaete, Telde, Tejeda, etc.

L. minutolii Bolle. Stem and leaves woolly. Leaves lanceolate, 1-pinnatifid; lobes oblanceolate. Inflorescence bracts oval equalling or longer than the calyx. **(243)**

GRAN CANARIA: Central and S. region of the island, Cruz de Tejeda, Barranco de Tejeda, Caldera de Tirajana, Tazarte, etc, 900–1600 m, locally abundant. Subsp. ***tenuipinna*** (Svent.) Bramwell. Robust, leaves with much longer, narrower lobes; inflorescence bracts apiculate.

TENERIFE: Valle de Masca, 1000 m, Barranco de Natero, 500–1000 m, very local.

L. pinnata L. fil. Like *L. minutolii* but the leaves pinnate with broad flat lobes and the hairs very short and dense. Calyx much longer than the subtending bract. **(82)**

TENERIFE: Locally abundant on the N. side of the island from Anaga to Teno, Taganana, Bajamar, Puerto de la Cruz, Icod, Los Silos, El Fraile etc, Masca, Tamaimo, Guimar, etc, near sea-level to about 600 m; LANZAROTE: N. region, Famara cliffs, Haria; GOMERA.

NEPETA

Perennial herbs sometimes woody at base. Verticillasters in a spike-like inflorescence. Calyx tubular with 5 subequal teeth, 15-veined. Corolla 2-lipped, the upper lip erect, 2-lobed; lower 3-lobed. Corolla-tube curved.

N. teydea Webb & Berth. Tall pubescent herb up to 1·5 m. Leaves opposite, lanceolate-oblong, obtuse; the margins coarsely toothed. Inflorescence often branched at base. Bracts and calyx teeth purplish. Corolla blue-purple or occasionally white; the middle lobe of the lower lip longer than the laterals. **(240)**

TENERIFE: Subalpine zone of Las Cañadas 2000–2200 m. Locally frequent on rocky slopes and amongst volcanic debris, El Portillo, Arenas Negras, Montaña de Diego Hernandez, Llano de Ucanca, etc; LA PALMA: Cumbres de Garafia in the high mountain zone, 1900 m.

SOLANACEAE ———— Tomato Family

Shrubs or herbs. Leaves simple or pinnate. Flowers regular or zygomorphic, hermaphrodite. Calyx 5-lobed. Corolla more or less campanulate, 5-lobed. Stamens 5, sometimes unequal. Ovary superior. Fruit a berry.

1. Calyx enlarging to enclose the mature fruit, flowers yellowish green .. .. **Withania**
1. Calyx not as above, flowers purplish-blue or whitish .. **Solanum**

SOLANUM

Shrubs. Stems with or without spines. Corolla deeply 5-lobed, the lobes spreading to reflexed. Anthers 5, one much longer than the other 4. Fruit a two-celled tomato-like berry. 3 sp.

1. Erect to more or less procumbent shrubs; stems with at least a few small spines.
 2. Leaves ovate-rhomboidal; stems and petioles densely spiny .. .. **S. vespertilio**
 2. Leaves narrowly oblong-lanceolate; stems and petioles remotely spiny .. .. **S. lidii**
1. Climbing or scrambling shrub; stems without spines .. **S. nava**

S. vespertilio Aiton. Erect shrub up to 1·5 m. Stems and petioles densely spiny. Leaves ovate to ovate-rhomboidal, 5–15 cm long, yellowish felted beneath. Flowers in clusters of 5–10, bluish-purple. Fruits up to 2 cm in diameter, red. ***Rejalgadera.*** **(258)**

TENERIFE: N. coast regions from Icod to Anaga, above 400 m, Valle de San Andrés, Igueste de San Andrés; GRAN CANARIA: many old records and herbarium specimens exist from the former forest regions of the N. coast of the island between Arucas and Moya but this species is now extremely rare in this area.

S. lidii Sunding. Like *S. vespertilio* but more or less procumbent, the stems and petioles only remotely spiny, the leaves narrowly oblong-lanceolate and the fruits smaller (1 cm) and orange when ripe. **(257)**

GRAN CANARIA: S. mountain slopes near Termisas in the Tirajana valley 600 m, rare.

S. nava Webb & Berth. Climbing or scrambling shrub with glandular-hairy stems and leaves. Leaves ovate, entire or toothed, occasionally with two basal leaflets. Flowers in axillary clusters. Fruits orange.

TENERIFE: Laurel forests on the N. coast, very rare; GRAN CANARIA.

WITHANIA

Shrubs up to 2 m. Leaves entire. Flowers in axillary clusters. Calyx broadly campanulate, 5-dentate, enlarging in fruit to surround the berry.

W. aristata (Aiton) Pers. Stems subglabrous. Leaves up to 14 cm, ovate, often unequally cuneate at base. Corolla yellowish-green. Calyx with 5 long-aristate teeth, inflating to surround the blackish berry in fruit. ***Orobal.*** **(256)**

ALL ISLANDS: Frequent in dry barranco beds in the lower zone, 0–600 m.

SCROPHULARIACEAE ——— Figwort Family

Shrubs or herbs. Leaves alternate or opposite. Flowers zygomorphic, usually in terminal spikes or axillary. Calyx 4- to 5-lobed. Stamens 2 or 4. Ovary superior, 2-locular. Fruit a capsule, usually opening by 2 valves or pores. Seeds small, numerous.

1. Flowers solitary, axillary, long-spurred, yellow .. .. **Kickxia**
1. Flowers in terminal inflorescences, orange, reddish, green or purplish, not long-spurred.
   2. Flowers purplish-blue, stamens 2.. .. .. .. **Campylanthus**
   2. Flowers orange, reddish or greenish, stamens 4.
      3. Stems square in cross-section, corolla-tube more or less globular .. .. **Scrophularia**
      3. Stems round in cross-section, corolla more or less tubular.
         4. Leaves spathulate, more or less opposite, densely woolly, corolla ephemeral .. .. **Lyperia**
         4. Leaves lanceolate, alternate, subglabrous, corolla persistent, bright orange-red .. .. **Isoplexis**

SCROPHULARIA

Stems square in cross-section. Leaves opposite. Flowers in small cymes forming a loose terminal panicle, dull purple, red-orange or greenish. Corolla more or less globular with broad, short lips. Fertile stamens 4; 1 sterile staminode forming a scale under the upper corolla-lip. Fruit a 2-celled capsule.

A rather problematical genus in the Canaries. The species limits are difficult to define. The commonest species are included in the key.

1. Small annual with conical, beaked capsule .. .. **S. arguta**
1. Robust biennials to perennials with ovoid or subglobose capsules.
   2. Leaves with 3 leaflets, corolla large (1 cm or more), orange-red .. .. **S. calliantha**
   2. Leaves more or less entire, corolla less than 0·7 cm, green, brownish or purple.
      3. Corolla green .. .. **S. smithii**
      3. Corolla with at least the lips brownish or purple.
         4. Inflorescence with axillary branches, stems subglabrous, capsules ovoid (Tenerife, subalpine) .. **S. glabrata**
         4. Inflorescence usually without axillary branches, stems woolly hairy, capsules subglobose .. .. .. **S. langeana**

S. arguta Soland. ex Aiton. Small annual up to 25 cm. Leaves ovate; margins double-dentate. Corolla lobes reddish-brown. Capsule conical, beaked.

TENERIFE: S. coast region, Guimar, Adeje, Granadilla, Las Raices, Masca, Teno, etc, dry rocks in the lower zone and in pine forest, 300–1400 m; GOMERA: N. coast, Agulo-Vallehermoso, rather rare; HIERRO: Frequent in the lower zone of El Golfo, Valverde; GRAN CANARIA, LANZAROTE.

S. calliantha Webb & Berth. Ascending to erect perennial up to 1·5 m. Leaves with 3 leaflets, the terminal larger than the others. Corolla large, more than 1 cm long, the throat yellow, the lobes bright orange to red. (**264**)

GRAN CANARIA: Local, N. coast in old laurel forest relicts, Los Tiles de Moya, Arucas, cliffs in Pinar de Tamadaba, Rincón de Tenteniguada, etc, 500–1000 m.

S. glabrata Aiton. Dense, much-branched shrub. Leaves entire, subglabrous the margins double-serrate. Sepals rounded. Corolla with dark red to purplish lobes, small. Capsules more or less ovoid. **(83)**

TENERIFE: Las Cañadas de Teide, locally very common and characteristic of the subalpine vegetation, La Fortaleza, Ucanca, Los Roques, El Portillo, Cumbres de Pedro Gil, Vilaflor, Pinar de la Guancha, 1600–2400 m, occasionally found in lower regions where seeds seem to have been washed down from the high mountains; LA PALMA: Pine forest regions of the N. and Cumbre Vieja, sporadic.

S. langeana Bolle (incl. *S. teucrium* Christ?). Perennial herb, sometimes woody at the base. Stems woolly-hairy. Leaves entire, hairy on both surfaces, the margins once or twice dentate. Inflorescence usually unbranched. Corolla lobes reddish-purple or brownish. Capsules subglobose. **(262)**

TENERIFE: N. coast region in laurel woods, Sierra Anaga, Taganana, Agua Garcia, Orotava valley etc, frequent in forest relicts, Icod, Barranco Ruiz, etc, 500–800 m; GOMERA: Forests of El Cedro, Barranquillos de Vallehermoso, Bosque de la Haya; LA PALMA: Laurel forests of the N., Barlovento, Barranco Franceses, Cueva de la Zarza; HIERRO: El Golfo.

S. smithii Hornem. Perennial herb, often woody at the base. Leaves more or less glabrous, the margins irregular-dentate. Inflorescence lax. Corolla green to whitish. **(263)**

A form with glandular hairs beneath the leaves and pale flowers has been separated as a distinct species *S. anagae* Bolle.

TENERIFE: Laurel forests of the Anaga region, locally frequent in shady situations, Vueltas de Taganana, Las Mercedes, El Bailadero, Chinobre, 600–800 m, occasionally in forests of the N. coast.

CAMPYLANTHUS

Shrubs with alternate, linear, succulent leaves. Calyx 5-parted. Corolla 5-lobed, the lobes round, flat, patent; corolla-tube narrow, long curved. Stamens 2, the filaments short. Capsule 2-valved, many seeded.

C. salsoloides Roth. Up to 2 m. Leaves green, fleshy, linear. Inflorescence terminal, lax, often curved. Corolla bluish-purple, pink or white, the tube pubescent. ***Romero marino.*** **(265)**

GRAN CANARIA: Locally frequent in the lower zone, Santa Lucia de Tirajana, Arguineguin, Mogan, Andén Verde, Cuesta de Silva, etc; TENERIFE: Coasts of the Anaga region, Bufadero, N. coast, Barranco Ruiz, Teno, Guimar, Adeje, Masca; GOMERA: Barranco de la Villa, Argaga, rocks and cliffs from 10–600 m.

LYPERIA

Small shrublets. Lower leaves more or less opposite, broadly spathulate; margins crenate. Racemes short. Calyx 5-parted. Corolla ephemeral; tube curved; lobes 5. Stamens 4, included. Capsule dry, 2-valved, the valves bifid. Seeds numerous.

L. canariensis Webb & Berth. Dwarf shrublet. Lower leaves opposite to fasciculate, broadly spathulate to rounded, obtuse, densely glandular-pubescent; margins coarsely toothed. Inflorescences branched. Flowers with long pedicels. Capsule light brown. **(266)**

GRAN CANARIA: Sporadic, cliffs and rocks in the lower zone, Degollada de Tazartico, Andén Verde, Agaete, Juncalillo, Tenteniguada, Arguineguin, etc, 300–700 m.

ISOPLEXIS

Small shrubs with simple, alternate, coriaceous leaves. Flowers in terminal, bracteate racemes. Calyx 5-lobed, shorter than the corolla tube. Corolla 2-lipped, the upper longer than the lower; lower lip with one ventral and 2 lateral more or less triangular lobes. Stamens 4.

1. Leaves lanceolate-ovate, corolla about 3 cm long (Tenerife, Gomera, La Palma) .. .. **I. canariensis**
1. Leaves narrowly lanceolate, corolla less than 2 cm long (Gran Canaria).
   2. Leaves pubescent beneath, inflorescence lax, pedicels more or less equalling calyx .. .. **I. chalcantha**
   2. Leaves more or less glabrous beneath, inflorescence dense, peduncles much shorter than the calyx .. .. **I. isabelliana**

I. canariensis (L.) Loud. Shrub up to 1·5 m. Leaves lanceolate to ovate, glossy, sparsely pubescent beneath, the margins serrate. Inflorescences dense. Corolla bright orange-red. Capsule longer than calyx. ***Cresto de gallo.*** **(260) (261)**

TENERIFE: Locally frequent in the laurel woods of Sierra Anaga, Las Mercedes, Vueltas de Taganana, El Bailadero, Valle de San Andrés, 500–800 m, Los Realejos, Icod, Guia de Isora, etc. Also reported in older literature from GOMERA and LA PALMA but not apparently found recently.

I. isabelliana (Webb & Berth.) Masf. Like *I. canariensis* but with narrowly lanceolate leaves and much smaller, darker reddish flowers. **(259)**

GRAN CANARIA: San Mateo, Cueva Grande, Tamadaba, pine forest cliffs, now rather rare, 800–1000 m.

I. chalcantha Svent. & O'Shan. Differs form the previous species in having densely hairy undersides to the leaves and a lax spike of long-pedicellate, copper-coloured flowers.

GRAN CANARIA: Laurel woods, Los Tiles de Moya, Barranco de la Virgen, forest relict near Galdar, 600–800 m, very rare.

KICKXIA

Perennial herbs or small shrubs. Leaves simple. Flowers solitary, axillary. Calyx of 5 more or less equal segments. Corolla zygomorphic, long-spurred, the palate hairy. Fruits capsulate, dehiscing by 2 spical pores.

1. Petioles short, not twining, plants shrubby, erect or pendant.
   2 Erect, sparsely branched shrubs of dry slopes .. .. **K. scoparia**
   2. Pendant, much-branched cliff plants .. .. .. **K. pendula**
1. Petioles long, at least some twining, scrambling or prostrate herbs.
   3. At least some leaves hastate at base .. .. .. **K. heterophylla**
   3. All leaves more or less attenuate at base .. .. **K. urbanii**

K. heterophylla (Schousb.) Dandy (*K. sagittata* (Poir.) Rothm.). Glabrous perennial, scrambling or prostrate. Leaves linear to oblong, at least some hastate or sagittate at base; petioles usually twining. Corolla yellow, the spur somewhat curved at the tip. Capsule ovoid. Seeds tuberculate. **(268)**

FUERTEVENTURA: La Oliva, Cotillo, Jandia, Moro Jable, Gran Tarajal, etc, locally very common in dry rocky places and in *Euphorbia* scrub; LANZAROTE: Arrieta, Playa Famara, Haria, locally frequent.

K. urbanii (Pitard) Larsen. Like *K. heterophylla* but much more compact and sometimes almost cushion-like. Leaves attenuate at base, never hastate. Seeds densely tuberculate.

GRAN CANARIA: N. coast, Las Palmas to Cuesta de Silva, Punta de Arinaga, locally frequent; TENERIFE: Montaña Roja near El Medano, rare.

K. scoparia (Brouss.) Kunkel & Sunding Erect, sparsely branched perennial. Leaves linear, often caducous, more or less sessile. Corolla yellow, spur more or less straight. **(267)**

TENERIFE: Dry rocky slopes of the W. & S., Barranco del Infierno, Tamaimo, Punta de Teno, Guimar, locally common, 50–600 m; GOMERA: Barranco de la Villa, La Laja, Vallehermoso, Valle de Argaga; GRAN CANARIA: S. region, Tirajana, Santa Lucia, Arguineguin, Termisas, Fataga, locally very common up to 600 m.

K. pendula Kunkel. Like *K. scoparia* but much more densely branched and leafy. Branches arcuate to pendulous. Leaves broader. Corolla spur somewhat shorter.

GRAN CANARIA: Basalt cliffs near Termisas and in Barranco Guayadeque, ca. 600 m, rare.

GLOBULARIACEAE

Shrubs or shrublets. Leaves alternate. Inflorescence a dense head with an involucre of bracts. Flowers hermaphrodite, zygomorphic. Calyx tubular, 5-lobed. Corolla 2-lipped, the upper lip almost absent, the lower with 3 long lobes. Fruit a 1-seeded nut.

GLOBULARIA

Shrubs with alternate, entire leaves. Flowers in globular heads. Calyx tubular, 5-lobed. Corolla blue or whitish with a slender tube, 2-lipped, the upper short, the lower with 3 long lobes. Stamens 4.

1. Inflorescence terminal, heads about 1·5 cm across.
   2. Flowers blue, peduncle long (5–6 cm), leaves small (2 cm), obovate .. .. **G. sarcophylla**
   2. Flowers whitish, peduncle short (1–2 cm), leaves long (5–10 cm), broadly lanceolate .. .. **G. ascanii**
1. Inflorescences axillary, heads less than 1·5 cm across .. **G. salicina**

G. salicina Lam. (*Lytanthus salicinus* (Lam.) Wettst.) Shrub up to 1·5 m. Leaves narrowly to broadly lanceolate, entire, erect to erectopatent. Inflorescences axillary, often crowded towards the tips of stems. Flowers small, less than 1·5 cm, pale blue or whitish. **(84)**

TENERIFE: Frequent on S. slopes in the upper xerophytic zone particularly in the Anaga region, San Andrés to Bailadero, Igueste, Taganana, Punta del Hidalgo, Los Silos, Teno, etc, 200–600 m; GOMERA: N. coast region particularly in the Vallehermoso area, Epina, Hermigua; LA PALMA: N.E. region, Cumbre Nueva, San Andrés y Los Sauces, Barlovento; GRAN CANARIA: Tirajana valley, rare.

G. sarcophylla Svent. Dwarf, pendulous shrub of basalt cliffs. Leaves small, obovate, fleshy, ca. 2 cm long. Flower heads solitary on long terminal peduncles. Flowers blue.

GRAN CANARIA: Mountain cliffs of the Tirajana region, very rare, 1600 m.

G. ascanii Bramwell & Kunkel. Small procumbent shrub resembling *G. sarcophylla* but with larger, lanceolate leaves, short peduncles and pale blue-white flowers. **(269)**

GRAN CANARIA: Tamadaba Massif, cliffs in pine forest zone, 1200 m, very rare.

ACANTHACEAE

Herbs or small shrubs. Leaves opposite, simple. Inflorescence cymose, bracteate. Flowers hermaphrodite, zygomorphic. Corolla bilabiate. Stamens 4 or 2. Ovary superior. Fruit an explosive capsule.

JUSTICIA

Leaves entire. Corolla 2-lipped, the upper lip notched. Stamens 2 with the two locules separated by a narrow membrane.

J. hyssopifolia L. Small shrubs 50–100 cm. Leaves opposite or in whorls of 4–6, entire, oblanceolate to ovate-lanceolate, about 4 cm long; apex obtuse to apiculate. Flowers axillary usually on the upper portions of the branches. Corolla whitish to cream. Calyx 5-lobed, the lobes lanceolate. Fruit about 1·5–2 cm, club-shaped with an apical projection, light brown. **(85)**

TENERIFE: Locally frequent in the lower zone to 500 m. Teno, Adeje, San Juan de la Rambla, Sierra Anaga.

PLANTAGINACEAE ——— Plantain Family

Annual to perennial herbs or small shrubs. Leaves spirally arranged or opposite. Flowers in small heads to spikes, hermaphrodite, regular, usually 4-parted. Perianth green or scarious, the lobes imbricate. Sepals fused at the base. Petals fused, scarious. Anthers large. Ovary superior. Capsule splitting into an upper and a lower portion. Seeds 2-many.

PLANTAGO

Annual or perennial; herbs with a basal rosette or small, much branched woody shrubs. Flowers in dense heads or spikes, regular, subtended by scale-like bracts. Calyx tubular with 4 lobes. Corolla 4-lobed, papery. Fruit a capsule opening by an apical cap. Seeds flattish-concave, black or dark brown.

1. Small shrubs with linear leaves and oblong-oviod heads.
 2. Leaves patent to ascending, pubescent, heads ovoid.
  3. Leaves succulent, linear-lanceolate, flat (Lanzarote) .. **P. famarae**
  3. Leaves linear, more or less filiform .. .. .. **P. arborescens**
 2. Leaves erect, velutinous, heads spherical (Tenerife, Palma, subalpine zone, Gran Canaria) .. .. .. **P. webbii**
1. Herbs with leaves in basal rosette.
 4. Leaves linear, usually entire, spike very slender, 8–15 cm long .. .. **P. asphodeloides**
 4. Leaves more or less lanceolate, margins toothed, spike narrowly cylindrical 1–6 cm long .. .. .. **P. aschersonii**

P. arborescens Poiret. Small shrub up to 60 cm, branches ascending. Leaves fasiculate, densely crowded towards tips of stems, patent or more or less ascending, finely pubescent, the margins ciliate. Peduncles short, 3–5 cm. Heads ovate, few-flowered.

TENERIFE: Common in Sierra Anaga, 400–700 m, N. Coast to Teno, Guimar, sporadic in the S. region; LA PALMA: Cumbrecita, Barranco Las Angustias, S. region nr. Fuencaliente, etc, locally abundant; GOMERA: Rocky places in the forest zone and below, Agulo, Vallehermoso, Epina, etc, frequent; GRAN

CANARIA: San Mateo, Moya, sporadically in rocky places from the lower zone to the mountains.

P. webbii Barn. Like *P. arborescens* but with erect greyish, sericeous to velutinous leaves appressed to the stem and the corolla tube much shorter. **(86)**

TENERIFE: Common in the subalpine zone of Las Cañadas amongst rocks and boulders, La Fortaleza, El Cabezon, Portillo, etc; LA PALMA: locally abundant in the high mountains surrounding the Caldera, Pico de la Cruz, Roque de los Muchachos, 1800 m; GRAN CANARIA: Central region, Cruz Grande-Paso de la Plata, 1400–1600 m, rare.

P. famarae Svent. Like *P. arborescens* but more compact with broader, flat, fleshy leaves. **(270A)**

LANZAROTE: Famara region, coastal cliffs 300–500 m, rather rare.

P. asphodeloides Svent. Small herb with a tuft of silvery, linear basal leaves and long slender inflorescences. **(270)**

TENERIFE: S. coast region, locally frequent, maritime rocks below Candalaria and Guimar, extending S. to below Arico and to El Medano; GRAN CANARIA: S. region, Valle de Ayaguare, Mogan, up to 400 m, rare.

P. aschersonii Bolle. Herb with basal rosette of lanceolate, hairy leaves with toothed margins. Inflorescences cylindrical.

ALL ISLANDS: Common in coastal regions on rocks and along roadsides, Teno, Taganana on TENERIFE, Andén Verde on GRAN CANARIA 0–200 m.

CAPRIFOLIACEAE ———— Elder Family

Shrubs. Leaves opposite. Flowers in a dense, umbel-like cyme. Flowers 5-parted. Corolla with a short tube or campanulate Ovary inferior. Fruits drupaceous.

1. Leaves pinnate .. .. **Sambucus**
1. Leaves simple .. .. **Viburnum**

VIBURNUM

Shrubs or small trees. Leaves simple. Flowers in umbel-like cymes. Corolla more or less campanulate with rounded lobes. Ovary 1-celled, fruit a drupe.

V. rigidum Vent. Shrub up to 5 m. Branches brown to reddish below, erect. Leaves ovate to suborbicular, hairy on both surfaces, acute to acuminate; margins entire. Umbels large, about 10–15 cm across. Corolla white, about 7 mm across. Fruits subglobular or oval, dark brown, purple or black, about 6–7 mm in diameter. ***Follao.*** **(271)**

TENERIFE, LA PALMA, HIERRO, GOMERA, GRAN CANARIA: Frequent in forest zones and *Erica* heaths, common in the Anaga region of Tenerife, 400–1200 m. Probably a marker species of old, natural regions of laurel forest.

SAMBUCUS

Shrubs or small trees. Stems pithy. Leaves pinnate. Flowers in dense umbels. Corolla with short tube and flat limb. Ovary 3- to 5-celled.

S. palmensis Chr. Sm. Shrub up to 5 m. Branching arching; bark corky. Leaves with three pairs of lateral leaflets and a larger terminal one; margins dentate to serrate; lower surface hairy. Umbels about 10 cm across. Corolla white, 5 mm across. Fruit small, about 6 mm, subglobose, brownish-black. ***Sauco.*** **(87)**

TENERIFE: Forests of Anaga region; LA PALMA: Monte de los Sauces, in laurel woods; an extremely rare and endangered species.

DIPSACACEAE —— Scabious Family

Herbs or shrubs with opposite or whorled leaves. Inflorescence usually a head with a calyx-like involucre of bracts. Flowers hermaphrodite, zygomorphic, each surrounded at the base by an epicalyx or 'involucel' of united bracteoles. Calyx cup-shaped, cut into teeth or hairs. Corolla tube often curved, the lobes 4–5, sometimes forming 2 lips. Stamens 4 or 2. Ovary inferior. Fruit 1-seeded, enclosed in the involucel and usually surmounted by the persistent calyx.

PTEROCEPHALUS

Shrubs with hairy stems. Leaves entire, pubescent. Heads 'scabious-like', pink. Calyx short with at least 5 long, feathery bristles. Florets 5-lobed, the outer radiate.

1. Shrubs up to 1 m; leaves oblanceolate to obovate.
   2. Inflorescences 3- to 7-flowered.
      3. Shrubs about 1 m, leaves acute, petals pink (Tenerife, Gran Canaria) .. .. **P. dumetorum**
      3. Shrub 30–50 cm, leaves mucronulate, petals magenta (La Palma) .. .. **P. porphyranthus**
   2. Inflorescences usually single flowered (Tenerife, subalpine) .. .. **P. lasiospermus**
1. Woody-based cliff plant, leaves oblong (Tenerife, Anaga region) .. .. **P. virens**

P. dumetorum (Brouss.) Coult. Shrub up to 1·5 m. Stems brownish. Leaves crowded towards tips of stems, dull green, broadly lanceolate to ovate, acute, pubescent on both surfaces. Inflorescences 3- to 7-flowered. Petals pink. **(272)**

TENERIFE: Ladera de Guimar, locally abundant, 300–600 m; GRAN CANARIA: Abundant in mountain areas, up to 1600 m at Roque Nublo.

P. lasiospermus Link. Up to 1 m. Stems greyish. Leaves pale greyish-green, narrowly oblanceolate, acute, pubescent. Flowers borne singly on long peduncles. Petals pale pink. **(88)**

TENERIFE: Subalpine zone of Las Cañadas 2000–2200 m, locally common, Montaña de Deigo Hernandez, La Fortaleza, Parador Naciónal.

P. porphyranthus Svent. Differs from *P. lasiospermus* by its broader, mucronulate leaves and deep magenta pink flowers.

LA PALMA: Confined to high mountain regions and upper pine forest zone, up to 1900 m, very rare.

P. virens Berthelot. Dwarf cliff plant with woody rootstock and dense rosette of leaves. Leaves oblong, dark green, rounded at tip. Flowers borne singly on short peduncles. Petals pink.

TENERIFE: Confined to coastal rocks between sea-level and 200 m on the N. coast of Anaga between Bajamar and Roques de Anaga, rare.

CAMPANULACEAE —— Bellflower Family

Herbs. Leaves simple. Calyx tube fused to ovary; teeth or lobes 5 small.

Corolla inserted within the calyx-lobes, bell-shaped or tubular. Stamens 5. Ovary inferior. Style lobed. Fruit a fleshy berry or a capsule.

1. Small annuals, flowers blue or white .. .. .. **Laurentia**
1. Robust perennial herbs, flowers orange-red .. .. **Canarina**

CANARINA

Glabrous, glaucous, scrambling herb with thick tuberous root. Stems renewed each year from the tuber, hollow, latex present. Leaves opposite, petiolate. Flowers axillary, solitary, bell-shaped, orange.

C. canariensis (L.) Vatke. Stems up to 3 m. Leaves triangular, hastate, the margins dentate. Corolla 3–6 cm long, orange (darkening when dried). Fruit an ovate, fleshy berry, reddish to black when ripe, edible. ***Bicacaro.*** **(273)**

TENERIFE: Frequent in laurel forests and forest margins, Anaga region, N. coast from Orotava to Los Silos 300–1000 m, local in the S. of the island; GRAN CANARIA: Los Tiles de Moya, frequent in the laurel woods, very depleted in other localities near San Mateo, Santa Brigida and Teror; LA PALMA, GOMERA: Rare in the forest regions.

LAURENTIA

Small, annual herbs up to 10 cm, found in wet places. Leaves oblanceolate, subcrenate. Flowers blue or white, on long pedicels. Calyx tube campanulate; lobes long and narrow.

L. canariensis DC. Rare plant of wet areas below cliffs.

TENERIFE: Valle de Masca; GRAN CANARIA: Tenteniguada.

Two small non-endemic annuals, *Campanula erinus* and *Wahlenbergia lobelioides* are common in dry areas of the lower zone.

COMPOSITAE ——————— Daisy Family

Flowers (florets) small, aggregated into heads simulating a single, larger flower surrounded by a calyx-like involucre of 1 – several rows of bracts. Receptacle of the head expanded, flat, concave, convex or conical, with or without receptacular scales subtending the florets. Florets all similar or the outer ones differing. Calyx represented by bristles, hairs or scales (pappus), sometimes absent. Corolla tubular (disc florets) or with a strap-like ligule (ray florets). Ovary inferior with a single ovule. Style bifid. Fruit a hard achene (cypsela) crowned by the pappus, sometimes with a slender 'beak' between pappus and seed.

A very large and variable family, *Lactucosonchus*, *Heywoodiella*, *Sventenia*, *Vieraea*, *Allagopappus*, *Schizogyne*, *Gonospermum* and *Lugoa* are endemic to the Canary Islands and *Argyranthemum* occurs only in the Canaries and on Madeira.

1. Disc florets present, ray florets present or absent, plants without milky latex.
 2. Disc-florets only present.
  3. Leaves not spiny.
   4. Pappus absent or nearly so.
    5. Receptacle without scales, involucral bracts with a papery margin.. .. .. **Artemisia**
    5. Receptacle with short scales, involucral bracts without a papery margin .. .. **Gonospermum**
   4. Pappus present.

6. Leaves sticky, heads in a dense, flat corymb .. **Allagopappus**
6. Leaves not sticky, heads not densely flat-corymbose.
7. Receptacle without scales.
8. Pappus consisting of hairs and a crown of scales (shrub of coastal regions) .. .. **Schizogyne**
8. Pappus of simple hairs.
9. Involucral bracts very large, papery, shining, heads in a dense inflorescence .. .. .. **Helichrysum**
9. Involucral bracts smaller, only the margin papery, peduncles usually with a single, terminal head .. **Phagnalon**
7. Receptacle with scales.
10. Outer involucral bracts with a distinct, dentate or digitate terminal appendage, shrubs .. .. **Centaurea**
10. Outer involucral bracts with a wide, expanded papery margin, robust herbs .. .. .. **Rhaponticum**
3. Leaves spiny.
11. Stems with spiny wings, flowers purplish or whitish.
12. Receptacle without scales, pappus hairs feathery .. **Onopordum**
12. Receptacle with scales, pappus hairs simple .. **Carduus**
11. Stems without spiny wings.
13. Flowers purplish .. .. **Carduus**
13. Flowers yellow .. .. **Carlina**
2. Disc- and ray-florets present.
14. Leaves spiny (dwarf shrublets of exposed coasts) .. **Atractylis**
14. Leaves not spiny.
15. Pappus of scales, bristles or hairs.
16. Pappus of simple hairs, involucral bracts in a single row .. .. **Senecio**
16. Pappus of hairs and scales, involucral bracts in several rows.
17. Leaves glabrous, succulent (W. Tenerife) .. **Vieraea**
17. Leaves densely white-hairy, not succulent (E. islands) .. .. **Pulicaria**
15. Pappus absent or reduced to a small, narrow crown.
18. Herbs with leaves in a basal rosette (Tenerife, N. coast of Anaga) .. .. **Lugoa**
18. Shrubs.
19. Cypselas all similar.
20. Ray-florets yellow, outer involucral bracts leafy **Asteriscus**
20. Ray-florets white, outer involucral bracts scale-like .. .. **Tanacetum**
19. Outer cypselas large, 3-ribbed or winged, inner small .. .. **Argyranthemum**
1. Ray-florets only present, plants with milky latex.
21. Receptacle with scales or long hairs between the florets, involucral bracts with dense, whitish-glandular or black hairs.
22. Involucral bracts densely white-glandular tomentose **Andryala**
22. Involucral bracts with black hairs .. .. .. **Heywoodiella**
21. Receptacle without scales or long hairs between the florets, involucral bracts not densely white-glandular or with black hairs.
23. Cypselas grooved.

24. Leaves entire, white-waxy, inflorescence with rusty-glandular-hairs (W. Gran Canaria) .. .. .. **Sventenia**
24. Leaves pinnatisect, green, inflorescence more or less glabrous (La Palma) .. .. **Lactucosonchus**
23. Cypselas ribbed or transversely rugose.
25. Stems spiny .. .. **Launaea**
25. Stems without spines.
26. Cypselas beaked.
27. Cypselas cylindrical, pappus of many rows of hairs **Crepis**
27. Cypselas flattened, pappus of 2 rows of hairs .. **Lactuca**
26. Cypselas not beaked.
28. Pappus of a few long bristles with scales at their base .. .. **Tolpis**
28. Pappus of hairs or sometimes hairs and bristles.
29. Heads slender with about 6 florets, pappus of 1 or 2 rows of simple hairs .. .. .. **Prenanthes**
29. Heads with at least 8 florets and usually many more, pappus of more than 2 rows of hairs or with hairs and bristles.
30. Involucral bracts with a broad scarious margin, pappus of many rows of soft hairs .. .. **Reichardia**
30. Involucral bracts without a broad scarious margin, pappus with soft, silky fasciculate hairs and stiffer deciduous bristles .. .. .. **Sonchus**

PULICARIA

Heads with ray and disc florets, solitary, involucral bracts in several rows. Receptacle without scales. Pappus a row of fused scales forming a toothed crown and an inner row of simple hairs. Cypselas not beaked.

1. Leaves linear to narrowly oblanceolate, heads about 5–8 mm across .. .. **P. burchardii**
1. Leaves broadly oblanceolate, heads up to 3 cm across .. **P. canariensis**

P. canariensis Bolle. Small subshrub, woody at the base. Leaves broadly oblanceolate, densely woolly; margins entire or remotely sinuate-dentate. Heads large, up to 3 cm across. Outer involucral bracts oblanceolate, densely hairy, the inner linear. Ray florets about 7 mm long, golden yellow. **(316)**

FUERTEVENTURA: Jandia region on mountain cliffs, rare; LANZAROTE: Famara region, Montaña de los Helechos, Playa de Famara, coastal rocks, rare.

P. burchardii Hutch. Prostrate, silvery-farinose shrub. Leaves entire, linear to narrowly oblanceolate, obtuse or rounded. Heads small, about 5–8 mm across, solitary at ends of lateral branches. Involucral bracts linear. Ray florets very short, pale yellow.

FUERTEVENTURA: Confined to the coastal region of the S. tip of Jandia, near sea-level, sporadic.

SCHIZOGYNE

Like *Pulicaria* but ray-florets completely absent and the pappus with an entire, more or less unlobed ring of scales and a row of simple hairs. Heads in groups of 5–6.

1. Plants densely white-sericeus, leaves 3–5 cm long, flat .. **S. sericea**
1. Plants glabrous, green, leaves succulent, about 2 cm long, filiform .. .. **S. glaberrima**

S. sericea (L. fil.) Sch. Bip. Shrub up to 1·5 m. Stems white-sericeus. Leaves linear, 3–5 cm long, flat, obtuse, greyish white. Inflorescences dense. Heads 5–6 mm long and broad. Involucral bracts scarious, lanceolate, the outer small, triangular. Ray florets absent; disc florets yellow. ***Salado, dama.*** **(92)**

ALL ISLANDS: Locally very common on rocks in coastal regions, halophyte; N. coast of TENERIFE, Teno to Anaga; GRAN CANARIA: S.E. near Arinaga, N. coast.

S. glaberrima DC. Like *S. sericea* but very dense with linear, succulent, glabrous leaves and very compact inflorescences with pale yellow heads.

GRAN CANARIA: Confined to the coastal regions of the island, locally common Maspalomas, Cuesta de Silva, coastal rocks and dunes.

ASTERISCUS

Flower heads usually solitary. Disc and ray florets present. Outer involucral bracts leaflike, obtuse. Receptacle with scales. Pappus of very small, inconspicuous scales.

1. Leaves very densely sericeus with long silky hairs.
    2. Leaves linear, heads about 2 cm across (Gran Canaria) .. **A. stenophyllus**
    2. Leaves oblanceolate or spathulate, heads 2·5–3·5 cm across.
        3. Stems silvery, leaves oblanceolate to spathulate (Lanzarote) .. .. **A. intermedius**
        3. Stems dark grey to blackish, leaves broadly oblanceolate (Fuerteventura) .. .. **A. sericeus**
1. Leaves shortly pubescent to subglabrous.
    4. Leaves succulent, spathulate, margins ciliate, remotely toothed, ray florets pale yellow to whitish (Fuerteventura, Lanzarote) .. .. **A. schultzii**
    4. Leaves not succulent, linear to lanceolate, margins of lower leaves rarely toothed, upper entire, ray florets golden yellow (Gran Canaria) .. .. **A. odorus**

A. sericeus L. fil. Tough, strongly scented shrub. Older stems dark grey to blackish. Leaves crowded towards apex of stems, 3–5 cm, broadly oblanceolate, very densely silvery-sericeus on both surfaces. Flower heads 3–3·5 cm across. Ray florets yellow, with 3 teeth at tip of the ligule. **(274)**

FUERTEVENTURA: Locally common in rocky and mountain areas, Jandia, Betancuria, N. coast area, sea-level to 700 m.

A. intermedius Webb. Like *A. sericeus* but the stems silvery, the rather narrower and longer leaves not crowded towards the apex of stems. **(91)**

LANZAROTE: N. region on the Famara Massif and towards the coast of Jameo del Agua, 100–700 m, locally frequent.

A. stenophyllus Link. Small, densely branched shrublet 15–40 cm. Stems greyish. Leaves very densely clothing stem, linear, densely grey-hairy. Heads about 2 cm across. Ray florets pale yellow to golden. **(275)**

GRAN CANARIA: S. region on dry rocky slopes, Tirajana to Mogan, 50–500 m, locally common.

A. odorus Sch. Bip. Like *A. stenophyllus* but more lax and more or less dichotomously branched. Leaves broader, linear to narrowly lanceolate, densely and shortly pubescent, the lower occasionally very remotely toothed. Flower heads small, about 1·5 cm.

GRAN CANARIA: Common particularly on the dry east and south slopes, 50–600 m.

A. schultzii Bolle. Dwarf shrub. Leaves succulent, spathulate, sparsely hairy, the margins toothed and ciliate. Heads about 2–2·5 cm across. Ray florets pale yellow-white.

FUERTEVENTURA: N. coast between Toston and Corralejo 20–100 m, rare; LANZAROTE: Coastal rocks at Playa de Famara, 10 m, very local.

ALLAGOPAPPUS

Shrubs with glandular-sticky leaves. Heads less than 0·5 cm across, in terminal corymbs, homogamous without ray florets. Disc florets mostly hermaphrodite. Involucral bracts in several series, imbricate, narrow, the outer shorter than the inner which have subdentate, ciliate margins. Cypselas scarcely ribbed.

1. Leaves linear, margins entire .. .. **A. viscosissimus**
1. Leaves narrowly lanceolate to oblong, margins toothed .. **A. dichotomus**

A. dichotomus (L. fil.) Cass. Shrub up to 1·25 m. Stems rusty-brown. Leaves 2–3 cm long, narrowly lanceolate to oblong; margins toothed; apex obtuse to mucronate. Heads in dense, flattish corymbs, sometimes with small secondary corymbs projecting above the main one. Florets yellow. **(89)**

TENERIFE, LA PALMA, GOMERA, GRAN CANARIA: Rocky slopes, cliffs and lava flows in the lower zone, sea-level to 600 m, locally abundant particularly in the S. of TENERIFE.

A. viscosissimus Bolle. Like *A. dichotomus* but with very narrow, linear leaves with entire margins.

GRAN CANARIA: Locally frequent, N. region, Agaete, S. coasts, Arguineguin, Puerto de Mogan, Caldera de Tirajana.

VIERAEA

Like *Allagopappus* but leaves succulent; heads much larger (2 cm across); ray-florets present.

V. laevigata Webb & Berth. Shrub up to 1 m. Stems greyish. Leaves up to 5 cm, fleshy, ovate to ovate-lanceolate, strongly toothed towards tip, light green or glaucous. Inflorescence of 5–10 heads. Pedicels bracteose. Flowers yellow. ***Amargosa.*** **(277)**

TENERIFE: Known only from the W. mountain regions bounded by Los Silos Teno and Masca, on basalt cliffs, 50–300 m, rare.

PHAGNALON

Dwarf shrubs, leaves simple, entire. Involucral bracts imbricate, in up to 5 rows. Receptacle without scales. Ray florets absent. Cypselas unbeaked. Pappus hairs in 1 or 2 rows, simple.

1. Leaves linear, strongly revolute, heads not dense umbels.. **P. purpurascens**
1. Leaves narrowly lanceolate, flat, heads in dense umbels .. **P. umbelliforme**

NOTE: the Mediterranean species *P. rupestre* with flat, oblanceolate dentate leaves and *P. saxatile* with an undulate appendage on the outer involucral bracts are also locally frequent in the Canaries.

P. purpurascens Sch. Bip. Dwarf shrub up to 40 cm. Stems white-lanate. Leaves 2–3 cm, linear, strongly revolute, densely lanate beneath. Heads on long, slender peduncles, solitary. Involucral bracts slender, the inner deep purple. ***Mecha.***

ALL ISLANDS: Locally common up to 500 m particularly on the dry S. slopes of TENERIFE, Guimar to Masca, and GRAN CANARIA, Arguineguin, Mogan.

P. umbelliforme Sch. Bip. Erect shrub up to 75 cm. Stems yellowish-lanate.

Leaves up to 5 cm, linear to narrowly lanceolate, not strongly revolute. Heads small, in dense umbels of up to 40.

HIERRO: El Golfo region near Sabinosa, 200 m, frequent on N. facing slopes; LA PALMA: Frequent in the dry volcanic area from Fuencaliente to the S. coast, locally very common, 200 m; TENERIFE: Ladera de Guimar, dry S. slopes, now rather rare; GRAN CANARIA: Barranco de Guiniguada, possibly an escape from the Jardin Canario where it is cultivated.

HELICHRYSUM

Heads bell-shaped. Involucral bracts white, papery, overlapping. Ray florets absent. Receptacle without scales.

H. gossypium Webb. White, woolly, cliff plant with woody stems up to 30 cm. Leaves lanceolate or spathulate, entire, obtuse. Inflorescences dense, corymbose. Heads about 6 mm long, with many large, papery bracts. Florets brownish. **(90)**
LANZAROTE: Endemic to the Famara region on dry mountain cliffs, 300–600 m. Riscos de Famara, Montaña de los Helechos. Rare.

H. monogynum Burtt & Sunding. A recently discovered red-flowered species similar in habit to *H. gossypium*, is also endemic to LANZAROTE in the Famara region, Peñitas de Chache. **(276)**

ARTEMISIA

Small aromatic shrubs. Leaves alternate, sometimes fasciculate. Heads small, in paniculate or racemose inflorescences. Involucral bracts in few rows. Florets all tubular. Pappus absent.

1. Leaves fasciculate, sessile, very short (0·5 cm) .. .. **A. reptans**
1. Leaves alternate, petiolate, at least 1·5 cm long or more.
   2. Capitula ovoid, leaves with linear, cylindrical lobes .. **A. ramosa**
   2. Capitula globose, leaves with flat lobes .. .. .. **A. canariensis**

A. reptans Chr. Sm. Small shrub 12–30 cm. Leaves fasciculate, small, 0·5 cm, simple to 3- to 5-sect, greyish tomentose. Heads yellow, in bracteate racemes or panicles, shortly pedunculate. Involucral bracts greenish, pubescent, the margins scarious.

GRAN CANARIA: Coastal regions from the N., Barranco de Bañadero, to the S.E., Punta Melenara, up to 50 m, locally common.

A. ramosa Chr. Sm. Aromatic greyish tomentose perennial up to 60 cm. Base woody. Branches usually erect. Leaves up to 4 cm, 2-pinnatisect with more or less linear lobes, long petiolate. Heads pale yellow, very small (4 mm long), ovoid, in lax to dense panicles. Involucral bracts densely tomentose with wide scarious margins.

GRAN CANARIA: Dry slopes in the deep valleys of the southern region, Arguineguin, Mogan, etc, 100–600 m, locally abundant; TENERIFE: Los Cristianos.

A. canariensis Less. Grey, much branched shrub up to 1 m. Leaves variable, about 3–7 cm long, 2-pinnatisect; lobes flat, linear to oblanceolate, obtuse. Inflorescences dense, elongate. Heads golden to brownish-yellow, about 4 mm across. Involucral bracts tomentose, the inner with scarious margins. ***Incienso.***
ALL ISLANDS: Frequent shrub in dry lower zone, 50–700 m.

GONOSPERMUM

Shrubs. Leaves pinnatisect to pinnatifid. Flowers in dense yellow corymbs. Involucral bracts linear, usually without a white papery margin. Receptacle with a scale at the base of each floret. Ligulate florets absent. Pappus reduced to inconspicuous scales. Cypselas small, grooved.

1. Heads 4–5·5 mm across.
 2. Leaves pinnatifid; segments with serrate, obtuse lobes; heads at least 5 mm across .. .. **G. fruticosum**
 2. Leaves pinnatisect; segments scarcely lobed; heads 4–4·5 mm across .. .. **G. gomerae**
1. Heads 3 mm or less.
 3. Leaves 2-pinnatifid, ultimate lobes dentate, rounded .. **G. elegans**
 3. Leaves pinnatifid; lobes serrate, acute or obtuse .. **G. canariense**

G. fruticosum Less. Small shrub up to 1·5 m. Stems brown. Leaves sparsely hairy with woolly hairs, pinnatifid; segments with serrate, obtuse lobes. Inflorescence a flattish, dense corymb. Heads subglobose, about 5 mm across. ***Corona de la Reina, Faro.***

TENERIFE: Frequent in the lower and laurel forest zones especially on cliffs 100–700 m, Ladera de Guimar, Taganana, Teno.

G. gomerae Bolle. Differs from *G. fruticosum* by its less divided leaves with broad, irregularly toothed segments and narrower heads. **(94)**

GOMERA: Rare, on cliffs along the N. coast between Agulo and Vallehermoso 300–500 m and sporadically in other parts of the N. and C. regions of the island.

C. elegans Cass. Shrub up to 2·5 m. Stems greyish. Leaves with very short hairs, 2-pinnatifid; the tips of the ultimate lobes rounded to apiculate. Inflorescence a dense corymb of narrow (ca. 3 mm) heads.

HIERRO: Cliffs and rocks in the lower zone and forest regions between 50 and 700 m, locally frequent in the region of El Golfo between Frontera and Sabinosa.

C. canariense Less. Distinguishable from *G. elegans* by its less divided leaves and very narrow heads (2 mm). ***Faro.*** **(278)**

LA PALMA: Common in the lower and forest zones (200–1000 m) of the N. part of the island, the village of Roque del Faro on the N. side is so-called because of the local abundance of this species.

LUGOA

Like *Gonospermum* but woody only at the base; involucral bracts with papery margins; outer florets with long white ligules.

L. revoluta DC. Perennial herb with woody stock. Leaves pinnatifid, hairy, the margins revolute. Inflorescence few-flowered. Heads up to 2 cm across. **(93)**

TENERIFE: coastal rocks between Punta del Hidalgo and Taganana (up to 300 m), rare.

ARGYRANTHEMUM ——— (Chrysanthemum)

Shrubs, 40–150 cm. Leaves more or less entire or dissected. Inflorescence lax, corymbose. Heads with ray and disc florets. Cypselas heteromorphic, the outer trigonous with 1–4 wings or turbinate, sometimes fused into groups of 2–9; inner obconical, terete, 4-angled or laterally compressed, with 1–2 wings or wingless. Pappus a marginal ridge, crown-like to absent.

1. Outer cypselas wingless, pappus a marginal ridge or absent.
 2. Stems and leaves hispid to tomentose .. .. .. **A. canariense**

2 Stems and leaves glabrous.
3. Outer cypselas not fused into groups .. .. .. **A. broussonetii**
3. Outer cypselas fused into groups of 2–8.
4. Leaves sessile .. .. **A. canariense**
4. Leaves petiolate.
5. Leaves oblong, segments filiform, stems slender, heads small (S. Gran Canaria) .. .. .. **A. filifolium**
5. Leaves obovate, segments more or less flat, stems robust, heads at least 3 cm.
6. Primary leaf-lobes obtriangular in outline, narrowing to the base, cypselas usually tinged with purple (S.W. & W. Gran Canaria) .. .. .. **A. escarrei**
6. Primary leaf-lobes ovate to obovate in outline, cypselas creamish (Tenerife, Gomera) .. .. **A. broussonetii**
1. Outer cypselas with 1–4 wings, pappus crown-like.
7. Leaves more or less entire or with broad, flat lobes.
8. Ray-florets pale yellow (Lanzarote) .. .. .. **A. ochroleucum**
8. Ray-florets white or cream (N.W. Tenerife) .. .. **A. coronopifolium**
7. Leaves deeply lobed.
9. Leaf-lobes filiform to linear-lanceolate, less than 3 mm wide.
10. Leaves 2- to 3-fid, stems slender (S. Tenerife) .. **A. gracile**
10. Leaves 4- to 6-fid, stems robust.
11. Inflorescences 1-flowered, leaves glaucous, crowded (S.W. Tenerife) .. .. **A. foeniculaceum**
11. Inflorescence corymbose, leaves green, generally not crowded.
12. Leaves with 2–6 primary lobes .. .. .. **A. frutescens**
12. Leaves with 6–14 primary lobes.
13. Heads at least 1·1 cm across (La Palma) .. **A. haouarytheum**
13. Heads usually less than 1 cm across (Hierro) .. **A. sventenii**
9. Leaf-lobes flat, more or less ovate, up to 8 mm wide.
14. Outer cypselas fused into groups of 2–5.
15. Leaves with 2–5 primary lobes (Hierro) .. .. **A. hierrense**
15. Leaves with 6–14 primary lobes (Gomera).. .. **A. callichrysum**
14. Outer cypselas not fused.
16. Inner involucral bracts triangular-lanceolate (Tenerife, subalpine) .. .. **A. teneriffae**
16. Inner involucral bracts obovate-spathulate with expanded papery apices.
17. Leaves with 2–6 primary lobes.
18. Peduncles with dense hairs towards the apex (Tenerife, N. part of Sierra Anaga) .. .. **A. lemsii**
18. Peduncles glabrous.. .. .. **A. frutescens**
17. Leaves with 6–18 primary lobes.
19. Heads up to 2 cm across (Fuerteventura, Jandia) **A. winteri**
19. Heads about 1 cm across.
20. Leaf-lobes acute, ray-florets yellow or white (Gomera).. .. .. **A. callichrysum**
20. Leaf-lobes obtuse, ray-florets white (W. Gran Canaria) .. .. **A. lidii**

A. frutescens (L.) Webb ex Sch. Bip. Globose shrubs to creeping prostrate mats, 20–80 cm. Leaves up to 8 cm, linear-lanceolate to obovate, 1- to 2-pinnatisect, petiolate, coriaceous to succulent; lower surface sometimes slightly hispidulous; lobes linear-lanceolate to lanceolate, obtuse or acuminate. Inflorescence corymbose. Heads 4–30, up to 2 cm across. Ray-florets white. Outer cypselas 3-winged, the two lateral wings expanded, the ventral small. Inner cypselas laterally compressed, 1-winged. Pappus an irregularly dentate crown. **(95) (288) (289)**

TENERIFE, GRAN CANARIA, LA PALMA, GOMERA: Widespread species of coastal regions, confined to the N. on GRAN CANARIA and very rare on LA PALMA, sea-level to 700 m.

A. lemsii Humphries. Like *A. frutescens* but with broad, obspathulate leaf-lobes and hairy peduncles.

TENERIFE: S. slopes of Sierra Anaga, Chamorga, Igueste, 400 m.

A. sventenii Humphries & Aldridge. Like *A. frutescens* but usually taller, with very narrow leaf-lobes and small heads.

HIERRO: Dry S. slopes of the island near Taibique, rare.

A. gracile Webb. Like *A. frutescens* but very slender with filiform, trilobed leaves and broad corymbose inflorescences with small heads. **(96) (279)**

TENERIFE: S. coast region, Guimar, Granadilla, Arico, El Medano, Adeje, Guia de Isora, Tamaimo, etc, locally very common up to 600 m.

A. haouarytheum Humphries & Bramwell. Slender shrubs up to 1 m. Leaves up to 16 cm, obovate, 2-pinnatisect, petiolate, glabrous; lobes lanceolate, acuminate, coriaceous or occasionally fleshy. Inflorescence corymbose with up to 50 heads. Heads about 1·5 cm across. Ray-florets white. **(282)**

LA PALMA: Locally common in pine forests of the W. side, Tijarafe, La Cumbrecita, Barranco de las Angustias, Fuencaliente, frequent in coastal regions of the dry zone below Fuencaliente, 100–1500 m.

A. winteri (Svent.) Humphries. Like *A. haouarytheum* but smaller, compact with shorter leaves and broad leaf-lobes. **(287)**

FUERTEVENTURA: Confined to the summit region of Pico de la Zarza in the Jandia region, rare.

A. lidii Humphries. Like *A. winteri* but scarcely shrubby and with few-flowered corymbs; leaf-lobes obtuse. **(286)**

GRAN CANARIA: W. coast region near Andén Verde, 200–400 m, very local and rare.

A. callichrysum (Svent.) Humphries. Resembling *A. haouarytheum* but with much larger leaves and usually yellow or occasionally cream ray-florets.

GOMERA: Central forest region, El Cedro, Hermigua Valley, Roque Cano, Roque de Agando, 500–1200 m, locally frequent.

A. foeniculaceum Webb ex Sch. Bip. Dense candelabra-shrub up to 1 m. Leaves densely crowded 2- to 3-pinnatisect, glabrous, grey-glaucous; lobes acute. Inflorescences 1- to few-headed. Heads up to 2 cm across. Ray-florets white. **(281)**

TENERIFE: S. and S.W. regions, usually on dry cliffs, Masca, Tamaimo, Arafo, Adeje, etc, 200–1100 m, a few places in the high mountains above Vilaflor, Cumbre de Pedro Gil, 1800 m, locally abundant.

A. teneriffae Humphries (*A. anethifolium* Webb non Brouss.) Small shrublet up to 50 cm, branching from the base, more or less globose. Leaves up to 6 cm, oblong to obovate, 1- to 2-pinnatisect, hispidulous; Petiole long, thick, parallel-sided. Inflorescence a lax corymb with up to 12 heads. Heads 7–15 mm across. Ray-florets white. Inner cypselas wingless or with a small ventral wing. Pappus reduced to a marginal ridge. **(284)**

TENERIFE: Subalpine zone of Las Cañadas, locally very common, Cañada del Portillo, Montaña Blanca, Parador Nacional, etc, 1900–2300 m.

A. ochroleucum Webb ex Sch. Bip. Small, slender shrub, 40–70 cm. Leaves up to 8 cm, spathulate to obovate, pinnatifid, more or less sessile, cuneate at the base, glabrous, glaucous. Inflorescence subcorymbose, the peduncles usually short. Heads up to 1·5 cm across. Ray-florets pale yellow. **(285)**

LANZAROTE: Famara region on coastal cliffs, Peñitas de Chache, Riscos de Famara, Haria, etc,. also occasionally found in inland rocky areas between Arrecife and the Famara Massif, 500–650 m.

A. coronopifolium (Willd.) Webb ex Sch. Bip. Prostrate, procumbent or ascending shrubs, 40–120 cm. Leaves up to 9 cm, more or less obovate, 1- to 2-pinnatifid, glabrous, fleshy. Inflorescence with 1–8 heads. Heads up to 2·2 cm. Ray-florets white. Outer cypselas trigonous, 1- to 4-winged. Inner cypselas laterally compressed, usually 2-winged.

TENERIFE: N. coast on wet, N. facing cliffs in the Teno and Buenavista region up to 200 m; once common in the Bajamar area but now very rare.

A. broussonetii (Pers.) Humphries. Dense, robust shrub up to 120 cm. Leaves large, up to 16 cm, obovate-elliptical, 2-pinnatifid or rarely 2-pinnatisect, glabrous or sometimes sparsely hairy on the midrib. Inflorescence corymbose with up to 30 heads. Heads 1·2–2·2 cm across. Ray-florets white. Outer cypselas strongly 3-winged or wingless. Inner cypselas obconical, usually 2-winged. Pappus crown-like. **(280)**

TENERIFE: Open clearings in laurel forest areas, Sierra de Anaga, Las Mercedes, Cumbres de Taganana, El Bailadero, Orotava Valley, Agua Mansa, Icod el Alto, 600–1200 m; GOMERA: Cumbre de Carboneros, El Cedro, locally abundant. A very spectacular species which is often cultivated as an ornamental in parks and gardens.

A. webbii Sch. Bip. Like *A. broussonetii* but smaller, with petiolate leaves, fewer heads and the cypselas much smaller with a marginal pappus.

LA PALMA: Laurel forest regions of the E. coast, Barranco del Rio, La Galga, Los Tilos, etc, locally frequent from 500–800 m.

A. hierrense Humphries. Small shrub up to 80 cm. Leaves ovate, 2- to 3-pinnatisect, glabrous, petiolate. Inflorescence densely corymbose with up to 50 heads. Heads generally small, about 1 cm across. Ray-florets usually white.

HIERRO: Locally abundant in the El Golfo region, coastal cliffs to the laurel forests, 100–650 m. Las Casitas, Cuestas de Sabinosa, Frontera, etc.

A. filifolium (Sch. Bip.) Humphries. Up to 80 cm, forming slender, wiry shrubs branching only at the base. Leaves up to 11 cm, 1- to 2-pinnatisect with few filiform lobes, glabrous. Inflorescence corymbose with numerous very small heads. Heads usually less than 1 cm across. Ray-florets white. Outer cypselas usually fused into groups of 2–8, the wings very small or absent. Inner cypselas obconical, small, often fused to the outer. Pappus more or less absent.

GRAN CANARIA: Locally dominant in the dry, scrub vegetation of the S. of the island, Fataga, Mogan, Arguineguin, San Augustin, Maspalomas, etc, up to 500 m.

A. escarrei (Svent.) Humphries. Like *A. filifolium* but smaller, the leaf-lobes broader, cuneate and the florets much larger. **(283)**

GRAN CANARIA: W. and S.W. regions, mountain regions near Tazarte, Degollada de Tazartico, Aldea de San Nicolas, Tirma, Andén Verde, locally abundant.

A. canariense (Sch. Bip.) Humphries. Very variable shrub up to 1 m. Leaves up to 8 cm, obovate to rhomboidal, 2- to 3-pinnatifid or pinnatisect, sessile, subglabrous to densely tomentose. Inflorescence corymbose with 5–20 heads. Heads about 1 cm across. Ray-florets white. Outer cypselas trigonous, usually fused into groups, more or less wingless. The inner cypselas obconical, usually sterile. Pappus usually absent.

GRAN CANARIA: Very common in the mountain regions and pine forests of the C. part of the island, Cumbres de San Mateo, Cruz de Tejeda, Paso de la Plata, Tamadaba, Artenara, etc, 350–1950 m; TENERIFE: Pine forest zone of the C. region, Pinar de la Espèranza, Fuente de Joco, Vilaflor, Agua Mansa, Los Organos, 1200–1900 m, locally frequent, outer area of Las Canadas, Cumbres de Pedro Gil, Izana, etc; HIERRO: Forest regions of Riscos de Jinamar, 800 m, rare. A broad-leaved form from the N. coast of GRAN CANARIA is sometimes separated as *A. jacobifolium* Bornm.

TANACETUM

Like *Argyranthemum* but the heads very small, with few ray florets, the cypselas all similar, obconical with 5–10 ribs and the pappus a membranous rim.

1. Leaves 1- to 2-pinnatisect, green .. .. **T. ferulaceum**
1. Leaves at least 2- to 3-pinnatisect, silvery-grey .. .. **T. ptarmaciflorum**

T. ferulaceum (Webb & Berth.) Sch. Bip. Shrub up to 50 cm. Leaves 1- to 2-pinnatisect, subglabrous or pubescent, green. Heads in very dense corymbs. Involucral bracts sparsely hairy. Ray-florets white. **(291)**

GRAN CANARIA: S. and S.W. regions, in the lower zone, Caldera de Tirajana, Santa Lucia, Termisas, 300–600 m, Riscos de Guayedra, locally frequent.

T. ptarmaciflorum (Webb & Berth.) Sch. Bip. Like *T. ferulaceum* but the leaves at least 2- to 3-pinnatisect, very densely silver-grey tomentose. Heads in dense corymbs. Involucral bracts densely hairy. **(97) (290)**

GRAN CANARIA: S. region in the high mountain zone, Risco Blanco de Tirajana, Paso de la Plata, very rare.

SENECIO

Perennial herbs or shrubs with rhizomes. Basal leaves orbicular, deltate or broadly lanceolate; or with linear to lanceolate, succulent stem leaves. Inflorescences corymbose. Flower heads usually with ray and disc florets. Involucral bracts in a single row. Cypselas ribbed. Pappus white, caducous.

1. Flowers pink, white or mauve; stems and leaves not succulent.
 2. Shrubs with densely white-felted stems .. .. .. **S. appendiculatus**
 2. Perennial herbs with glabrous or pubescent stems.
  3. Inflorescence of few (1–4) heads.
   4. Base woody, leaves and stems silvery hairy, flowers large, usually solitary .. .. **S. heritieri**
   4. Base herbaceous, leaves glabrous to pubescent, flowers small in groups of 2–4 .. .. .. **S. tussilaginis**
  3. Inflorescence many-headed (up to 45), densely corymbose.
   5. Leaves orbicular, entire, margins wavy, the upper leaves sheathing the young corymbs (Gran Canaria) **S. webbii**
   5. Leaves deltate, suborbicular or broadly lanceolate, usually shallowly dentate-lobed, the upper usually not sheathing the young corymbs.

6. Involucral bracts with 10–20 dark tubercles or echinules, central florets usually yellow (Tenerife).. **S. echinatus**
6. Involucral bracts glabrous, pubescent or with one or two small tubercles, central florets usually mauve-purple.
   7. Basal leaves without auricles, ray-florets usually 8 (La Palma) .. .. **S. papyraceus**
   7. Basal leaves with auricles, ray-florets usually about 10–12.
      8. Heads small, corymbs dense, up to 45-flowered (Hierro) .. .. **S. murrayi**
      8. Heads large, corymbs lax (10–30 flowered).
         9. Leaves pink-carmine beneath, ray-florets pink-mauve (Tenerife, Gran Canaria) .. .. **S. cruentus**
         9. Leaves white-greenish beneath, ray-florets usually white or pale pink (Gomera) .. .. .. **S. steetzii**

1. Flowers yellow; stems or leaves succulent.
   10. Stems thick, succulent, leaves linear, entire .. .. **S. kleinia**
   10. Stems woody, thin, leaves succulent, lobed ot dentate.
      11. Leaves with narrow, cylindrical lobes (Gomera) .. **S. hermosae**
      11. Leaves lanceolate, margins toothed (La Palma, Tenerife).. .. .. **S. palmensis**

S. hadrosomus Svent. A tall robust species of the *S. cruentus* group from the Tenteniguada region of Gran Canaria is now thought to be extinct.

S. heritieri DC. Ascending to procumbent perennial. Stems and lower surface of leaves densely pubescent. Leaves suborbicular, about 3 cm across with 5–7 shallow lobes. Flower-heads more or less solitary, 3–5 cm across with 10–12 mauve ray-florets and a purple centre. ***Palomera.*** **(295)**

TENERIFE: S. coast region from Guimar to Masca, rocky slopes, cliffs, walls, 200–650 m, usually N.-facing, common.

S. tussilaginis (L'Her.) Less. Tuberous-rooted perennial up to 50 cm. Basal leaves suborbicular, crenately lobed, 6–10 cm across, pubescent above, white-tomentose beneath. Cauline leaves sessile, amplexicaul, lanceolate. Inflorescence few-flowered. Heads 3–5 cm across with 10–14 carmine ray-florets and a yellowish centre. Involucral bracts usually glabrous.

TENERIFE: N. coast region from Anaga to Icod de los Vinos, forests, cliffs, wayside scrub in the lower zone, usually in moist conditions, Bajamar, Las Mercedes, Icod el Alto, etc, 200–800 m.

S. webbii Sch. Bip. Erect, sturdy very variable perennial. Basal and cauline leaves orbicular, entire with wavy margins about 12 (6 – 14) cm across; petioles expanded at base, amplexicaul, with 2–8 sessile appendages; upper cauline leaves auriculate, sheathing the young corymbose inflorescences. Heads 10–20, about 1–2 cm across with 8–14 white to carmine ray-florets, the centre purple. Involucral bracts more or less glabrous, occasionally with a few tubercles. **(294)**

GRAN CANARIA: Locally very common in the N. sector of the island from the coastal region to the mountains 50–1600 m.

S. appendiculatus (L. fil.) Sch. Bip. Ascending shrub up to 1 m. Stems and petioles white-lanate. Leaves ovate-dentate, cordate at base, shallowly 7- to 9-lobed, upper surface glabrous, green, glossy, the lower white-lanate; margins dentate. Inflorescence 5- to 30-flowered. Heads about 1·3 cm across with 9–12 white ray-florets; centre yellow, occasionally tinged purple. Involucral bracts usually glabrous. ***Mato blanco.*** **(99)**

TENERIFE, GOMERA, HIERRO, LA PALMA: Locally abundant in the laurel forest zones 600–1000 m, Las Mercedes, El Cedro, Los Tilos, etc.

S. echinatus (L. fil.) DC. Erect perennial herb 30–50 cm. Leaves suborbicular, shallowly lobed with dentate margins; upper surface glabrous to pubescent, the lower lanate; petioles with up to 5 small appendages. Inflorescence of 5–15 heads. Heads up to 2·5 cm across with 11–14 pink ray-florets; centre yellow or mauve. Involucral bracts with up to 20 small dark tubercles or spines on the outer surface.

TENERIFE: N.W. region from the Orotava valley to Teno, locally abundant in *Cistus* scrub and rocky slopes on the N. coast, 50–400 m, San José, Los Silos, El Fraile, very local in the S., Adeje.

S. papyraceus DC. (incl. *S. hillebrandii* Christ). Like *S. echinatus* but with more or less orbicular, shallowly dentate glabrous to arachnoid-hairy leaves and the heads with 8–12 pink ray-florets and a purple centre. Involucral bracts usually glabrous. (**292**)

LA PALMA: Laurel and pine forests in most parts of the island, coastal region below Fuencaliente, 100–1600 m, locally frequent and very variable.

S. murrayi Hornm. Like *S. echinatus* but with the petiole-base expanded and amplexicaul. Inflorescence a dense corymb of 30–45 small (7 mm) heads. Ray-florets 11–13, white or pink; centre yellow. **(293)**

HIERRO: Common in forests and on dry slopes in the lower zone, El Golfo region, 50–700 m.

S. steetzii Bolle. Like *S. murrayi* but the petiole-base auriculate. Inflorescence of 7–25 heads. Ray-florets white to carmine. Involucral bracts with a few tubercles.

GOMERA: Forest regions, 600–1200 m, locally common between Roque Agando and El Cedro, Vallehermoso, forests between Epina and Arure.

S. cruentus DC. Erect pubescent perennial. Leaves orbicular, sinuate-dentate, pubescent above, densely pink-carmine tomentose beneath; petioles auriculate at base. Inflorescence 7- to 20-flowered. Ray-florets pink; centre purple. (**98**)

TENERIFE: Forest zones of the N. coast, Orotava valley, Agua Mansa, Agua Garcia, etc, very common 800–1500 m (this species is the ancestor of the florist's *Cineraria*).

S. kleinia (L.) Less. Stems succulent, glaucous. Leaves linear-lanceolate, acute, fleshy, glaucous. Inflorescences umbelliforme. Heads long, very slender, pale yellow. ***Verode.*** (**296**)

ALL ISLANDS: Very common in the lower zone usually in association with *Euphorbia* communities 50–1000 m.

S. palmensis Chr. Sm. Small shrub. Leaves fleshy, lanceolate, the margins dentate. Inflorescence dense, corymbose. Heads slender, golden yellow. ***Turgayte.***

TENERIFE: Subalpine zone of Las Cañadas, sporadic on the cliffs of the S. rim, Roca de Tauce, Montaña de Diego, Hernandez, etc, 2000 m; LA PALMA: abundant in the Caldera region, Barranco de las Angustias, 450–1800 m.

S. hermosae Pitard. Like *S. palmensis* but taller, more robust. Leaves 3-lobed, the lobes fleshy, linear (**100**)

GOMERA: Rare, Vallehermoso region, cliffs and rocks in the lower zone near Roque Cano.

ATRACTYLIS

Compact, small shrublets. Outer involucral bracts leaf-like, pinnatisect, spiny, the inner entire, with dark spiny tip. Cypselas with silvery hairs. Pappus of feathery bristles fused at the base.

A. preauxiana Webb & Berth. Small, woody cushion plant, 5–10 cm. Leaves entire, more or less rosulate, linear to narrowly oblanceolate, silvery. Inner involucral bracts woolly; margins brown, the tip blackish with a long spine. Outer florets ligulate, the ligules pinkish or cream to white. **(103)**

GRAN CANARIA: Rare, rocky coasts near the sea, Arinaga, Melenara, etc; TENERIFE: Similar habitats in the S. of the island between Guimar and El Medano.

Populations from N. Gran Canaria and Lanzarote with a much taller habit and broader leaves have been described as belonging to a distinct species *A. arbuscula* Svent. & Kahne.

CARLINA

Shrubs up to 1 m. Leaves entire, white-lanate at least beneath, the margins spiny. Flower heads yellow, ray florets absent. Receptacle with scales. Outer involucral bracts large, spiny at least at apex. Pappus hairs feathery, caducous.

1. Heads at least 2 cm across; inflorescences on short peduncles .. .. **C. salicifolia**
1. Heads less than 1·5 cm across; inflorescences on long, leafy peduncles.
  2. Leaves linear, margins with long spines; heads about 1 cm across .. .. **C. xeranthemoides**
  2. Leaves linear-lanceolate, margins remotely spiny; heads 1·3–1·5 cm .. .. **C. canariensis**

C. xeranthemoides L. fil. Shrub up to 50 cm, branching near base. Leaves linear, woolly on both surfaces; margins with long spines. Inflorescence branches bracteate. Flower heads 3–6, bright yellow, 1 cm across. ***Malpica.*** **(101)**

TENERIFE: Subalpine zone of Las Cañadas de Teide, pumice slopes and lava debris, locally frequent, 1900–2100 m.

C. canariensis Pitard. Like *C. xeranthemoides* but leaves linear-lanceolate; margins remotely spiny. Involucral bracts narrower. Heads larger, about 1·3–1·5 cm across.

GRAN CANARIA: Locally common in the S., Tirajana, Fataga, Tazartico, on dry cliffs 150–600 m.

C. salicifolia Cav. Shrub up to 1 m. Leaves narrowly lanceolate, densely hairy below, more or less glabrous above; margins spiny, sometimes very remotely so. Outer involucral bracts leaf-like, patent. Flower heads 1–3 yellow, 2–3·5 cm across. ***Cabezote, Cardo de Cristo.*** **(102)**

ALL ISLANDS: Frequent on cliffs in the upper xerophytic and forest zones 200–1600 m, very rare on LANZAROTE (Famara) and FUERTEVENTURA (Jandia).

Populations from the northern part of LA PALMA with curved, unarmed leaves and distinctive small bracts have been referred to a separate species *C. falcata* Svent. but are here included in *C. salicifolia*.

ONOPORDUM

Flower heads solitary, with short, spreading or reflexed, spine-tipped involucral bracts; ray florets absent. Recpetacle without scales. Pappus feathery.

O. nogalesii Svent. Perennial herb. Basal and lower stem leaves white-felted, sinuately lobed, the lobes with spiny, dentate margins. Flowering stems winged and spiny. Heads up to 5 cm across. Involucral bracts purplish, with long apical spines. Florets purplish. Cypselas transversely striate-rugose, yellowish.

FUERTEVENTURA: Dry slopes of the Jandia mountains, locally frequent on the dry S. slopes below Pico de la Zarza, 300 m. (*O. carduelinum* Bolle, a very rare species from the Tenteniguada region of GRAN CANARIA is said to differ from the above species by its smaller leaves, slightly larger flower heads and pink florets. It has not been found recently.)

CARDUUS

Annuals with spiny, winged stems. Ray florets absent. Involucral bracts with spiny tips. Receptacle with scales. Pappus of several rows of simple hairs.

1. Leaves shallowly sinuate-pinnatifid, the upper linear to narrowly lanceolate.
 2. Flowers white, apical spines of involucral bracts short .. **C. baeocephalus**
 2. Flowers purplish, apical spines of involucral bracts long **C. bourgaeanus**
1. Leaves ovate, entire, the upper ovate-lanceolate .. .. **C. clavulatus**

C. baeocephalus Webb & Berth. Unbranched, robust herb. Leaves shallowly sinuate-pinnatifid, arachnoid-tomentose; margins spiny; lower leaves obovate-lanceolate, the upper linear to narrowly lanceolate. Inflorescences few-headed. Heads small, shortly pedunculate, erect. Involucral bracts in 4–5 series, the outer arachnoid-hairy. Flowers whitish to occasionally pale pink.

GRAN CANARIA: N. coast, Barranco de la Virgen, Cuesta de Silva; Agaete; HIERRO: Cuesta de Sabinosa, coastal malpais, 200–300 m.

C. bourgaeanus (Bolle) Sch. Bip. Like *C. baeocephalus* but with larger heads; longer apical spines on the involucral bracts and purplish florets.

FUERTEVENTURA: Mountains of Jandia, N. coast region above Playa de Barlovento, locally abundant.

C. clavulatus Link. Small herbs. Stems scarcely winged. Leaves more or less ovate, entire, sparsely pubescent, the upper ovate-lanceolate. Heads few. Florets deep pink to mauve.

TENERIFE: Widespread on the N. side of the island from Taganana to Teno, Las Mercedes, Guimar, 50–1000 m; GRAN CANARIA: N. region, forest zone, San Mateo, Tafira, Moya, Barranco de Balbuzanos, etc; LA PALMA: Barlovento, La Galga, Fuencaliente; HIERRO: Riscos de Jinamar, Sabinosa, from coasts to forest zone, 50–800 m.

Two European weed species *C. pycnocephalus* L. and *C. tenuiflorus* Curt. are also widespread as field weeds and along roadsides in the Canaries.

CENTAUREA

Shrubs. Heads ovoid-subconical. Involucral bracts imbricate, in several series, each with a terminal, laciniate, pectinate or subentire appendage. Florets tubular, hermaphrodite or the outer row neuter. Receptacle with bristles. Cypselas cylindrical to compressed, smooth. Pappus of bristles and scales.

1. Heads white, cream or yellow.
 2. Appendage almost as long as bract (Tenerife near Taganana) .. .. **C. tagananensis**
 2. Appendage much shorter than bract or more or less absent.
  3. Appendage more or less absent, leaves linear (Tenerife, Las Cañadas) .. .. **C. arguta**
  3. Appendage present, leaves lanceolate to ovate.
   4. Appendage very small (2 mm), lacerate and deflexed, not decurrent on the bract .. .. **C. webbiana**

4. Appendage more than 3 mm, fimbriate or pectinate.
  5. Leaves ovate, coarsely toothed (Hierro) .. .. **C. durannii**
  5. Leaves lanceolate, margins finely dentate.
    6. Appendage evenly long-fimbriate, scarcely decurrent on bract (Gomera) .. .. **C. sataratensis**
    6. Appendage laciniate-pectinate, decurrent on bract (La Palma) .. .. **C. arborea**

1. Heads deep pink or purple.
 7. Shrub up to 3·5 m, leaves always entire (Gran Canaria) .. .. **C. arbutifolia**
 7. Shrub up to 1·5 m, leaves usually pinnately lobed (occasionally entire but then plants less than 1 m).
  8. Leaves with linear lobes less than 0·4 mm wide.
   9. Bracts of involucre with wide, scarious margins, appendages fimbriate (Tenerife) .. .. .. **C. canariensis**
   9. Bracts without wide scarious margins, appendages laciniate (Gomera) .. .. **C. ghomerytha**
  8. Leaves with lanceolate lobes more than 0·4 mm wide (La Palma) .. .. **C. junoniana**

C. arguta Nees. Shrub up to 1·5 m. Leaves entire, linear-lanceolate, coarsely toothed, sticky. Flower-heads solitary, conical. Florets pale yellow to cream. Involucral bracts without a distinct appendage, the margins fimbriate.

TENERIFE: Subalpine zone of Las Cañadas, locally common, 2000–2200 m, on dry rocky slopes and screes, La Fortaleza, Montaña de Diego Hernandez and other cliff areas of the S. rim of Las Cañadas.

C. tagananensis Svent. Small, little-branched shrub with oblong, entire, dentate, glossy leaves. Inflorescences few-flowered. Florets cream. Involucral bracts with a large terminal appendage as long as the bract.

TENERIFE: Very rare in the N. coast region of Anaga in the Taganana area, Roque de las Animas, 100 m.

C. webbiana Sch. Bip. Like *C. tagananensis* but the leaves oblanceolate and the involucral bracts with very small lacerate, deflexed appendages.

TENERIFE: N. coast between Orotava and Garachico, very sporadic, 100–400 m, San Juan de la Rambla, El Rincón near Icod de los Vinos.

C. durannii Burchd. Like *C. tagananensis* but the leaves ovate, coarsely toothed and the involucral bract appendage pectinate.

HIERRO: El Golfo region, on the cliffs in the lower zone between Frontera and Sabinosa, 100–300 m, rare.

C. sataratensis Svent. Like *C. tagananensis* but leaves lanceolate with finely denticulate margins and the involucral bract appendages long fimbriate, not decurrent on the bract. **(104)**

GOMERA: S.W. region in the mountains S. of Vallegranrey, Pico de Satarata, Barranco de Cabrito, 250–600 m, very rare.

C. arborea Webb & Berth. Tall branched shrub with lanceolate, dentate leaves. Inflorescence with leaf-like bracts. Florets cream-white. Involucral bracts with more or less decurrent, deflexed, laciniate-pectinate appendages.

LA PALMA: Barranco de las Angustias, 300 m, extremely rare in the lower zone, also recorded from the E. side of the island.

C. arbutifolia Svent. Shrub up to 3 m with tough, woody stems. Leaves entire, lanceolate, somewhat leathery and often sticky; margins denticulate to serrate.

Florets pinkish purple. Involucral bracts with a small decurrent, laciniate appendage. **(298)**

GRAN CANARIA: Very rare, N.W. zone of the island, Guayedra, Agaete valley, Berrezales, on basalt cliffs, 400–800 m.

C. canariensis Willd. Shrub up to 1·5 m. Leaves pinnately lobed with 4–5 pairs of linear-lanceolate lobes or occasionally subentire. Florets pale mauve. Involucral bracts with a large, broadly decurrent fimbriate appendage. **(105)**

TENERIFE: Very rare, known only form the valley of Masca, 450 m, on basalt cliffs. A more or less entire-leaved from is locally frequent in the Teno region (var. *subexpinnata* Burchd.), El Fraile, Punta de Teno; GRAN CANARIA.

C. ghomerytha Svent. Like *C. canariensis* but the leaves pubescent on both surfaces and the involucral bracts narrow with a small, pectinate appendage.

GOMERA: N.W. coast at Punta de San Marcos, 150 m, rare.

C. junoniana Svent. Small shrub up to 1 m. Leaves pinnately lobed or entire (var. *isoplexiphylla* Svent.), the lobes lanceolate, obtuse. Florets pale mauve. Involucral bracts with a large fimbriate appendage. **(297)**

LA PALMA: S. region, on old basalt outcrops in the coastal zone S. of Fuencaliente, Roque de Teneguia, 300 m, locally abundant.

RHAPONTICUM

Heads very large, solitary. Involucral bracts expanded into a broad, toothed papery margin and tip. Receptacular scales present. Cypselas ribbed. Pappus of several rows of rough, scabrid hairs.

R. canariensis DC. (*Serratula canariensis* Webb & Berth.) Robust perennial. Leaves large, pinnatisect, white-woolly beneath. Heads 6–8 cm across, pink-purplish. Involucral bracts with white, papery margins.

TENERIFE: Subalpine zone of Las Cañadas, known only from Llano de Maja and La Fortaleza, 2000 m, very rare and endangered species.

ANDRYALA

Leaves lobed or toothed to subentire. Involucral bracts in a single row, usually densely hairy on outer surface. Flowers yellow. Receptacle with long, hair-like scales. Pappus of simple, deciduous hairs. Cypselas with 8–10 ribs.

1. Leaf-margins undulate-crispate, peduncles 4–5 cm .. **A. cheiranthifolia**
1. Leaf-margins not undulate-crispate, peduncles 3 cm or less .. .. **A. pinnatifida**

A. pinnatifida Aiton. Very variable, polymorphic species. Perennial herb, sometimes woody at the base. Stems and lower leaf-surface white-floccose. Leaves lanceolate to ovate, sinuate-pinnate to dentate or subentire. Inflorescence usually dense, flat-topped, up to 20-flowered. ***Estornudera.***

ALL ISLANDS: In all zones from coasts to subalpine, forest forms are often very slender and few-flowered. A very robust, tall form with a single stem from the high mountains of TENERIFE is usually distinguished as var. *teydea* Webb and an almost entire-leaved form with pale yellow flowers from the laurel woods on the N. side of LA PALMA can be separated as var. *webbii* (Sch. Bip.) Christ.

A. cheiranthifolia L'Her. Like *A. pinnatifida* but very densely white-pubescent, with undulate, crispate margins to the lanceolate, pinnate leaves. Flower-heads with long peduncles. **(319)**

TENERIFE: Sierra Anaga, Igueste de San Andrés; LANZAROTE: N. region, Haria,

Las Mesas, up to 600 m; FUERTEVENTURA: Betancuria, Jandia, Pico de la Zarza.

HEYWOODIELLA

Stems with translucent, pale-yellowish latex. Receptacle with long scales. Outer involucral bracts with black hairs. Pappus hairs sparsely scabrid.

H. oligocephala Svent. & Bramwell. Rosette cliff plant with short woody stock. Leaves lanceolate, subglabrous, glossy, margins undulate, coarsely dentate. Inflorescences 1- to 2-flowered. Heads about 1·5 cm long. Cypselas 5-ribbed, rugose. Pappus with 12–14 unequal bristles.

TENERIFE: Known only from one small area on the N.W. coast of the island, 100–200 m, crevices on basalt cliffs, very rare.

TOLPIS

Perennial with woody stock. Basal leaves rosetted, pinnatifid to pinnatisect, Cauline leaves pinnate to subentire. Inflorescences lax, corymbose to paniculate. Peduncles 1- to 3-headed. Involucral bracts in 2 to 3 rows, linear-lanceolate, tomentose, the margins scarious, the outer often very small and scale-like or filiform. Receptacle without scales. Cypselas ribbed. Pappus a single row of scabrid hairs with scales between.

1. Stock short, branching at ground level, lower leaves suberect, linear-lanceolate, pinnatifid; lobes unequal (Tenerife, subalpine zone) .. .. **T. webbii**
1. Plants with short, woody stem, branching above ground level, lower leaves rosulate, lanceolate to ovate, subentire to 2-pinnatisect, lobes more or less equal.
  2. Leaves 2-pinnatisect, greyish tomentose.
    3. Lobes linear to lanceolate, petioles and involucral bracts not densely white floccose .. .. .. **T. laciniata**
    3. Lobes lanceolate-ovate, petioles and involucral bracts densely white floccose .. .. **T. proustii**
  2. Leaves pinnate to entire with dentate margins, usually glabrous or subglabrous.
    4. Cliff plant with glabrous, more or less fleshy leaves .. **T. crassiuscula**
    4. Forest plant with thin, subglabrous leaves .. .. **T. lagopoda**

T. webbii Sch. Bip. Rootstock thick, woody, covered with the remains of old leaf-bases, branching at ground level. Basal leaves suberect, linear to linear-lanceolate, dentate to unequally pinnatifid, pubescent, stem leaves similar to the basal becoming smaller above. Inflorescence a lax corymb. Heads about 1 cm long. Outer bracts small, somewhat reflexed.

TENERIFE: Las Cañadas de Teide, 2000 m, subalpine rocks and pumice slopes, locally very common, Cañada del Portillo, Montaña Blanca, etc.

T. laciniata Webb & Berth. Very variable. Stem woody. Branches divaricate. Rosette leaves ovate-lanceolate, usually 2-pinnatisect, 3–8 cm long, glabrescent or minutely pubescent on both surfaces; cauline leaves few, the lower similar to the basal, the upper linear. Inflorescence corymbose, lax. Heads 1–5 cm long. Outer involucral bracts shorter than the inner, hairy. (**314**)

TENERIFE: Locally frequent in forest and upper xerophytic zones, 300–800 m, Barranco de San Andrés, Guia de Isora, San Juan de la Rambla, etc; LA PALMA: Common in the S. below Fuencaliente and in the forest zones of the N. region, Barranco de Las Angustias, El Paso, Caldera de Tabouriente, Barranco Carmen etc, *T. calderae* Bolle, a rare species from the Caldera de Tabouriente appears to be a local relative of this species; GOMERA: Forests of the El Cedro region in the C. of the island.

T. proustii Pitard. Like *T. laciniata* but with larger leaves with very broad lobes; stems and petioles white woolly to floccose.

HIERRO: Common in the El Golfo region up to 500 m, on dry cliffs and rocks; GOMERA: Locally frequent in many areas of the island up to 1000 m, Barranco de La Laja, Roque Agando, Barranquillos de Vallehermoso, Monteforte, etc.

T. crassiuscula Svent. Rosettes flat. Leaves lanceolate with dentate or crenate margins or shallowly pinnatifid, fleshy, glabrous. Inflorescences lax, few-flowered. Heads large, up to 2 cm long. **(106)**

TENERIFE: Apparently confined to two localities, sea-cliffs W. of Buenavista 50–200 m where it is locally frequent and Barranco del Infierno de Adeje where it is found only on a single cliff face.

T. lagopoda Chr. Sm. Small herb with somewhat woody stock. Leaves small, subentire or crenate, thin, subglabrous or pubescent. Inflorescences few-flowered. Heads small, about 1 cm long.

TENERIFE, GRAN CANARIA, LA PALMA: Locally frequent on pine-forest cliffs on TENERIFE (Agua Mansa) and LA PALMA, rather rare on GRAN CANARIA, 800–1600 m.

CREPIS

Perennial herbs with thick, fleshy caudex. Involucral bracts in several series. Receptacular scales absent. Cypselas shortly beaked, 10-ribbed. Pappus of white, soft hairs.

C. canariensis (Sch. Bip.) Babc. 15–30 cm. Basal leaves rosulate, up to 18 cm long, obovate-lanceolate or elliptic, glabrous, irregularly dentate at margin; petiole winged; upper stem leaves triangular. Inflorescence corymbiform, the ultimate branches with 1–4 heads. Peduncles finely glandular-hairy. Heads 1–1·5 cm across. Florets yellow. Cypselas light brown, rough.

LANZAROTE: Common in the N. region, Peñitas de Chache, Haria, from coastal rocks to 600 m; FUERTEVENTURA: Common in the C. and N. regions on the coast and inland, Betancuria, La Oliva, Montaña Atalaya, etc; GRAN CANARIA.

LACTUCOSONCHUS

Herbaceous, glabrous, perennial. Stem cylindrical, hollow, leafy. Lower leaves lyrate, sessile, the upper linear. Heads many, with 12–16 florets. Involucral bracts in several series, more or less imbricate. Receptacle without scales. Cypselas with 3 longitudinal grooves; the pappus white, denticulate, caducous.

L. webbii (Sch. Bip.) Svent. Perennial with long tuberous root. Leaves in a basal rosette, very variable, pinnatisect to almost entire. Inflorescence corymbose, with many small heads. Heads up to 5 mm across with 12–16 florets.

LA PALMA: N. coast region in pine forests of the Cumbres de Garafia, Roque del Faro, 400–600 m, rather rare.

SVENTENIA

Glabrous perennials with a basal rosette of leaves and an almost leafless flowering stem. Involucral bracts in several series. Receptacle without scales. Florets yellow. Cypselas shallowly grooved; the pappus of 2 types of hairs, some deciduous and stiff, others persistent and soft.

S. bupleuroides Font Quer. Up to 30 cm. Leaves rosulate at the top of a short stem, entire, more or less lanceolate, glabrous, pruinose, obtuse. Flowering stem glandular-hairy in the upper part, the hairs reddish-brown. Heads about 1 cm across. **(317)**

GRAN CANARIA: W. region on the high cliffs of Barranco de Guayedra below the Tamadaba Massif, 600–800 m, on wet, vertical rocks, very rare.

PRENANTHES

Involucre long, cylindrical, outer bracts much shorter than the inner. Heads with 5–6 yellow florets, all ligulate. Cypselas shallowly ribbed, not beaked. Pappus hairs simple.

P. pendula (Webb) Sch. Bip. Cliff plant, stems woody, often pendulous. Leaves pinnately lobed, the lobes lanceolate to triangular, acute, glaucous, often only the lower pair cut to midrib. Inflorescence dense. Heads long, slender, 2 mm across. **(315)**

GRAN CANARIA: Mountain cliffs, generally on the S. and W. sides of the island from 200–800 m or more, Agaete, Fataga, etc, locally common.

SONCHUS

Perennial herbs or shrubs up to 3 m. Leaves denticulate to pinnatisect. Heads few–many. Involucral bracts in at least 3 rows, imbricate. Receptacular scales absent. Florets yellow. Cypselas not beaked, compressed, tapering at both ends, the faces with 1–4 ribs. Pappus hairs of 2 types, rough bristles and softer, fasciculate hairs. ***Cerraja, Cerrajon.***

1. Herbaceous perennial with a tuberous root (W. Tenerife) .. .. **S. tuberifer**
1. Shrubs or shrublets with a tap-root or rhizome.
 2. Shrublets with basal rosette(s) on a short, woody stem.
  3. Rosettes large, up to 1 m, leaves not pruinose.
   4. Leaf-lobes rounded (La Palma) .. .. .. **S. bornmuelleri**
   4. Leaf-lobes acute (Tenerife, Gran Canaria) .. .. **S. acaulis**
  3. Rosettes small, less than 50 cm, leaves usually pruinose.
   5. Stems very short, rosettes flat.
    6. Involucral bracts not extending down the peduncle.
     7. Heads 1–2 cm across (N. Tenerife) .. .. .. **S. radicatus**
     7. Heads 3–5 cm across (Gomera) .. .. .. **S. gonzalezpadronii**
    6. Involucral bracts extending down the peduncle (S.W. Tenerife) .. .. **S. fauces-orci**
   5. Stems up to 15 cm, rosettes not flat.
    8. Leaf-lobes overlapping (Tenerife, S. Anaga) .. **S. tectifolius**
    8. Leaf-lobes not overlapping (Tenerife, Guimar) .. **S. gummifer**
 2. Shrubs with terminal leaf-rosettes on the branches.
  9. Leaves with few lobes, inflorescence of 1–6 heads (N. & N.W. Gran Canaria) .. .. **S. brachylobus**
  9. Leaves with at least 4 pairs of lobes, inflorescence of more than 6 heads.
   10. Heads very narrow, 5 mm across or less, leaf-lobes very slender.
    11. Leaf-lobes linear-filiform, less than 1 mm wide.
     12. Heads very narrow, 1–2 mm, leaves more or less ascending, leaf-lobes few, stiff (S. Tenerife, Guimar) .. .. **S. microcarpus**

12. Heads up to 3 mm, leaves pendant, leaf-lobes in at least 3 pairs, not stiff.
13. Leaf-lobes flat, heads with up to 20 florets (N. Tenerife, Gran Canaria) .. .. .. **S. leptocephalus**
13. Leaf-lobes not flat, heads with less than 15 florets.
14. Heads very narrow, with 8–10 florets (Gomera) **S. filifolius**
14. Heads broader with 10–15 florets (S. Tenerife) .. **S. capillaris**
11. Leaf-lobes linear, flat, at least 2 mm wide.
15. Heads 4–5 mm across (Hierro) .. .. .. **S. gandogeri**
15. Heads less than 3 mm.
16. Leaf-lobes up to 5 mm wide, stems slender, usually about 1 m tall (Gomera) .. .. **S. regis-jubae**
16. Leaf-lobes usually 2–3 mm wide, stems thick, usually at least 1·5 m tall .. .. .. **S. arboreus**
10. Heads broad, at least 1 cm across, leaf-lobes broad.
17. Heads very large, at least 4 cm across.
18. Involucral bracts very broadly ovate, large, leaves densely pruinose with acute lobes (Gran Canaria) **S. platylepis**
18. Involucral bracts ovate-lanceolate, leaves glabrous, green with deltoid-rounded lobes (Gomera) .. **S. ortunoi**
17. Heads less than 4 cm across.
19. Heads densely white-tomentose, inflorescences congested.
20. Leaves tomentose .. .. **S. hierrensis**
20. Leaves glabrous.
21. Peduncle with bracts, heads at least 2 cm across **S. congestus**
21. Peduncle without bracts, heads less than 2 cm across .. .. **S. abbreviatus**
19. Heads glabrous or very sparsely tomentose, inflorescences not congested.
22. Leaf-lobes broad, acute, peduncle with 2–5 small bracts (Lanzarote, Fuerteventura).. .. .. **S. pinnatifidus**
22. Leaf-lobes narrow, obtuse-rounded, peduncle without bracts.
23. Heads narrow with up to 50 florets (La Palma) **S. palmensis**
23. Heads broad with at least 80 florets (Tenerife, Gran Canaria) .. .. **S. canariensis**

S. tuberifer Svent. Small, herbaceous perennial with tuberous root. Stems up to 30 cm. Leaves mostly basal, pinnatifid or coarsely serrate; lower surface glabrous, the upper finely pubescent; apex acute; petiole auriculate, amplexicaul at the base. Inflorescence a small corymb of 3–4 heads. Peduncles bracteate. **(310)**
TENERIFE: Valle de Masca, El Retamal, Tamaimo, Teno, crevices on shady cliffs, 150–1000 m, locally frequent.

S. brachylobus Webb & Berth. Small shrublet. Stem simple or branched, 30–80 cm. Leaves lyrate-pinnatifid, glabrous; terminal deltoid lobe larger than the rounded laterals; margins denticulate. Peduncles short. Heads few. **(308) (309)**
GRAN CANARIA: Locally frequent along the N. coast, usually on cliffs, Cuesta de Silva, coast below Moya, Galdar, Barranco de Guayedra, Andén Verde on the W. coast. Populations with pointed leaf-lobes are referable to var. *canariae* (Pitard) Boulos.

S. acaulis Dum.-Cours. Perennial. Base woody. Leaves in a single, large basal rosette up to 1 m across; pinnatifid, tomentose, acute. Scape up to 1·5 m, without bracts. Inflorescence umbellate. Peduncles with 1–several heads, tomentose. Involucral bracts broad, very densely white-tomentose. Heads about 2·5 cm across. **(305)**

TENERIFE: Widespread in forest and xerophytic zones, Sierra Anaga to Teno particularly along the N. coast; GRAN CANARIA: Montane regions, Los Tiles de Moya, Tenteniguada, Cruz de Tejeda, Roque Nublo, etc, 500–1600 m, locally frequent.

S. bornmuelleri Pitard. Perennial with woody stock. Leaves in a basal rosette, pinnatifid with rounded lobes; margins subspinose. Scape up to 80 cm, with a few small bracts. Inflorescence a dense corymb of up to 20 heads. Heads densely tomentose. **(303)**

LA PALMA: Cliffs S. of Santa Cruz de la Palma, coastal rocks near San Andrés y Los Sauces, up to 150 m, rare.

S. radicatus Aiton. Perennial. Base woody. Leaves in a basal rosette, lyrate-pinnatisect, pruinose. Inflorescence bracteate, with up to 10 heads. **(312)**

TENERIFE: N. coast regions from Sierra Anaga to Punta de Teno, locally very common on cliffs, Taganana, Bajamar, Barranco Ruiz, San Juan de la Rambla, Icod, Los Silos, Buenavista, etc, up to 400 m.

S. gonzalezpadronii Svent. Like *S. radicatus* but usually with a longer stem, the leaves pinnatifid-runcinate, amplexicaul at the base. Inflorescence with sessile bracts. Peduncles and head tomentose.

GOMERA: N. coast, Barranco de la Villa to Vallehermoso, Roque Cano, S. and S.W. region, Epina, Arure, Vallegranrey, Barranco de Argaga, Fortaleza de Chipude, 400–1200 m, locally frequent.

S. gummifer Link. Resembling *S. radicatus* but the leaves pinnatifid with triangular lobes, the inflorescence bracteate with pinnatifid bracts, few-headed. Heads small.

TENERIFE: S. region, Ladera de Guimar, Candelaria, Escabonal, etc, on cliffs up to 500 m, frequent only at Guimar.

S. tectifolius Svent. Like *S. gummifer* but the leaves more or less pendant, the lobes overlapping, densely pruinose and the heads slightly smaller. **(313)**

TENERIFE: S. slopes of Sierra Anaga, Valle de San Andrés, Igueste de San Andrés, very local.

S. hierrensis (Pitard) Boulos. Robust shrub up to 1·5 m. Leaves deeply lobed, the lobes up to 2 cm wide. Inflorescence densely branched with numerous heads. Peduncles up to 8 cm. Heads 2–3 cm across. **(311)**

HIERRO: Widespread throughout the island, El Golfo, Frontera, Sabinosa, etc, up to 1000 m; GOMERA: N.E. region, Barranco de la Villa, Hermigua, S. region, Roque de Agando, Benchijigua, 300–800 m; LA PALMA: S. region to the S. of Santa Cruz, Mazo, Fuencaliente, W. region, Barranco de las Angustias (var. *benehoavensis* Svent.).

S. ortunoi Svent. Robust shrub up to 1 m. Leaves rosulate at the tips of stems, pinnatisect, subglabrous, the lobes deltoid-rounded. Inflorescence subcorymbose with few heads. Heads large, 4–5 cm across. **(307)**

GOMERA: Locally frequent in the C. and S.W. regions, El Cedro, Vallegranrey, N. coast, Barranco de Vallehermoso, Roque Cano, 200–1000 m.

S. congestus Willd. Shrub up to 1·5 m. Leaves lanceolate, pinnatifid, the lobes acuminate; margins serrulate. Inflorescence, dense, subumbellate, the flowering stem short. Heads 2–2·5 cm across, tomentose. **(304)**

TENERIFE: Widespread in the forest zones of Sierra Anaga and the N. coast Orotava Valley, La Rambla, La Guancha, Icod el Alto, Buenavista, 150–800 m; GRAN CANARIA: C. and N. reions, San Mateo, Tafira.

S. abbreviatus Link in Buch. Similar to *S. congestus* but the leaves more deeply pinnatifid with a long, slender terminal lobe, the lateral lobes narrow. Inflorescence very densely congested.

TENERIFE: Forest zones of Sierra Anaga, Vueltas de Taganana, El Bailadero, 550–700 m, in laurel woods; GRAN CANARIA: Forest of Los Tiles de Moya, Teror.

S. pinnatifidus Cav. Shrub up to 2 m. Leaves rosulate at tips of stems, oblong-lanceolate, pinnatifid, the lobes ovate-triangular. Inflorescence corymbose with few heads. Peduncles bracteate, swollen at the bases of the heads. Heads sparsely tomentose, up to 2 cm across.

LANZAROTE: N. region, Riscos de Famara, Barranco de Pocela, 500–700 m, rare; FUERTEVENTURA: Montaña de Cardones, N. side of Macizo de Jandia. (S.W. Morocco).

S. fauces-orci Knoche. Glabrous shrub. Leaves rosulate at the tip of a short stem, lanceolate-obovate, pinnatifid-lyrate with about 8, ovate-triangular, pointed lobes. Inflorescence subumbellate with 5–20 heads. Peduncles with numerous small bracts which merge into the involucral bracts. **(108)**

TENERIFE: S.W. region, Barranco del Infierno de Adeje, Guia de Isora, Tamaimo, Masca, 300–600 m, local.

S. platylepis Webb & Berth. Leaves rosulate at the tips of stems, pinnatifid, with linear lobes and broad, rounded sinuses, densely pruinose. Inflorescence of a few very large heads. Involucral bracts very large, broadly ovate. **(306)**

GRAN CANARIA: Mountains of the C. region, Cruz de Tejeda, Barranco de Tejeda, Paso de la Plata, Roque Nublo, Artenara, Juncalillo, Santa Brigida, etc, locally very common, 800–1600 m.

S. capillaris Svent. Tall, slender shrub, 1–2 m. Leaves crowded at stem apex, lanceolate, pinnatisect, the lobes in 5–9 pairs, filiform. Inflorescence a loose, corymbose panicle with many heads. Heads very narrow, less than 2 mm with 10–15 florets. **(107)**

TENERIFE: S. and S.W., Punta de Teno, Barranco de la Cueva, Masca, Barranco de Tejina, Guia de Isora, Barranco de las Hiedras, 200–700 m, locally frequent; GRAN CANARIA: S. region.

S. microcarpus (Boulos) Aldridge. Like *S. capillaris* but the leaves shorter, with fewer lobes and the flower-heads narrower (1·2 mm).

TENERIFE: Ladera de Guimar, very local, rare, 400–500 m.

S. leptocephalus Cass. Tall, slender shrub up to 1·5 m. Leaves crowded at the tips of stems, glabrous, the lobes linear, flat. Inflorescence a dense panicle. Heads up to 3 mm across with 12–20 florets. ***Balillo.*** **(301)**

TENERIFE: Very frequent along the N. coast from Taganana to Garachico, Barranco de Ruiz, San José, Cuevas Negras de los Silos, etc, 10–600 m; GRAN CANARIA: N. and N.W. regions, Tafira, Barranco de Guiniguada, Cuesta de Silva, S. region, Barranco de Tejeda, Fataga, locally frequent.

S. filifolius Svent. Similar to *S. leptocephalus* but with very long leaves, usually with fewer lobes. Heads very slender with 8–10 florets. ***Balillo.***

GOMERA: W. and S.W. of the island, Epina, Lomo de Carretón, Vallegranrey, Argaga, Chipude, La Fortaleza, Vallehermoso, etc, 200–1000 m, locally abundant.

S. arboreus DC. Tall shrub up to 2·5 m. Leaves rosulate at the tips of branches, pinnatisect with linear, flat lobes up to 5 mm wide, inflorescence a dense corymb. Heads with 15–20 florets. **(300)**

TENERIFE: N.W. region, La Guancha, Punta de Teno, etc, very rare; LA PALMA: N.W. region between Tijarafe and Garafia, 300–600 m, rare; recent records of *S. regis-jubae* and *S. heterophylla* from LA PALMA and GRAN CANARIA probably refer to forms of this species.

S. regis-jubae Pitard. Like *S. arboreus* but up to 1·5 m, with broader leaf-lobes and rather smaller inflorescences.

GOMERA: Roadsides and cliffs from Barranco de la Villa to Vallehermoso, Riscos de Agulo, Roque Cano, 200–600 m, locally frequent.

S. palmensis (Sch. Bip.) Boulos. Tall shrub up to 2 m. Leaves pinnatisect with 10–15 pairs of equally spaced lobes, the lobes 6–40 mm wide. Inflorescence very large and dense, compound-corymbose. Heads 3–4 mm across with 30–50 florets. **(299)**

LA PALMA: Widespread especially along the E. coast, Fuencaliente, Mazo, Santa Cruz, Las Breñas, Puntallana, La Galga, Los Tilos, etc, W. coast, Los Llanos, etc, up to 1000 m in the lower and forest zones.

S. canariensis (Sch. Bip.) Boulos. Like *S. palmensis* but up to 3 m, with less dense inflorescences and larger heads up to 1·5 cm across with about 100–150 florets. **(302)**

TENERIFE: S. region, Barrancos near Adeje, Valle Seco, Barranco Tinadaya, up to 600 m, very rare; GRAN CANARIA: Local, Barranco de la Virgen, Barranco de la Angostura, Andén Verde, Aldea de San Nicolas, 200–800 m, rare.

S. gandogeri Pitard. Shrub up to 1·5 m. Leaves pinnatisect with long, slender lobes about 5 mm wide. Inflorescence a corymb of about 20–60 heads. Heads 4–5 mm across.

HIERRO: El Golfo region on cliffs N.E. of Frontera, Las Casitas, Los Llanillos. Hybrids of this species and *S. hierrensis* are very frequent in the El Golfo region between Frontera and Los Llanillos and a number of these have been given binomials (*S. lidii* Boulos, *S. pitardii* Boulos).

LACTUCA

Like *Sonchus* but the outer involucral bracts short and broad and the cypselas compressed with a short or long beak.

1. Florets pale blue, leaves pinnate with deflexed lobes (La Palma) .. .. **L. palmensis**
1. Florets yellow, leaves more or less entire, ovate with toothed margin .. .. **L. herbanica**

L. palmensis Bolle. Small slender herb up to 40 cm. Lower leaves pinnate with deflexed linear lobes, glaucous, the veins with short, stiff hairs. Heads few, 2–3 mm across, cylindrical. Florets pale blue.

LA PALMA: Caldera de Tabouriente, 2000 m, Cumbre Nueva, Cumbre Vieja, forest regions, 1200–1500 m, locally frequent.

L. herbanica Burchd. Robust perennial herb with ovate, more or less entire leaves; margins deeply toothed. Heads in groups of 5–7 at tips of branches, about 5 mm across. Florets yellow.

FUERTEVENTURA: Rocky cinder slopes near La Oliva in the N. of the island, 300 m, rare.

LAUNAEA

Spiny shrubs or herbs. Leaves few, basal. Involucral bracts in several rows, imbricate; margins scarious. Receptacle without scales. Florets yellow. Cypselas slightly compressed, ribbed, not beaked. Pappus of several rows of simple hairs.

L. arborescens (Batt.) Murb. Shrub up to 7 cm or more. Stem densely spiny. Leaves pinnatisect with narrowly linear lobes. Heads about 1 cm across. Cypselas attenuate at the base. ***Aulaga.*** **(318)**

ALL ISLANDS: Locally dominant in very dry areas especially on S. slopes, up to 700 m. (S.W. Mediterranean region, N. Africa.)

REICHARDIA

Annual to perennial herbs. Leaves entire to deeply pinnatisect. Heads few. Involucral bracts in several rows, at least the outer with broad, scarious margins. Receptacle without scales. At least the outer cypselas 4- to 5-angled, transversely rugose. Pappus of several rows of soft hairs.

1. Leaves densely white-papillose, plants with a basal rosette and short scapes .. .. **R. crystallina**
1. Leaves more or less glabrous.
   2. Leaves pinnatisect with spiny margins, scapes usually long, with several heads .. .. **R. ligulata**
   2. Leaves more or less entire, scapes short, usually with a single head (Lanzarote, Famara) .. .. .. **R. famarae**

Three Mediterranean species, all small annual to perennial herbs also occur in the Canaries: *R. tingitana* (L.) Roth, *R. picroides* (L.) Roth and *R. intermedia* (Sch. Bip.) Hayek.

R. crystallina (Sch. Bip.) Bramwell. Perennial herb of coastal areas. Leaves in a basal rosette, pinnatisect, densely white-papillose. Scapes usually with a single head. Ligules with a reddish stripe on the back.

TENERIFE: S. coast region below Guimar, Playa de Viuda, Punta de Teno, local.

R. ligulata (Vent.) Aschers. Stock short, woody. Leaves pinnatisect, glabrous or with a few white papillae, the margins more or less spiny. Scapes long, bracteose. Heads usually 3–4.

ALL ISLANDS: Locally abundant in the lower zone on rocks and cliffs up to 800 m.

R. famarae Bramwell & Kunkel. Small cliff-plant with woody stock and dense rosettes. Leaves more or less entire, obovate to broadly spathulate. Scapes short. Heads usually solitary. **(109)**

LANZAROTE: Riscos de Famara, 50–400 m, cliff crevices, rare.

MONOCOTYLEDONES

Grasses, sedges, palms, orchids, lilies, snowdrops, daffodils, etc.

1. Trees.
   2. Leaves pinnately lobed .. .. **Palmae**
   2. Leaves entire, sword-shaped .. .. .. **Liliaceae (Dracaena draco)**
1. Shrubs, climbers or herbs.
   3. Inflorescence in the formof a spadix surrounded by a spathe .. .. **Araceae**

3. Inflorescence not as above.
  4. Perianth entirely scarious and reduced to scales or bristles.
    5. Flowers arranged in flat or cylindrical spikelets, each flower subtended by a membranous bract (glume), ovary with 1 locule.
      6. Stems triangular in cross-section .. .. .. **Cyperaceae**
      6. Stems circular in cross-section .. .. .. **Gramineae**
    5. Flowers in heads, not subtended by a membranous bract, ovules 3–many .. .. **Juncaceae**
  4. Perianth not reduced to scales or bristles, usually petaloid (in *Tamus* very small), in 1 or 2 whorls.
    7. Ovary superior .. .. **Liliaceae**
    7. Ovary inferior.
      8. Unisexual climbers, usually with cordate leaves .. **Dioscoreaceae**
      8. Not as above.
        9. Flowers more or less regular or sometimes irregular, stamens 3 or 6.
          10. Stamens 6 (sometimes 3 + 3 staminodes) .. **Amaryllidaceae**
          10. Stamens 3 (staminodes absent) .. .. .. **Iridaceae**
        9. Flowers strongly irregular, stamen 1 .. .. **Orchidaceae**

LILIACEAE —— Lily Family

Trees, shrubs, climbers or herbs. Leaves alternate, Flowers regular, 3-merous. Perianth usually of 2 petaloid whorls. Stamens in 2 whorls. Ovary superior, usually 3-celled, rarely 1-celled. Fruit a capsule or berry.

1. Trees with sword-shaped leaves .. .. **Dracaena**
1. Shrubs, climbers or herbs.
  2. Leaves with tendril-like stipules .. .. **Smilax**
  2. Stipules not tendril-like or absent.
    3. Herbs, leaves normally developed.
      4. Flowers violet-bluish, in a spike-like inflorescence .. **Scilla**
      4. Flowers white, in a compact, sessile inflorescence .. **Androcymbium**
    3. Shrubs or climbers, leaves scale-like with leaf-like cladodes in the axils.
      5. Cladodes needle-like, linear, not bearing the flowers .. **Asparagus**
      5. Cladodes leaf-like, ovate or lanceolate, bearing the flowers in the middle of the lower surface or on the margins .. .. **Semele**

DRACAENA

Trees. Leaves entire. Inflorescences paniculate. Tepals and stamens 6. Fruit a berry.

D. draco L. Trunk silvery-grey. Branches dichotomous. Leaves sword-shaped, glaucous, coriaceous, reddish at the base, borne in dense rosettes at the summits of branches. Inflorescences terminal, branched. Tepals greenish-white. Fruits globose, up to 1·5 cm across, red-orange. ***Drago.*** **(322)**

The dragon-tree is commonly cultivated in parks and gardens in most parts of the islands. Wild specimens are, however, now extremely rare.

TENERIFE: Cuevas Negras de los Silos, Barranco del Infierno, Roque de las

Animas, Punta del Hidalgo, Masca, Barranco del Fraile, on vertical cliffs, 50–500 m; LA PALMA: N. region, Barlovento, Tijarafe, Garafia, Las Breñas; GRAN CANARIA: Barranco de Arguineguin.

SMILAX

Woody liana. Stems branched, thorny. Leaves alternate, the nerves more or less parallel. Stipules tendril-like. Flowers dioecious. Inflorescences more or less umbelliform. Fruit a 1- to 2-seeded, globose berry.

S. canariensis Willd. Lower stems thorny. Leaves ovate, cuneate to truncate at base; margins not spiny. Flowers in more or less simple umbels. Fruits reddish. ***Zarzaparrilla.***

TENERIFE: Los Silos, Barranco del Agua, Los Organos de Agua Mansa, Teno, Sierra Anaga, Vueltas de Taganana; LA PALMA: Barlovento, Los Tilos, La Galga; GOMERA: Arure, Chorros de Epina, laurel forests and below, 400–1300 m, locally frequent.

S. mauritanica Poir. With cordate leaves with spiny margins is frequent in the lower zone and on rocks in the forests of the W. islands (N. Africa, Iberian Peninsula).

ANDROCYMBIUM

Bulb tunicate. Stem hypogeous. Inflorescence subsessile. Perianth of 2 rows of 3 subequal, free tepals. Stamens 6. Ovary sessile, 3-locular. Styles 3.

A. psammophilum Svent. Bulb elongate-ovoid. Leaves linear, 8–15 cm long, acute, glaucous. Inflorescence 3- to 6-flowered. Flowers subsessile. Tepals white. Seeds rugose, black.

FUERTEVENTURA: N. coast region, in dunes and maritime sands nr. Corralejo, locally abundant; LANZAROTE: El Jable nr. La Caleta in the N.W., locally frequent.

SCILLA

Bulb tunicate. Leaves all basal, lanceolate or oblong. Flower stem simple, leafless. Flowers small, in clusters, blue-purple. Perianth segments free. Stamens 6 attached to the base of the perianth segments.

1. Large, robust plant, leaves broad, inflorescence many-flowered .. .. **S. latifolia**
1. Small plant up to 15 cm, leaves lanceolate, inflorescence few-flowered .. .. **S. haemorrhoidales**

S. latifolia Willd. (*S. iridifolia* Webb & Berth.) Large, robust, glaucous plant with a large bulb. Flowering stems up to 50 cm. Leaves 3–6 in a basal rosette, broadly lanceolate, up to 40 cm. Inflorescence dense, many-flowered, occasionally branched. Flowers blue-lilac, about 4 mm; anthers violet.

TENERIFE: Teno region, San Andrés, La Rambla, rocky slopes in *Euphorbia* scrub; LANZAROTE: Famara region, 150–500 m, rather rare (Morocco).

S. haemorrhoidales Webb & Berth. Like *S. latifolia* but very much smaller in all its parts, leaves lanceolate. Flowering stems up to 15 cm.

TENERIFE: Coastal zone of the Anaga region, Santa Cruz, Igueste, San Andrés, Taganana, Bajamar, Teno, Guimar, Badajoz, Orotava valley, near sea-level to nearly 1600 m at Vilaflor; GRAN CANARIA: Tafira Alta, Santa Brigida, Agaete, etc, locally common; GOMERA: Roque Agando, Barranco de la Laja, Vallehermoso region; LANZAROTE: Teguise, Las Nieves; HIERRO, LA PALMA.

ASPARAGUS

Erect or scrambling shrubs with linear, needle-like cladodes in fascicles. Flowers small, in axillary clusters. Stamens 6. Fruit a reddish berry.

1. Stems thorny .. .. **A. pastorianus**
1. Stems unarmed.
  2. Cladodes less than 3 cm, needle-like or channelled.
    3. Scrambling plant, cladodes channelled .. .. .. **A. umbellatus**
    3. Shrub, cladodes not channelled.
      4. Branches erect or patent, cladodes very fine, needle-like .. .. **A. scoparius**
      4. Branches arching, pendulous, cladodes capillary-like **A. plocamoides**
  2. Cladodes longer than 3·5 cm, fleshy .. .. .. **A. arborescens**

A. pastorianus Webb & Berth. Procumbent, scrambling shrub with thorny stems. Cladodes densely crowded in the axils of the thorns. ***Espina blanca.*** **(323)**

TENERIFE: Coastal and xerophytic regions especially in the S., San Andrés, Bajamar, Garachico, Teno, Guimar, Candelaria, Granadilla, Adeje, etc; GRAN CANARIA: Tafira Alta, Guia, Moya, Tirajana valley, Santa Lucia, etc; GOMERA: Puerto de Vallehermoso; locally very common 50–800 m.

A. umbellatus Link. Scrambling shrub. Stems thornless. Cladodes flat, channelled, short. Inflorescence a simple, pendulous umbel.

TENERIFE: Locally very common, Sierra Anaga, Orotava valley, Guimar, Bandas del Sur; GRAN CANARIA: Caldera de Bandama, Tafira, Telde, Guia, Agaete; LA PALMA: Los Tilos, Barranco de las Angustias, La Galga, etc; GOMERA: N. coast, Agulo, Roque Cano de Vallehermoso; HIERRO: Cuesta de Sabinosa, cliffs and rocks in the lower and laurel forest zones.

A. arborescens Willd. Tall erect shrub. Cladodes long, more or less patent, fleshy. **(110)**

TENERIFE: Dry regions of the S. and W. of the island up to 1000 m, Guia de Isora, Candelaria, Arico, Guimar, etc, Teno: GOMERA: Puerto de Vallehermoso; GRAN CANARIA: Bañaderos, Playa de Jinamar, Telde, Barranco Goteras.

A. plocamoides (Bolle) Svent. Like *A. arborescens* but the branches arching and pendulous and the cladodes very fine, needle-like and short.

TENERIFE: Agua Mansa in pine woodland, S. region, Arico, Vilaflor, 1200–2000 m; GRAN CANARIA: S. region, Caldera de Tirajana, Rincón de Tenteniguada; LA PALMA: Lomo de Vizcaino.

A. scoparius Lowe. Erect shrub. Cladodes needle-like, 2 cm or less, very densely crowded on alternate, ascending branches. Inflorescence many-flowered.

TENERIFE: Common in the lower zone, Los Silos, Cuevas Negras, El Fraile, Garachico, Masca, Valle de Santiago del Teide, Chio; GRAN CANARIA: Barranco de Moya, Guiniguada, Atalaya, Santa Lucia de Tirajana, up to 800 m; LA PALMA: Nr. Santa Cruz de La Palma, Barranco Carmen, Los Sauces, etc.

A. fallax Svent. from Vueltas de Taganana, TENERIFE differs from the above species in being densely leafy and having few-flowered inflorescences.

SEMELE

Climber with leaf-like, alternate cladodes. Inflorescences borne on the margins or towards the centre of the lower surface of the cladodes. Anthers forming a more or less continuous ring. Fruit a 1-seeded berry.

S. androgyna (L.) Kunth. Liana. Cladodes glabrous, lanceolate to ovate, the margins sometimes more or less sinuately lobed. Flowers small, in fascicles of

2–6 on the margins or centre of the cladode or both. Tepals small, greenish. Fruit globose, greenish to black, about 1 cm across. ***Gibalbera, Alcacan.*** **(324)**

TENERIFE: Laurel forest regions, Sierra Anaga, Vueltas de Taganana, Los Silos, Guimar, etc, 400–1200 m, often on cliffs; GRAN CANARIA: Los Tiles de Moya var. ***gayae*** (Webb & Berth) Burchd. with the inflorescence usually in the centre of the cladode); LA PALMA: Frequent in the forests of the N.W. region, Los Sauces, Barranco de los Tilos, La Galga, Barlovento, Cumbre Nueva; GOMERA: El Cedro, Arure, Riscos de Agulo, etc; HIERRO: El Golfo, Frontera, Miradero. Often grown as an ornamental.

AMARYLLIDACEAE — Snowdrop and Daffodil Family

Bulbous herbs with basal leaves. Flowers solitary or umbellate, usually regular, 3-merous. Perianth petaloid, in 2 whorls, often with a tube-like corona. Stamens 6 in 2 whorls often inserted on the perianth segments. Ovary inferior. Style simple, capitate or 3-lobed. Fruit a capsule.

PANCRATIUM

Bulbs tunicate. Leaves strap-shaped with parallel sides. Corolla with a broad tube and an outer, free perianth. Capsule subglobose, 3-locular. Seeds angular, black, glossy.

P. canariensis Ker.-Gawl. Flowering stems up to 80 cm. Leaves fleshy, obtuse, somewhat glaucous. Flowers in umbellate clusters of 5–10. Perianth white, strongly scented. **(111)**

TENERIFE: N. coast, frequent from the Orotava valley to Teno, Adeje, San Andrés; FUERTEVENTURA: Frequent, Puerto Rosario, La Oliva, Gran Tarajal; LANZAROTE: Arrecife, Haria, Peñitas de Chache, dry slopes in the lower zone.

DIOSCOREACEAE

Climbers with swollen rootstock. Leaves cauline, cordate. Flowers in a raceme, unisexual, regular, small. Ovary inferior. Fruit a berry.

TAMUS

Dioecious liana with annual stems and large, underground, stem-tubers. Leaves alternate. Flowers in axillary racemes. Stamens 6. Style with 3, 2-lobed stigmas. Fruit a berry.

T. edulis Lowe. Leaves entire, broadly ovate, deeply cordate at base; veins numerous, curving, more or less parallel. Female inflorescences shorter than the male. Berries reddish-orange, more or less globose or somewhat elongated. **(112)**

TENERIFE: Laurel forest and lower zone, Sierra Anaga, Taganana, Igueste, N. coast, Los Silos, Masca, etc, 200–600 m; GRAN CANARIA: Barranco de Moya, Santa Brigida, Tafira Alta, Bandama; LA PALMA: Cumbre Nueva, La Galga, Los Tilos, Cumbre Vieja, Mazo, up to 1200 m; GOMERA: El Cedro, Hermigua valley, Roque Monte Forte; HIERRO: Forests of El Golfo.

IRIDACEAE — Iris Family

Herbs with rhizomes, corms or bulbs. Flowers 3-merous, hermaphrodite, usually regular, usually with 1 or 2 bracts forming a spathe at their base. Perianth in 2 whorls. Stamens 3. Ovary inferior. Style 3-lobed. Fruit a capsule opening by valves.

ROMULEA

Bulb small, tunicate. Basal leaves linear, acute; cauline leaves similar, shorter. Flowers more or less sessile in the spathe. Ovary inferior. Perianth tubular. Anthers sagittate. Capsule subglobose.

R. columnae Seb. & Maur. var. ***grandiscapa*** Gay (incl. *R. hartungii* Parl.) Bulb ovoid. Flower-stem very short, overtopped by the long, narrow leaves. Spathes ovate-lanceolate, about 1 cm. Flower purple with a yellow throat. Seeds brown-orange, subglobose.

ALL ISLANDS: Locally frequent in the lower zone and occasionally in forests, 150–800 m.

JUNCACEAE —————— Rush Family

Herbs. Leaves mostly basal, sheathing. Inflorescences cymose or paniculate, sometimes condensed to a head. Flowers bisexual, regular. Perianth 6, scarious. Stamens 6, rarely 3. Ovary 3-locular, superior. Fruit a capsule.

LUZULA

Tufted perennials with flat or channelled, white-hairy leaves. Flowers in a dense or lax cluster. Perianth segments papery. Stamens 6. Fruit a capsule with 2–3 seeds.

1. Inflorescences dense, parianth segments silvery-white .. **L. canariensis**
1. Inflorescences lax, perianth segments brownish-purple .. **L. purpurea**

L. canariensis Poiret. Robust perennial. Leaves flat, somewhat reddish and with long white hairs near the base. Flowering stem leafy. Inflorescence dense, flat-topped to somewhat elongate. Perianth segments silvery-white. **(113)**

TENERIFE: Sierra Anaga, in laurel forests, Las Mercedes, Vueltas de Taganana, Cruz de Afur, El Bailadero; GOMERA: El Cedro forest, Cumbre de Carboneros 600–850 m, locally frequent.

L. purpurea Link. Small herb. Leaves flat, sparsely white-hairy at base. Inflorescence large, lax with long peduncles. Perianth segments brownish-purple.

TENERIFE: Locally common in forests and in the lower zone, Guimar, Agua Mansa, Los Realejos, Cumbre de Taganana, Las Mercedes, La Esperanza; GRAN CANARIA: Barranco de Tirajana, Los Berrezales, Caldera de Bandama, etc; LA PALMA: Cumbre Nueva, El Paso, Barranco de las Angustias, La Galga; GOMERA: Hermigua valley, El Cedro, Roque Agando, Laguna Grande; HIERRO: Fuente de Tinco, Riscos de Jinamar, Tinor, 300–1200 m.

PALMAE —————— Palm Family

Trees. Leaves large, pinnately divided. Inflorescences panicles or spikes. Perianth 6. Stamens 6. Ovary 1 to 3, locular, superior. Fruit a drupe.

PHOENIX

Unbranched trees. Leaves in a large terminal rosette, pinnate. Flowers dioecious. Fruit a one-seeded, pale orange-yellow, fleshy berry.

P. canariensis Chabaud. Up to 15 m. Old leaves and leaf-bases persistent. Leaves green, arched, numerous (up to 200), 5–6 m long; rachis spinose at base, the pinnules linear, acute. Fruiting peduncles pendulous. Fruits about 2 cm long, ellipsoid. ***Palma*** or ***Palmera.*** **(320) (321)**

ALL ISLANDS: Common in the lower zones and frequently cultivated; GOMERA: Vallehermoso, Vallegranrey; GRAN CANARIA: Maspalomas; FUERTEVENTURA: Pajara; LANZAROTE: Haria.

GRAMINEAE —— Grass Family

Leaves entire, often ligulate at the base. Flowering stems round, hollow. Flowers borne between a membranous bract (lemma) and a bracteole (palea), arranged in spikelets subtended by 1 or 2 empty bracts, the glumes. Styles feathery.

A species of bamboo (*Arundo donax L.*) is commonly cultivated and widely naturalized in the Canaries.

1. Spikelets of 1 floret.
  2. Inflorescence slender, compound .. .. **Agrostis**
  2. Inflorescence simple, dense, oval or cylindrical .. .. **Phalaris**
1. Spikelets of 2 or more florets.
  3. Upper floret club-shaped, sterile .. .. **Melica**
  3. Upper floret not club-shaped, usually fertile.
    4. Glumes much larger than florets, awn bent .. .. **Avena**
    4. Glumes equalling or shorter than florets or if longer then only a single plume present.
      5. Spikelets borne singly on the ends of long peduncles .. **Tricholaena**
      5. Spikelets aggregated into a dense inflorescence or long spike.
        6. Glume 1 in lateral spikes .. .. **Lolium**
        6. Glumes 2.
          7. Glumes more or less equal, spikelets crowded into 1-sided masses at tips of panicle branches .. **Dactylis**
          7. Glumes unequal, spikelets not crowded in dense masses.
            8. Spikelets very long, flattened .. .. .. **Brachypodium**
            8. Spikelets short, more or less terete .. .. **Festuca**

AGROSTIS

Perennials. Inflorescences slender, compound. Spikelets small, 1-flowered. Glumes papery, 1-veined. Lemma oval, blunt, 3- to 5-veined with a short awn originating near the base or absent.

A. canariensis Parl. Rhizome short. Culms up to about 1 m but usually shorter. Leaves flat or inrolled. Panicle long, narrow, branched near base. Lemma scabrid, 3-nerved, the nerves usually excurrent with aristate projections. Glumes scabrid.

TENERIFE: N. coast, Mesa de Mota, La Guancha, Icod, 600–800 m, San Andrés, Sierra de Anaga, S. region, very sporadic; GRAN CANARIA: Caldera de Tirajana, Valle de Mogan; GOMERA: N.E., Barranco de Cabrito, Barranco de la Laja.

PHALARIS

Inflorescence dense, oval or cylindrical. Spikelets appearing 1-flowered, with

2 small sterile florets and a single, terminal fertile one. Glumes large, papery, keeled, 3- to 7-veined. The 5-veined lemma subtended by 2 sterile ones at base.

P. canariensis L. Annual up to 1·5 m. Inflorescence dense, oval, up to 6 cm. Spikelets broad, overlapping, whitish with green veins.

ALL ISLANDS: Locally frequent in dry areas particularly on LANZAROTE and FUERTEVENTURA: Native of N. Africa and Macaronesia, widely introduced into Europe as the seeds are used in bird-seed mixtures.

FESTUCA

Inflorescence a compact or spreading panicle. Spikelets several-flowered. Glumes unequal, keeled. Lemmas convex, shortly awned.

F. bornmuelleri Hack. Perennial with robust, densely tufted culms up to 50 cm. Leaves inrolled. Inflorescence erect, branched; branches more or less erect. Awns very short.

TENERIFE: Las Mercedes, Agua Mansa, Cumbres de Pedro Gil, pine and laurel forests 700–1600 m; LA PALMA: La Cumbrecita, Caldera de Tabouriente, rocks and slopes in pine woodland, 800–1400 m, locally frequent.

LOLIUM

Inflorescence simple, spike-like. Spikelets sessile, flattened. Glume solitary, 5- to 9-veined. Lemma 5-veined with short awns or awnless. Palea translucent.

1. Glumes much longer than spikelets .. .. .. **L. lowei**
1. Glumes equalling spikelets .. .. **L. canariense**

L. canariense Steud. (*L. gracile* Parl.) Slender annual. Glume more or less equalling spikelet in length. Lemma with awns up to 2 cm. The distribution of this species is very poorly known. It is probably far more widespread than indicated below.

TENERIFE: Local, Punta de Teno, Valle de Santiago del Teide, Adeje; FUERTEVENTURA: Jandia, Pico de la Zarza, Betancuria.

L. lowei Menzes. Erect annual up to 35 cm. Spikes erect, very rigid and stout, cylindrical. Glumes much longer than the spikelets. Lemma sometimes shortly awned.

GRAN CANARIA: Coastal region, Cuesta de Silva, etc.

AVENA

Inflorescence compound. Branches with 2–4 florets, drooping when mature. Glumes large, papery, longer than florets. Lemma thick, leathery, 7-veined, with a long-projecting, rough, bent awn on the back. Palea ciliate on the edges.

1. Spikelets about 3·5 cm, lemma with 2 small teeth at tip .. **A. canariensis**
1. Spikelets less than 3 cm, lemma with 2 long teeth at tip .. **A. occidentalis**

A. canariensis Baum, Rajh. & Samp. Erect annual with a lax, 1-sided, nodding inflorescence with several large (3·5 cm) open spikelets. Lemmas hairy; lower lemma with 2 small teeth at the apex.

LANZAROTE: N. region about 300 m, Montaña los Helechos; FUERTEVENTURA: Interior of the island, La Oliva, Betancuria, etc, dry slopes above 200 m, locally common; TENERIFE: Valle de Santiago del Teide, El Molledo, 900 m.

A. occidentalis Dur. Like *A. canariensis* but the spikelets much smaller and the lower lemma with 2 long teeth.

TENERIFE: N. coast, Cuevas Negras, Valle de Orotava, Santiago del Teide;

FUERTEVENTURA: La Oliva, central region, rather common, often a weed of cultivated ground; HIERRO.

TRICHOLAENA

Inflorescence lax. Spikelets borne singly on long peduncles, with one fertile and several rudimentary florets. The glume of one of the rudimentary florets often appears to form a 3rd glume in the fertile floret.

T. teneriffae (L. fil.) Link. Perennial. Culms up to 60 cm. Panicle of several filiform, flexuous branchlets. Spikelets finely hairy. Lower glume small.

ALL ISLANDS: Locally very common in dry rocky places in the lower zone especially on S. slopes. (N. Africa).

DACTYLIS

Inflorescence usually compound. Spikelets densely crowded in ovate to oblong clusters at the ends of side branches. Spikelets short-stalked, 2- to 5-flowered, flattened, the glumes 3-veined, keeled. Lemma 5-veined, keeled with a short awn.

D. smithii Link. Robust perennial, culms often lignified at base. Dead leaves persistent, often coiled. Spikes with lower branches spreading or patent. Spikelets often with a violet tinge. **(114)**

TENERIFE: Frequent along the N. coast, Punta del Hidalgo, Taganana, Icod el Alto, Barranco Ruiz, Los Silos, Teno, etc, 50–650 m; LA PALMA: Occasional, cliffs near Puntallana, La Galga, Barranco de Las Angustias, Caldera de Tabouriente, up to 800 m.

BRACHYPODIUM

Spikelets many-flowered, long, forming a single short spike. Lemmas shortly awned. Palea with shortly, sparsely ciliate margins.

B. arbuscula Gay ex Knoche. Very robust, glaucous perennial often with lignified bases to the culms. Leaves flat or inrolled, very stiff. Spikes erect, yellowish, with long, lateral spikelets. ***Pajonazco.*** **(115)**

TENERIFE: Between Buenavista and Punta de Teno, on coastal cliffs, rather rare, 80–200 m; GOMERA: N. coast region, locally frequent, Barranco de la Villa, Agulo, Vallehermoso, Roque Cano, up to 500 m, cliffs and steep rocky slopes in the lower zone; HIERRO.

MELICA

Inflorescence usually simple. Spikelets with 2–4 florets; the upper floret club-shaped, sterile.

1. Spikelets glabrous .. .. **M. teneriffae**
1. Spikelets with dense, silky hairs .. .. **M. canariensis**

M. teneriffae Hack. Perennial. Spikes long, slender, sometimes branched, somewhat arched and one-sided. Spikelets glabrous. **(325)**

TENERIFE: Icod, Garachico, Teno, El Palmar, Sierra Anaga, Batan, 200–600 m, dry slopes in the lower zone, locally frequent; LANZAROTE: Peñitas de Chache, 450–600 m, rare on dry cliffs; GOMERA: N. coast region, Hermigua to Vallehermoso.

M. canariensis Hemp. Glaucous perennial. Leaves usually inrolled, narrow. Spikes long, slender, dense. Spikelets with dense, silky hairs on the lemmas.

TENERIFE: Granadilla, Los Silos, Barranco del Agua, Los Azulejos, Bajamar, rather rare, 400–2000 m; GOMERA: Barranco de la Villa, 400 m; LANZAROTE: Riscos de Famara, Haria region.

ARACEAE —— Arum or Cuckoo-pint Family

Herbs. Flowers minute, borne on a spadix enclosed by a spathe. Fruit usually a berry.

DRACUNCULUS

Herbs. Leaves long-petiolate, pedately divided. Spathe with a long tube, not hooded. Spadix almost as long as spathe. Fruit a cluster of small berries.

D. canariensis Kunth. Up to 1·5 m. Leaves 5- to 7-lobed; the petiole 20–40 cm. Spathe greenish-white or cream. Spadix pale yellow. Berries reddish-orange. ***Tacarontilla.*** **(116)**

TENERIFE: In and below the forest zone, Taganana, San Andrés, Candelaria, Icod, Los Silos, Cuevas Negras, etc, 200–600 m, sporadic; GRAN CANARIA: Bandama, Moya, Tafira Alta, Agaete, Los Berrezales, etc; LA PALMA: Barranco del Rio, Mazo, Barranco del Carmen; HIERRO: Las Lapas, Los Lanillos, Mocanal, Taibique, up to 800 m.

CYPERACEAE —— Sedge Family

Herbs. Leaves usually basal, sheathing at the base. Stems round or triangular in cross-section. Flowers subtended by membranous bracts (glumes), arranged spirally in bisexual or unisexual spikelets which are usually aggregated into spike-like inflorescences. Stamens 3. Ovary 1. Fruit nutlike.

CAREX

Stem triangular in cross-section. Flowers unisexual, each subtended by a papery bract (glume), densely clustered into 1 or several spikes. Petals absent; male flowers with 2–3 stamens; female flowers with a single ovary surrounded by the sack-like perigynium which is usually beaked. Fruit a 1-seeded nutlet enclosed within the perigynium.

1. Leaves broad (0·8–1·2 cm), upper spikelets male .. .. **C. perraudieriana**
1. Leaves narrow (0·3–0·5 cm), spikelets all andryogynous .. **C. canariensis**

C. canariensis Kükenth. Densely tufted perennial with a short, somewhat woody, scaly rhizome. Up to 1 m. Stems sharply 3-angled. Leaves stiff, very rough, flat, 3–5 mm broad, glaucous. Inflorescences androgynous; spikelets short (1–2 cm), 3–10 per inflorescence. Female glumes papery, light brown, triangular.

TENERIFE: Laurel forests, in damp areas near streams and springs, Vueltas de Taganana, El Bailadero, Icod el Alto etc, 600–800 m; HIERRO: Forest regions of El Golfo, Jinamar to Frontera, Sabinosa; LA PALMA: Forests of La Galga, Barranco de Los Tilos, Barlovento, etc; GOMERA: Monte del Cedro, Riscos de Agulo, Laguna Grande.

The much smaller, widespread *C. divulsa* with very short spikelets is common in the W. and C. islands, (Europe, Mediterranean region).

C. perraudieriana Gay. Perennial. Rhizome short. Culms up to 120 cm. Leaves flat, broad (0·8–1·2 cm) with 2 prominent veins. Spikes unbranched or occasionally branched near the base, the upper portion slender, male; the lower broader, andryogynous.

TENERIFE: Laurel forests of Sierra Anaga, Vueltas de Taganana, El Bailadero, etc, sporadic.

ORCHIDACEAE —————— Orchid Family

Perennial herbs with rhizomes or tubers. Leaves entire, usually alternate, sometimes spotted. Inflorescence a spike. Flowers bilaterally symmetrical. Perianth segments usually petaloid, the inner directed downwards forming a lower lip or labellum which is sometimes spurred; the outer sometimes forming a helmet. Stamen 1. Stigmas 2, often confluent. Ovary inferior. Fruit a capsule.

1. Leaves spotted .. .. **Neotinea**
1. Leaves not spotted.
 2. Flowers pink with purplish markings .. .. .. **Orchis**
 2. Flowers greenish-yellow.
  3. Leaves basal, labellum deeply divided into 3 long, linear lobes .. .. **Habenaria**
  3. Leaves cauline, labellum briefly 3-lobed .. .. **Gennaria**

NEOTINEA

Slender. Leaves usually spotted. Flowers small. Upper perianth segments forming a small, white-pinkish helmet; lip 3-lobed, the lateral lobes narrow.

N. intacta (Link) Reichenb. fil. Leaves oval with purplish spots. Inflorescence a small dense spike. Flowers pale pink to greenish-white, smelling of vanilla.

TENERIFE: Forest regions, Agua Mansa, Lomo de Pedro Gil, Las Mercedes, Las Raices, etc, 800–1400 m; HIERRO: Forests between Frontera and Jinamar; LA PALMA: Cumbre Nueva; GOMERA: forest regions between Roque de Agando and Roque Ojilla.

HABENARIA

Upper perianth segments forming a helmet; the lower forming a lip with 3 very long, linear, deeply divided lobes. Long spur present.

H. tridactylites Lindl. Basal leaves 2, large, ovate. Flowering stem leafless. Inflorescence long, 10- to 30-flowered. Flowers greenish. **(118)**

TENERIFE: Forest cliffs and cliff-ledges in the lower zone, Sierra Anaga, Icod el Alto, Barranco de Masca, Agua Mansa, Los Silos, etc, locally common; GRAN CANARIA: Barranco de Moya, Los Tiles, Caldera de Bandama, San Mateo, Agaete, 200–800 m; GOMERA: Barranco de la Laja, Roque de Agando, Vallehermoso; LA PALMA: Barranco Herradura, La Galga, Las Breñas, Cumbre Nueva, Fuencaliente, Caldera de Tabouriente; HIERRO: Valverde region, forests of El Golfo, Las Lajas, locally abundant.

GENNARIA

Stem with 2 well-developed cordate leaves, the upper smaller than the lower. Flowers in a spike, small, greenish-yellow. Labellum 3-lobed. **(117)**

G. diphylla Parl. Cliffs and banks in laurel forests, TENERIFE, LA PALMA, HIERRO, GOMERA, locally frequent. (N. Africa, S. Spain, Macaronesia.)

ORCHIS

Flowers with the upper perianth lobes forming a helmet and the lower a large, 3-lobed lip. Spur present.

C. canariensis Lindl. Basal leaves ovate, not spotted. Flowers pink, in a rather short, lax spike. Lower lip of corolla with purple markings.

TENERIFE: Frequent in the upper xerophytic zone and in pine woodland, Cumbre de Masca, Teno, Los Organos de Agua Mansa, Arago, etc; GRAN CANARIA: Cumbre de San Mateo, Agaete, Tamadaba, etc; HIERRO: Miradero.

O. mascula L. with darker flowers and a spotted lip occurs at a single station in the pine woods of the Cumbrecita of LA PALMA. (Europe, Mediterranean region).

Some Canary Islands Ferns

The forests and valleys of the Canaries have a rich fern flora. The following species, some of which are illustrated, are amongst the most common.

Dry Rocks and Walls
Davallia canariensis Sm. (**119**)
Asplenium trichomanes L.
Notholaena marantae (L) R.Br. (**1**).
Rock Crevices
Adiantum reniforme L. (**120**)
Cheilanthes maderensis Lowe (**121**)
Notholeana vellea (Ait.) Desv.

Laurel Forest Regions
Woodwardia radicans (L.) Sm. (**1**)
Athyrium umbrosum (Ait.) Presl
Asplenuim onopteris L.
Asplenium hemionitis L. (**1**)
Blechnum spicant (L.) With.
Subalpine Regions
Cheilanthes guanchica Bolle

Glossary of Botanical Terms

Fig. i.

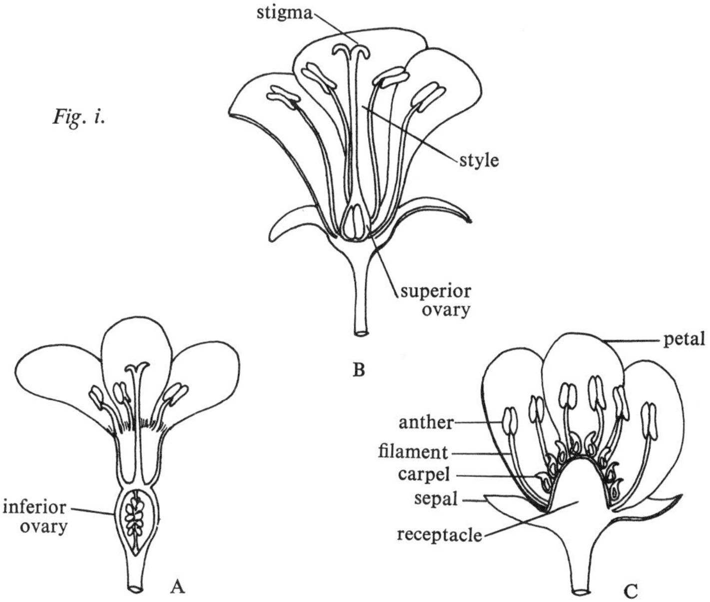

Achene: A small, dry, indehiscent, 1-seeded fruit.

Actinomorphic: Regular, radially symmetrical.

Acute: Sharp, ending in a point. *Fig. iii*, B.

Alveolate: Spongy.

Amplexicaul: Clasping or embracing the stem.

Androgynous: Having male and female flowers in the same inflorescence.

Anther: The pollen-bearing part of the stamen.

Apiculate: Having a short, sharp, pointed tip. *Fig. iii*, C.

Appressed: Closely and flatly pressed against an organ (as hairs on a leaf).

Arachnoid: With slender, entangled hairs.

Arcuate: Curved, bowed.

Aril: An appendage or outer covering of a seed.

Aristate: Provided with a stiff bristle. *Fig. iii*, A.

Attenuate: Long tapering. *Fig. iii*, J.

Auricle: An ear-shaped appendage.

Awn: A bristle-like appendage.

Axil: Upper angle formed by the union of stem and leaf.

Axillary: Situated in the angle formed by the base of a leaf or petiole and the stem.

Beak: A long prominent point, usually applied to a fruit appendage.

Berry: Pulpy, indehiscent, few- or many-seeded fruit with no true stone (e.g. tomato).

Bract: A reduced leaf associated with the inflorescence.

Bracteole: A secondary bract.

Caducous: Falling off prematurely.

Calyx: The whorl of sepals.

Campanulate: Bell-shaped.

Canescent: Grey-pubescent, hoary.

Capitate: In a head, aggregated into a compact cluster.

Capsule: A simple, dry fruit formed by the fusion of 2 or more carpels. *Fig. v*, H.

Carnose: Fleshy.

Carpel: An ovule-bearing unit of a simple ovary in the female flower, 2 or more carpels may be fused to form a compound ovary.

Carpophore: The stalk to which carpels are attached, in *Umbelliferae* a portion of the receptacle prolonged between the carpels.

Caruncle: A small appendage at or near the point of attachment of a seed.

Catkin: A hanging spike of small flowers, usually with scales between the flowers and often unisexual.

Caudate: With a tail-like appendage. *Fig. iii*, G.

Caudex: The woody base of a perennial plant.

Cauline: Belonging to the stem.

Chasmophyte: A plant dwelling in cliff crevices.

Ciliate: Bearing hairs on the margin. *Fig. ii*, J.

Cladode: A branch or stem simulating a leaf.

Coccus: One of the parts of a lobed fruit with 1-seeded cells.

Connate: United.

Cordate: Heart-shaped. *Fig. iii*, N.

Coriaceous: Leathery.

Corolla: The whorl of petals, often united into a tube.

Corymbose: In a more or less flat-topped inflorescence with the outer flowers opening first. *Fig. vi*, I.

Crenate: With shallow, blunt teeth, scalloped. *Fig. ii*, B.

Crustaceous: With a hard, brittle texture.

Culm: The jointed stem of grasses and sedges, usually hollow except at the nodes.

Cuneate: Wedge-shaped, triangular with the narrow end at the attachment point. *Fig. iii*, K.

Cupule: The cup of fruits such as the acorn.

Cyathium: The 'flower' of *Euphorbia*

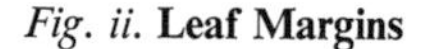
Fig. ii. **Leaf Margins**

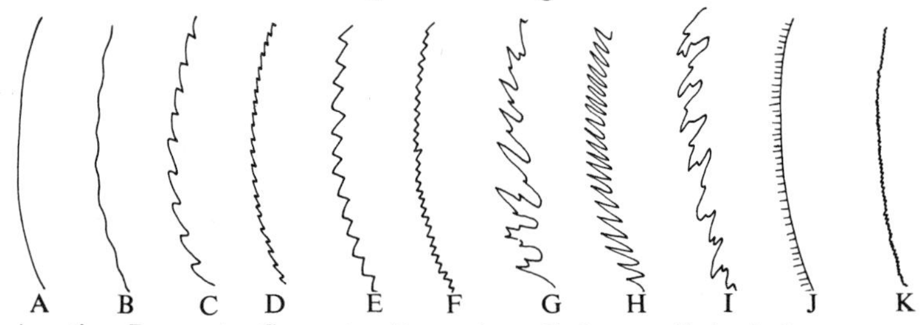

A, entire; B, crenate; C, serrate; D, serrulate; E, dentate; F, denticulate; G, lacerate; H, lacinate; I, runcinate; J, ciliate; K, erose;

consisting of a cup-like involucre containing the small, true flowers.

Cyme: Broad, more or less flat-topped inflorescence in which the central flowers open first. *Fig. vi*, C.

Cypsela: The single-seeded fruit of Compositae, derived from an inferior, unilocular ovary. *Fig. v*, J.

Decurrent: Extending down and fused to the stem or other organ.

Decussate: Leaves in pairs alternately crossing at right angles.

Deflexed: Reflexed or bent sharply downwards.

Deltoid: Triangular.

Dentate: With sharp teeth at right angles to the margin. *Fig. ii*, E.

Denticulate: Finely dentate. *Fig. ii*, F.

Dichasium: A cyme-like inflorescence with 2 lateral axes. *Fig. vi*, H.

Dichotomous: Branching by equally forking in pairs.

Dioecious: With male and female flowers on separate plants.

Disc-floret: The central, tubular flowers in the heads of some Compositae.

Discoid: Having only disc-florets.

Divaricate: Extremely divergent.

Drupe: A stone fruit, a single-seeded, indehiscent fruit with the seed having a stony endocarp (drupaceous). *Fig. v*, C.

Echinulate: Having tiny prickles.

Ellipsoidal: Solid with an elliptical profile.

Elliptical: Oval, tapering at both ends. *Fig. viii*, E.

Epicalyx: An extra calyx-like structure below the calyx.

Endocarp: The inner layer of the fruit wall.

Entire: With a continuous margin. *Fig. ii*, A.

Erose: With an irregularly toothed or eroded margin as if bitten. *Fig. ii*, K.

Farinose: Mealy.

Fascicle: A condensed close cluster.

Filament: The stalk of a stamen.

Filiform: Thread-like.

Fimbriate: Fringed.

Flexuous: Having a wavy zig-zag form.

Floccose: Covered with tufts of soft, woolly hairs.

Follicle: Dry dehiscent fruit splitting only the dorsal side. *Fig. v*, G.

Glabrescent: Almost glabrous or becoming glabrous with age.

Glabrous: Without hairs.

Glaucous: Covered with a white bloom.

Globose: Globe-shaped.

Glochidiate: Pubescent with barbed, stiff hairs.

Glomerule: A cluster of heads usually in a common involucre.

Fig. iii. **Leaf Bases and Tips**

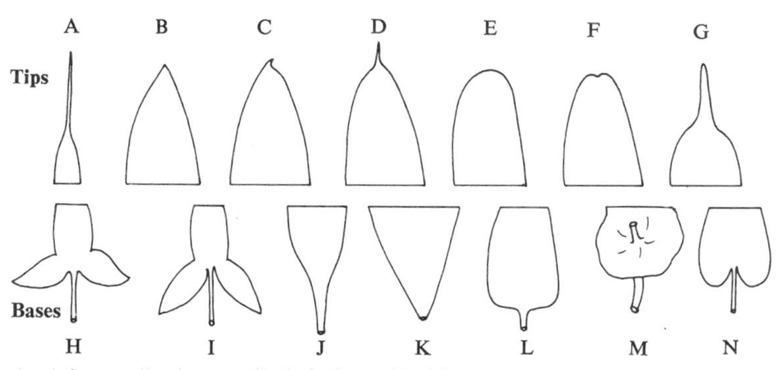

A, Aristate; B, Acute; C, Apiculate; D, Mucronate; E, Obtuse; F, Retuse; G, Caudate; H, Hastate; I, Sagittate; J, Attenuate; K, Cuneate; L, Truncate; M, Peltate; N, Cordate.

Glume: One of a pair of sterile bracts at the base of a grass spikelet.

Hastate: Arrow-shaped with the basal lobes turned outwards. *Fig. iii*, H.

Hemicryptophyte: Herbs with renewal buds at soil-level.

Hermaphrodite: With stamens and ovary in the same flower.

Hispid: With rough hairs or bristles.

Hispidulous: Minutely hispid.

Homogamous: With the simultaneous ripening of stamens and styles in the same flower.

Hypogeous: Underground.

Hypogynous: Inserted beneath the female organs.

Imbricate: Overlapping like roof-tiles.

Indumentum: A hairy or pubescent covering.

Inferior: Below, usually an ovary located below the perianth. *Fig. i*, B.

Involucel: A secondary involucre.

Involucre: A whorl of bracts subtending a flower of inflorescence.

Involute: Rolled in from the edges with the upper surface innermost.

Keel: The lower, united petals in a leguminous flower.

Labellum: The lip of an orchid flower.

Lacerate: Irregularly cut at the edge.

Laciniate: Deeply cut into narrow lobes separated by narrow, irregular incisions.

Lanate: Woolly, with interwoven curly hairs.

Lanceolate: Lance-shaped, tapering at both ends with the widest part below the middle. *Fig. viii*, C.

Leaf Arrangement *Fig. vii.*

Legume: A simple, superior fruit dehiscent into two valves. *Fig. v*, A.

Lemma: The lower of two bracts enclosing the grass flower.

Liana: Woody climber.

Ligulate: Strap-shaped (ligule).

Linear: Long and narrow with the margins more or less parallel. *Fig. viii*, A.

Lomentaceous: Having the form of a flat legume constricted between the seeds, the legume falling apart at the constrictions when mature to give a number of 1-seeded segments. *Fig. v*, B.

Lyrate: Pinnatifid with the terminal lobe enlarged. *Fig. viii*, K.

Medifixed: Of hairs fixed in the middle with either end free.

Mericarp: One-seeded part of an ovary split off at maturity. *Fig. v*, I.

Monoecious: Having separate male and female flowers borne on the same plant.

Mucronate: With a short, narrow point. *Fig. iii*, D.

Nectary: A nectar-secreting gland.

Oblanceolate: Lanceolate but broadest above the middle. *Fig. viii*, D.

Obovate: Ovate but with the distal end broader. *Fig. viii*, J.

Obtuse: Blunt or rounded. *Fig. iii*, E.

Orbicular: Flat with a circular outline. *Fig. viii*, H.

Ovate: Flat with an egg-shaped outline. *Fig. viii*, I. (Oval, *Fig. viii*, F.)

Palate: A rounded projection on the lower lip of some corollas which closes or almost closes the throat.

Fig. iv. **Leaf Types**

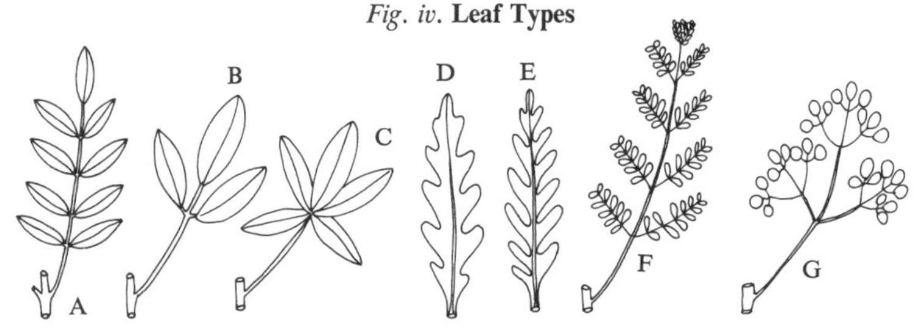

A, Pinnate; B, Trifoliate; C, Palmate; D, Pinnatifid; E, Pinnatisect; F, Bi-pinnate; G, Ternate.

Palea: The innermost bract of a grass floret.

Palmate: Divided in a hand-like manner. *Fig. iv*, C.

Panicle: A branched racemose inflorescence. *Fig. vi*, B.

Papillae: Minute, nipple-shaped projections.

Pappus: The various tufts of hairs or scales on the apex of Compositae fruits, generally considered to be a modified calyx.

Patent: Spreading.

Pectinate: Comb-like.

Pedate: Palmately lobed with the lateral lobes again divided.

Pedicel: The stalk of a single flower.

Peduncle: The stalk of a flower cluster.

Peltate: With the stalk arising towards the centre rather than at the margin (of a leaf). *Fig. iii*, M.

Perianth: Collective term for the whorls of petals and sepals.

Perigon: A perianth not differentiated into petals and sepals.

Perygynium: Papery sheath enclosing the achene in sedges (*Carex*).

Petaloid: Petal-like.

Pilose: Shaggy, with soft hairs.

Pinnate: Feather-like, as the leaflets of a compound leaf placed on either side of the rachis. *Fig. iv*, A, F.

Pinnatifid: Leaf divided in a pinnate way but not cut to the midrib. *Fig. iv*, D.

Pinnatisect: Leaf divided in a pinnate way but cut to the midrib. *Fig. iv*, E.

Polygamous: With unisexual and bisexual flowers on the same plant.

Procumbent: Trailing, lying flat along the ground without rooting at the nodes.

Pruinose: With a thick, white, flaky bloom.

Pubescent: Covered with fine hairs.

Pungent: Ending in a stiff, sharp, point.

Pyriform: Pear-shaped.

Quadrate: More or less square in outline.

Raceme: Simple elongated inflorescence with the oldest flower at the base. *Fig. vi*, A.

Radiate: Spreading from a common centre.

Radical: Arising from the root or crown.

Ray-floret: The outer, ligulate florets in a composite head.

Receptacle: The modified apex of a stem bearing the floral parts.

Reniform: Kidney-shaped.

Reticulate: With a netted appearance.

Retuse: Notched slightly at the apex. *Fig. iii*, F.

Revolute: Margin rolled downwards with the lower surface innermost.

Rhizome: Underground stem.

Rhomboidal: Diamond-shaped. *Fig. viii*, G.

Rotund: Almost circular.

Rugose: With a wrinkled surface.

Runcinate: Coarsely serrate with the teeth pointing towards the base.

Saccate: Bag- or pouch-shaped.

Fig. v. **Types of Fruit**

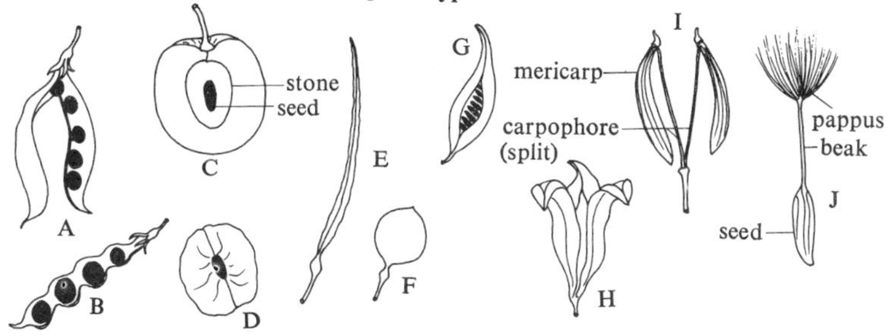

A, Legume; B, Loment; C, Drupe; D, Samara; E, Siliqua; F, Silicula; G, Follicle; H, Capsule; I, Schizocarp; J, Cypsela.

Sagittate: Arrow-shaped with the lobes pointing towards the base. *Fig. iii*, I.

Samara: An indehiscent winged fruit. *Fig. v*, D.

Scabrid: Rough.

Scabrous: With short, rough hairs.

Schizocarp: A dry, indehiscent fruit splitting into two valves. *Fig. v*, I.

Scorpioid: A coiled cluster in which the flowers are usually two-ranked. *Fig. vi*, G.

Sericeus: Silky.

Serrate: With saw-like teeth which point forwards.

Serrulate: Minutely serrate.

Sessile: Without a stalk.

Setaceous: Bristle-like, with a bristle.

Setose: Covered with bristles.

Setulose: With minute bristles.

Shrublet: A small shrub (perennial).

Silicula: Dry, dehiscent fruit of a Crucifer, not more than three times as long as broad. *Fig. v*, F.

Siliqua: Dry, dehiscent fruit of a Crucifer, at least three times as long as broad, with two valves and a papery septum. *Fig. v*, E.

Sinuate: With a pronounced wavy margin.

Spadix: Thick, fleshy spike bearing sessile flowers at the base and surrounded by a spathe (Aarceae).

Spathe: A large, sometimes coloured bract surrounding a spadix.

Spathulate: Spoon-shaped. *Fig. viii*, B.

Spike: An unbranched inflorescence of more or less sessile flowers with the oldest flowers at the base.

Spikelet: A secondary spike or the floral unit in a grass.

Staminode: A sterile stamen which may in some cases be petal-like.

Standard: The upper, broad petal of a leguminous flower.

Stipule: The basal appendage of a petiole.

Striate: With fine ridges, grooves or lines of colour.

Strigose: With sharp-pointed, appressed, stiff hairs.

Strophiole: An appendage at the point of attachment of certain seeds.

Sub-: Slightly, as sub-dentate, slightly toothed.

Subulate: Awl-shaped, tapering from base to apex.

Superior: Applied to an ovary situated above the perianth. *Fig. i*, C.

Tendril: A slender, often coiled extension of stem or leaf used for climbing.

Tepal: An undifferentiated perianth unit.

Terete: Circular in cross-section.

Ternate: Divided into threes. *Fig. iv*, G.

Thyrse: A compact panicle in which the flowers of the lateral branches open from the outside inwards.

Tomentose: Densely woolly or pubescent, with matted hairs.

Trifid: Divided into threes.

Trifoliate: Having compound leaves with three leaflets. *Fig. iv*, B.

Fig. vi. **Inflorescence Types**

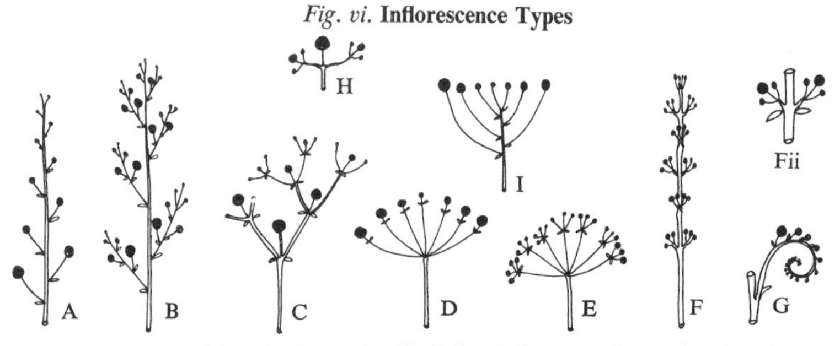

A, Raceme; B, Panicle; C, Cyme; D, Umbel; E, Compound umbel; Fi, Spike of verticillasters; Fii, Verticillaster; G, Scorpioid cyme; H, Dichasium; I, Corymb.

Fig. vii. **Leaf Arrangement**

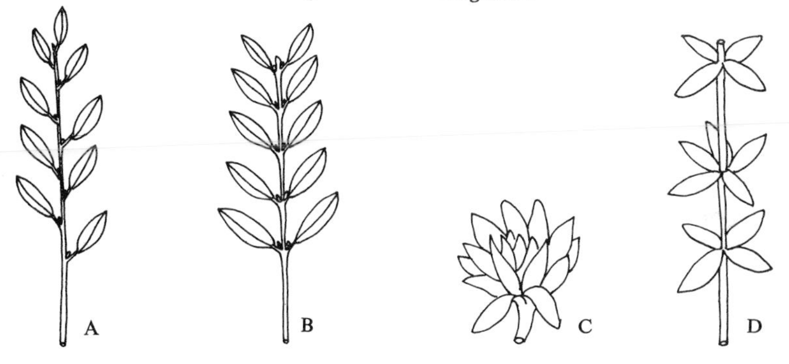

A, Alternate; B, Opposite; C, Rosulate; D, Whorled.

Truncate: Ending abruptly, the apex or base more or less straight across. *Fig. iii*, L.

Tuberculate: With small tubers or lumps on the surface.

Tunicate: Formed of concentric layers like the bulb of an onion, the outer loose coat is often referred to as a tunic.

Umbel: A flat-topped inflorescence characteristic of the Umbelliferae with the peduncles of more or less equal length and arising from a common point. *Fig. vi*, D, E.

Umbellule: A secondary umbel.

Valve: One of the segments into which a capsule or pod naturally splits on maturity.

Velutinous: With a velvety indumentum.

Verticillaster: A whorl-like structure composed of a pair of opposed cymes as in the inflorescence of Labiatae. *Fig. vi*, F.

Villous: With long, unmatted, silky, straight hairs.

Viscid: Sticky.

Wing: Dry, flat extension of some seeds or stems, the lateral petals of a leguminous flower.

Zygomorphic: Irregular or bilaterally symmetrical, divisible into two halves in one plane only.

Fig. viii. **Leaf Shapes**

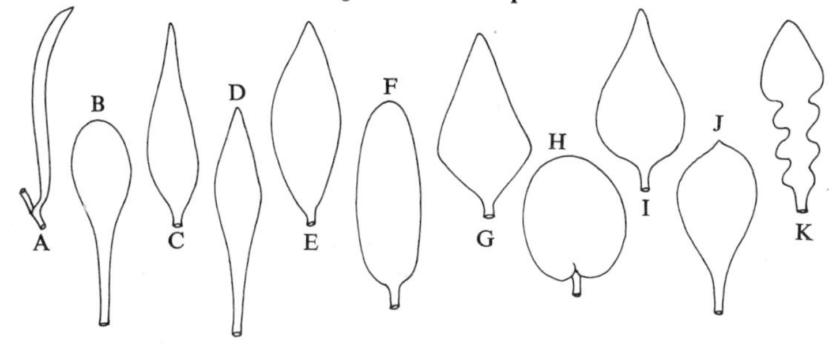

A, Linear; B, Spathulate; C, Lanceolate; D, Oblanceolate; E, Elliptical; F, Oval; G, Rhomboidal; H, Orbicular; I, Ovate; J, Obovate; K, Lyrate.

Index

In this index botanical names are printed in italics and local common names in bold type. Place names refer to the chapter on places of botanical interest. Numbers in bold type are illustrations, and other numbers refer to text pages.